IntechOpen

Supporting open minds since 2005

AI and Learning Systems - Industrial Applications and Future Directions
http://dx.doi.org/10.5772/intechopen.85833
Edited by Konstantinos Kyprianidis and Erik Dahlquist

Contributors
Weiwei Zhao, Tarek Hassan Mohamed, Gaber Salman, Hussein Abubakr Hussein, Mahmoud Hussein, Gaber Shabib, Andreas Roither-Voigt, Valdemar Lipenko, David Zelenay, Sebastian Nigl, Javad Khazaei, Dinh Hoa Nguyen, Moksadur Rahman, Örjan Larsson, Enislay Ramentol, Tomas Olsson, Shaibal Barua, Idelfonso B. R. Nogueira, Antonio Santos Sánchez, Maria. J. Regufe, Ana M. Ribeiro, Erik Dahlquist, Gladys Bonilla-Enriquez, Patricia Cano Olivos, José Luis Martínez Flores, Diana Sánchez Partida, Santiago Omar Caballero Morales, Karim Belmokhtar, Mauricio Higuita Cano, Jan Skvaril, Konstantinos Kyprianidis, Amare Desalegn Fentaye, Valentina Zaccaria, Ioanna Aslanidou

First published in London, United Kingdom, 2021 by IntechOpen
IntechOpen is the global imprint of INTECHOPEN LIMITED, registered in England and Wales, registration number: 11086078, 5 Princes Gate Court, London, SW7 2QJ, United Kingdom
Printed in Croatia

British Library Cataloguing-in-Publication Data
A catalogue record for this book is available from the British Library

Additional hard and PDF copies can be obtained from orders@intechopen.com

AI and Learning Systems - Industrial Applications and Future Directions
Edited by Konstantinos Kyprianidis and Erik Dahlquist
p. cm.
Print ISBN 978-1-78985-877-8
Online ISBN 978-1-78985-878-5
eBook (PDF) ISBN 978-1-83968-601-6

An electronic version of this book is freely available, thanks to the support of libraries working with Knowledge Unlatched. KU is a collaborative initiative designed to make high quality books Open Access for the public good. More information about the initiative and links to the Open Access version can be found at www.knowledgeunlatched.org

For EU product safety concerns:
IN TECH d.o.o., Prolaz Marije Krucifikse Kozulić 3, 51000 Rijeka, Croatia, info@intechopen.com or visit our website at intechopen.com.

Meet the editors

Prof. Konstantinos G. Kyprianidis is a Full Professor in Energy Engineering within the Future Energy Center at Mälardalen University in Sweden. He leads the SOFIA research group (Simulation and Optimization for Future Industrial Applications) and is the Head of Research Education for Energy & Environmental Engineering. He has been the Principal Investigator of a large number of national and international research projects related to automation in the energy and process industry. Among others, he has been the Chief Engineer for the 5.75mEuro project FUDIPO funded by the European Commission. Prior to coming to MDH, he worked for Rolls-Royce plc in the United Kingdom. He has co-authored over 140 peer-reviewed publications and currently supervises 15 doctoral candidates and is the Chair of the ASME/IGTI Aircraft Engine Committee.

Prof. Erik Dahlquist is a Senior Professor of Energy Technology and was the former Research Director of the Future Energy Center from 2000 to 2017 at Mälardalen University in Sweden. Among many other projects, he is the Project Coordinator Engineer for the 5.75mEuro project FUDIPO funded by the European Commission. Prior to joining MDH, he worked for 28 years at ABB Sweden in various senior technical and management roles. He has co-authored over 300 peer-reviewed publications and currently supervises 10 doctoral candidates.

Contents

Preface

Artificial Intelligence (AI) was a "hot" research topic within industrial automation in the early 1980s with developments focused on methods for diagnostics, simulation, model adaptation, and optimal control. Artificial neural networks were used extensively at the time within a number of industrial automation applications. Issues included robustness and available computational capacity for online and real-time applications.

It took the best part of the next 3 decades to develop practical solutions to these shortcomings, and gradually bring the technology to the level of "intelligence" required to leverage benefits compared to traditional approaches. As a result, interest in the industrial applications of AI and learning systems have surged anew over the last few years. Several powerful methods and digital platforms have been developed during the past decade, and the number of industrial applications has been growing exponentially. This is rapidly changing the way of doing things within the process and energy industry, as well as other industrial sectors.

This book covers recent developments and provides a broad perspective of the key challenges that characterize the field of Industry 4.0 with a focus on applications of AI. The target audience for this book includes engineers involved in automation system design, operational planning, and decision support. Computer science practitioners and industrial automation platform developers will also benefit from the timely and accurate information provided in this work.

The book is organized into two main sections comprising 12 chapters overall:

(i) Digital Platforms and Learning Systems

(ii) Industrial Applications of AI

The academic editor is indebted to all his colleagues from across the world that contributed to this book with their latest research, to Prof. Erik Dahlquist for joining this effort as co-editor, to several automation experts who volunteered as reviewers, as well as to IntechOpen Publishers for giving me the opportunity to work on this book and its members of staff for their constant support during its preparation.

Konstantinos G. Kyprianidis and Erik Dahlquist
Professor,
Mälardalen University,
Västerås, Sweden

Acknowledgments

This project has received funding from the European Union's Horizon 2020 research and innovation programme under grant agreement No 723523.

Section 1

Digital Platforms and Learning Systems

Chapter 1

AI Overview: Methods and Structures

Erik Dahlquist, Moksadur Rahman, Jan Skvaril and Konstantinos Kyprianidis

Abstract

This paper presents an overview of different methods used in what is normally called AI-methods today. The methods have been there for many years, but now have built a platform of methods complementing each other and forming a cluster of tools to be used to build “learning systems”. Physical and statistical models are used together and complemented with data cleaning and sorting. Models are then used for many different applications like output prediction, soft sensors, fault detection, diagnostics, decision support, classifications, process optimization, model predictive control, maintenance on demand and production planning. In this chapter we try to give an overview of a number of methods, and how they can be utilized in process industry applications.

Keywords: process industry, artificial intelligence (AI), learning system, soft sensors, machine learning

1. Introduction

During the 80th AI was a hot topic both in the academia and industries. Many researchers were working a lot with development of methods for diagnostics, simulation and adaptation of models. Artificial Neural Networks (ANN) were being implemented in real applications such as e.g. soft sensors to predict NOx concentration in exhaust gas from power plants. Still there was quite some “over-selling” and the enthusiasm for AI in the future was assumed to be useful tomorrow. But it took much longer to get the systems robust enough to be used and fast enough to be applicable in on-line applications. After year 2000, systems started to reach a more mature state and we got IBMs Watson, that could beat the Jeopardy master. Later the Google tool could beat the “Go-master”, a very complex Chinese game. This has changed the perception of AI. It is still similar type of tools as were developed during the 80th, but now they were refined a lot and hardwires has been developed dramatically. This has given us a much more positive perception of what can be done, and a lot is now being implemented. Still there is a risk for over-selling, as the tools are normally not that “intelligent” as we normally think of when we talk about Intelligence. But we are closing the gap day by day.

Concerning use of AI in process industry, we cannot just take the tools and hope they will fix everything. It is still important to identify “what is the problem to solve”? With Jeopardy the goal is to be good at Jeopardy, but what is the goal in process industry? It should be to increase production, reduce process variations, implement maintenance on-demand and give operator support. It also means to coordinate and optimize production lines as well as complete plants and later on complete corporations. It also means to adapt to changing customer demands, support in development of new products with production lines as well as handle new business models. These different functions demand quite different tools and thus we will not use only one but several. Often Machine learning is considered being "the tool", but often there is not data available to implement ML, especially not when starting a new production line. To implement new tools, it is also very important to pre-treat data. You have to sort data in "normal variations" or "anomalies". You may need to filter data with moving windows, but in different time perspectives. We need to do data reconciliation to handle drifting sensors. And you need to integrate all levels from orders to production planning down to coordinated and optimized production. In this chapter we will discuss a number of different methods as well as discuss integration between the different levels. Over the years many researchers have investigated different AI techniques for different process industrial application. A comprehensive review on different AI models applied in energy systems can be found in [1]. Applications of different AI tools based on simulation models in pulp and paper industry has been presented by researchers including Dahlquist [2–5]. Applications in power plants have been presented in many articles including Karlsson et al. [6–8]. In Karlsson et al. [9] a general discussion is made on how to make better use of data including pretreatment of data. Adaptation to degeneration in process models by time is discussed in Karlsson et al. [7]. [10] conducted an extensive review on different AI based soft sensors in process industries.

1.1 Similarities between AI and how the brain works

The mathematicians developing especially ANN have been looking a lot on how the brain works. In **Figure 1** we see a principal picture of a human.

Running in a forest: The brain stores many different factors locally by "tuning many soft sensors". During the night strength of connections are enhanced for the most important functions, while other less important connections are eliminated. Some information is used for direct control. Others is stored for use later on.

If it is rainy when you run there is a general feeling that "this was not so nice". Everything else happening in the forest then will be "colored" by this in your memory, aside of concrete thing like if you meet someone, like a friend, during the run.

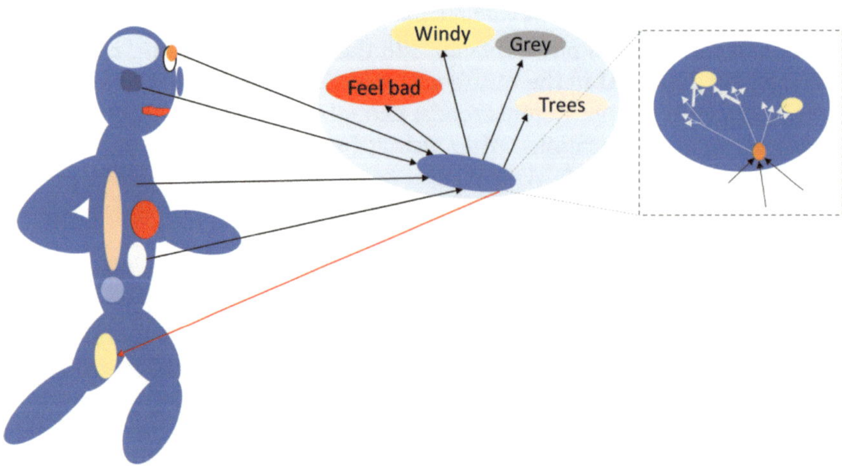

Figure 1.
How a human handle input from the surrounding.

Short term memory: Dorsolateral prefrontal cortex controls information stream from sensors. Skull lobe is for attention. Ventrolateral prefrontal cortex sort information into useful or not useful info. Supplementary motor area (SMA) repeat new memories all over.

Long term memory: Hippocampus and nearby areas in medial temple globe are essential for long term memory. Facts are stored. Small brain and basial ganglia contain procedural memory, like how to bike or swim.

A human may have approximately 120 billion nerve cells. Each connect to hundreds of other cells. Some connections enhance while other decrease signals. Very complex interactions where connections are established and broken continuously. No exact values or memories exist for control, but diffuse input give diffuse output, but with different feed-back mechanisms. The Swedish Nobel Prize winner Arvid Carlsson [11] found out the mechanism of how signals are transferred from the dendrite of one cell to the axon of the next, where complex feed-back mechanisms enhance a connection and thereby also enforced a memory by changing the easiness of transferring new signals. He explored how dopamine works as a signal substance, which we now know is of highest importance in the brain. By back-propagation in ANN we try to simulate this mechanism (**Figure 2**).

Input to the brain is sorted in Amygdala and hippocampus. Signals are sent to different part of the brain Here different signals are enhanced or decreased depending on previous experiences in many different "soft sensors", built up with tuning of Ca-channels working as parameters in a polynom. "= enhancement factors". The situation is triggering memory build up. All control is "diffuse" using many different "diffuse" measurements. Different individuals have different sensitivity and number of different sensors like sense for bitterness, sugar, pain etc. Soft sensors get input and react with output to other soft sensors. Signals are sent to direct different biochemical processes like when fear - increase production of Adrenalin and Cortisone. This in turn is affecting many other hormones and proteins etc. Also, microbiome in the stomach and skin send input to the brain on how these organs perform. When you run, the body feel good and e.g. endorphins are produced enhancing performance of stomach, muscles etc. Serotonin levels,

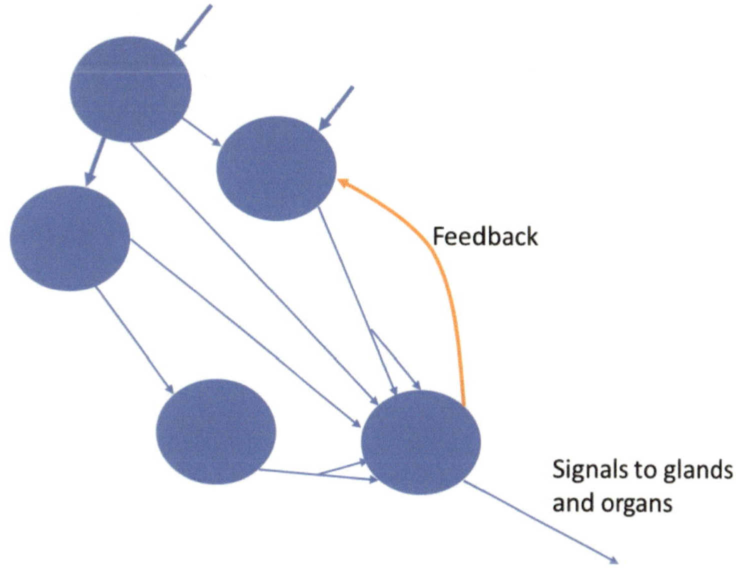

Figure 2.
Signals flow in the brain – Many connections and feed-back enhance learning.

gibberellins, insulin, cortisone etc. are interacting and tuning each other, but with influence "from the side" by other sensor inputs. The brain is interacting with all this. This is also the basic concept to mimic in "deep learning".

If we try to transfer this picture into a control system, it can look like below in **Figure 3**.

We start with sorting out "outliers" in pre-processing. This is what the brain does with information from the eye etc. The outliers can be used for anomaly detection. This is principally what is done in Amygdala. We then compare predictions from simulators and soft- sensors to measurements. We trend differences developed by time. Refined data are used for model building and adaptation of models. The models are used for soft sensors, diagnostics, control etc. We also make conclusions in decision a tree from previous experience and identify optimal action to take in different time perspectives. In the brain this is done by utilizing previous experience in a way where we try to "make sense". This means that we replace missing data with what is reasonable. In our computer system we do this by data-reconciliation using e.g. solving an equation system of physical models to get a best fit. We then take actions by control of many different functions more. In the body, this means e.g. control of sugar content in the blood, release of adrenalin to meet threats or melatonin to make you tired and go to sleep. We learn buy tuning soft sensors and decision trees with the new information just as the brain does, but where the brain is very much more complex than what we can handle today.

1.2 Market aspects

IndTech's market, i.e. Products and systems for industrial digitization and automation in the world are worth around USD 340 billion in 2016/2017 and have an average growth rate of 7–8 percent. The area can be divided into two parts: IT (industrial IT) and OT (operational technology). The share that can be categorized into industrial IT is about USD 110–120 billion. The remaining USD 220 billion is operational technology for the factory floors and in the field. It, in turn, is traditionally divided into discrete automation (about 45 percent) and process automation (about 55%). OT includes various types of industrial control systems (ICS) and field equipment such as instrumentation, analysis, drive systems, motors, robots and similar.

For the future of AI, we can see that this comes deeply into all these industrial market segments, but also far beyond as not only for industrial applications.

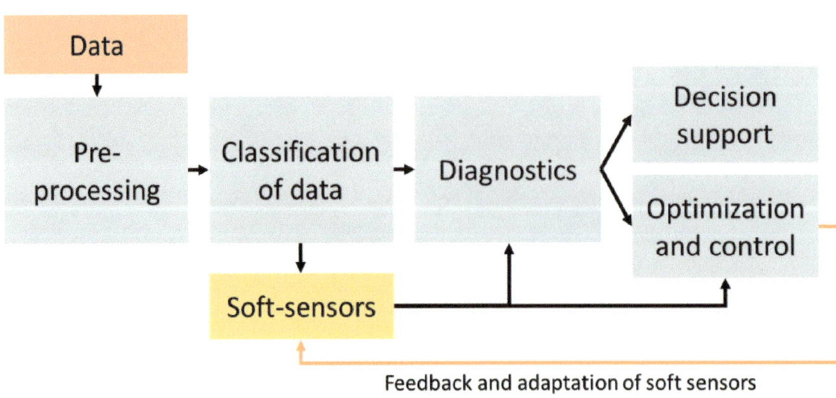

Figure 3.
Principal diagram of signal processing in a "learning system".

The tools thus will be developed for one application, but then will be used also for other applications most probable.

2. Different AI methods

There are many different methods developed. Some of them are very similar or aim to solve the same type of problems. If we look at Machine learning (ML), we have e.g. Regression. Artificial Neural Networks (ANN), Support Vector Machines (SVM), Principal Component Analysis (PCA), Partial Least Square regression (PLS) and etc. They both aim to sort different variables into group that correlate to different properties or faults.

PLS and ANN, both are very useful to create soft sensors. Deep learning is a sophisticated version of the ANN, but with the goal to produce models that can do much more than just be a soft sensor, which predicts one or more qualities. Examples of soft sensors is to predict strength properties of paper from e.g. NIR data and process variable values in paper machines, amount of different kind of plastics in Waste combustion plants or protein content in cereals in agriculture from NIR spectra. The deep learning on the other hand can be used to teach a robot to pick out machine components that are scrapped from a conveyor belt for instance. This then includes image pattern analysis from camera monitoring of the parts passing.

A selection of different tools is listed in **Table 1**.

2.1 Machine learning methods

Machine learning methods principally use a lot of process data measured preferably on-line, and identify correlation models from the data, which can be used for different purposes like soft sensors, anomaly detection and others.

There are several different machine learning methods. Some are correlating a specific property to process data. Reinforcement learning is described in e.g. Gattami Ather [12]. It is used in problems where actions (decisions) have to be made and each action (decision) affects future states of the system. Success is measured by a scalar reward signal and proceed to maximize reward (or minimize cost) where no system model is available. One example of this technique is deep reinforcement learning which was used in AlphaGo that defeated the World Champion in Go. Here a Q function is approximated with a deep neural network. Minimizing the loss function with respect to the neural network weights w is made as given below

$$l = (r(s,a) + \delta \sup Q(\acute{s}, \acute{a}, w_{-}) - Q(s,a,w))^2 \quad (1)$$

- Gaussian Process Regression (GPR)
- Partial Least Square (PLS) Regression
- Principal Component Analysis (PCA)
- Artificial Neural Networks (ANN)
- Support Vector Machines (SVM)
- Gray box models
- Physical models, MPC – model predictive control
- Bayesian networks (BN)
- Gaussian Mixture Model (GMM)
- Reinforcement Learning
- Google algorithm – search engines

Table 1.
A selection of different common AI-tools.

If the system is deterministic the model is given by

$$S_{k+1} = f_k(s_k, a_k) \quad (2)$$

If the system is stochastic the model is given by

$$P(s_{k+1}|s_k, a_k) \quad (3)$$

$f_k(s_k, a_k)$ is a scalar valued reward.

In Werbos Paul: A Menu of Design for reinforcement learning over time [13] reinforcement methods are described more generally.

2.2 Soft sensors

It is interesting to create soft sensors by creating models correlating process measurements on-line to quality measurements from samples analyzed at lab. The soft sensor then can be used to predict the quality property on-line from feeding the on-line measurements into the soft sensor model. There are several different methods for the regression, and a number of alternatives are given in **Figure 4** below.

In **Figure 5** we see how the data flow can look like for data collection, data pre-processing, model building and model validation. Here NIR measurements are correlated to properties like lignin content.

Soft sensors also can be built with other methods like using ANN, Artificial Neural nets. There are advantages and disadvantages with the different methods, but also commonalities. You need good data for building the models. This means that data need to be spread out in the value space in a good way. If we only have "white noise" the models will be unusable. We need to vary all variables in a systematic way to get useful data for model building.

2.3 Gaussian process regression model

Gaussian Process Regression takes more memory but gives better regression models than many other methods like (Nonlinear) System Identification, Neural Networks and Adaptive learning models. Can also be Combine with physics-based models. The method is presented in e.g. Fredrik et al. [14]. In **Figure 6** we see a first attempt to predict kappa number of pulps after a digester for two different wood

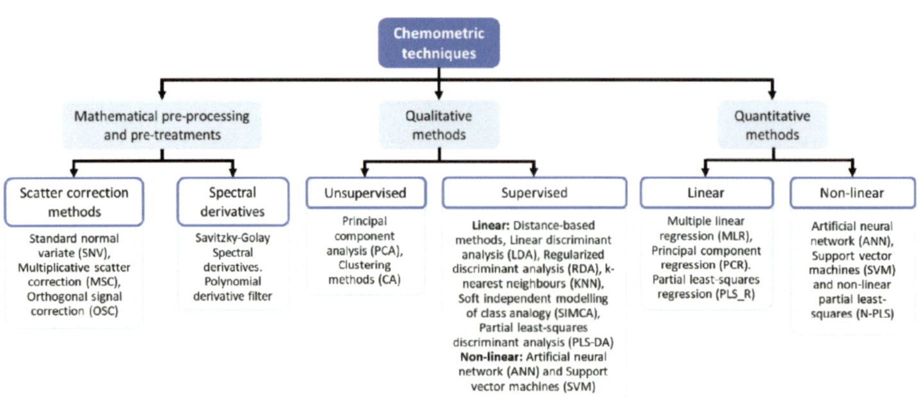

Figure 4.
A number of methods that can be used to develop soft sensor models from process data.

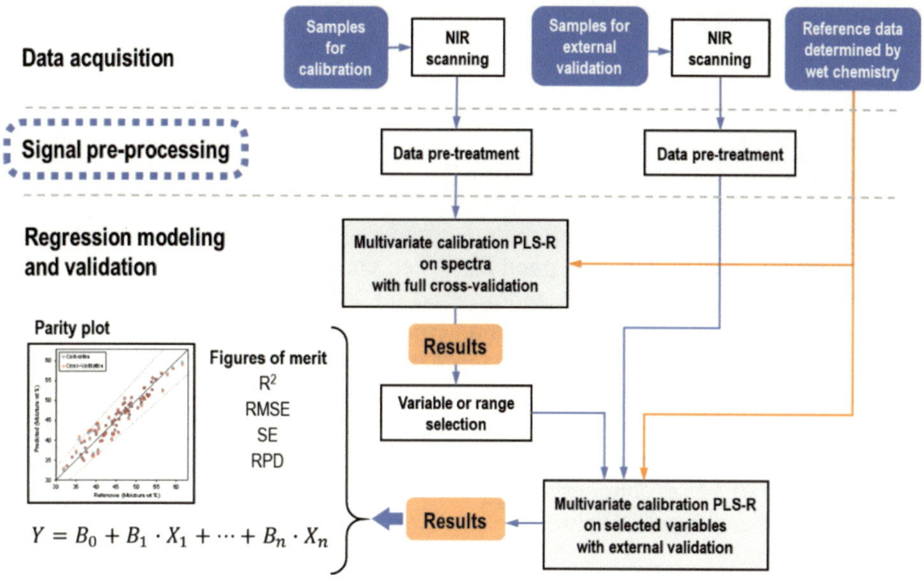

Figure 5.
Data flow for building and verification of soft sensors.

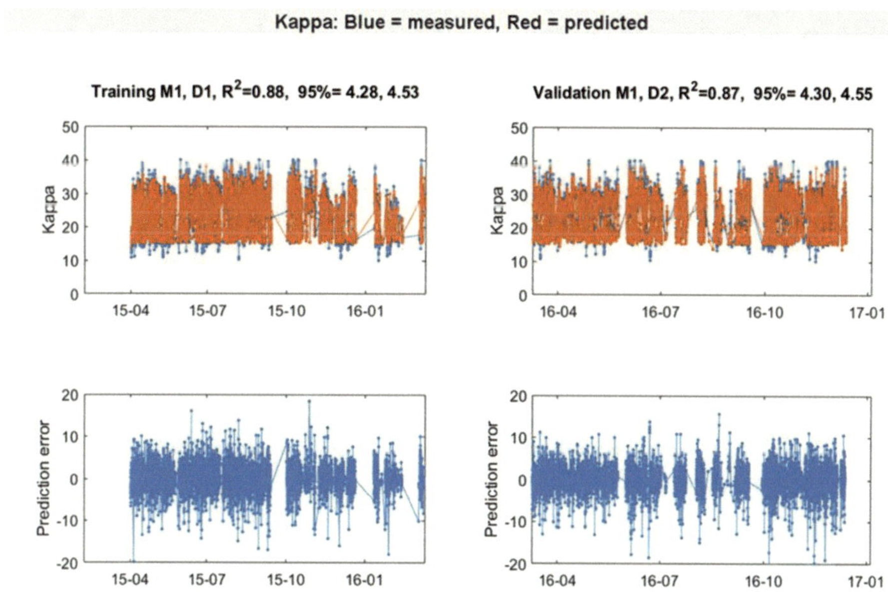

Figure 6.
Example of Gaussian process regression (GPR) for kappa prediction.

types, hardwood and soft wood. The training data fits quite well, while the predictions are less good. By using more data and fine-tune the estimation of residence time in the reactor the prediction power became significantly better. It went from $R^2 = 54$ to $R^2 > 90$.

2.4 Artificial neural nets, ANN

Artificial neural nets try to mimic the brain. In a simple way we can use the equation below to show how it is calculated:

$$\hat{y}(t|\theta) = a_1 * \kappa(\gamma_1 + \beta_{11}\varphi_1(t) + \beta_{21}\varphi_2(t)) + \alpha_2 * \kappa(\gamma_2 + \beta_{12}\varphi_1(t\ ) + \beta_{22}\varphi_2(t)) \quad (4)$$

In **Figure 7** we see three input variables to the left. Each variable is multiplied with a weight factor towards the two summa-nodes, where the products are summarized. Next these values can be treated to pass a threshold or only be passed on and multiplied with a second constant αi. The two products are summarized again, and we get a prediction of the value of a wanted property. When you build the net, you look at the difference between the measured and the predicted value and adjust the weight factors until you get a good fit. When you have been testing one set of input variables you go to the next and proceed for all data you have and try to get a fit that is the best for all input variables together. This is a simple net with only one "hidden layer", but you can have much more complex versions with many variables and many layers. If you have many layers the problem though can be that you get a good fit for the training data but it may also give risk for "over-fitting", which means less stable predictions.

An example of a first commercial application of ANN was for prediction of NOx in power plants. In **Figure 8** below we see a regression for the power boiler number four in Vasteras.

2.5 PLS, partial least square regression and factorial design of experiments

PLS is very popular to use for making prediction models after performing factorial designs of experiments. The basic idea is to start with a linear regression for a line, $y = a + b * x$, and adding non-linearity by $+c * x_2$ and if there are more than one variable the interaction between variable 1 and 2 by $d * x_1 * x_2$. The polynomial for a property like a strength property of a paper then becomes

$$y = A + B * x_1 + C * x_2 + D * {x_1}^2 + E * {x_2}^2 + F * x_{1*} x_2 \quad (5)$$

Here A-F are constants you get from fitting the experimental data to the model. If we use factorial design, it means that we try to expand the prediction space as much as possible within given borders. This means that we shall have a good distribution of experimental data in all parts of the space, and not only close to origo or in one part of the space. This means for example that you shall not make correlation for one variable at a time but vary all variables in a systematic way. In Ferreira et al. [15] the Box–Behnken design is described more in detail. In **Table 2** below we see an example for three variables:

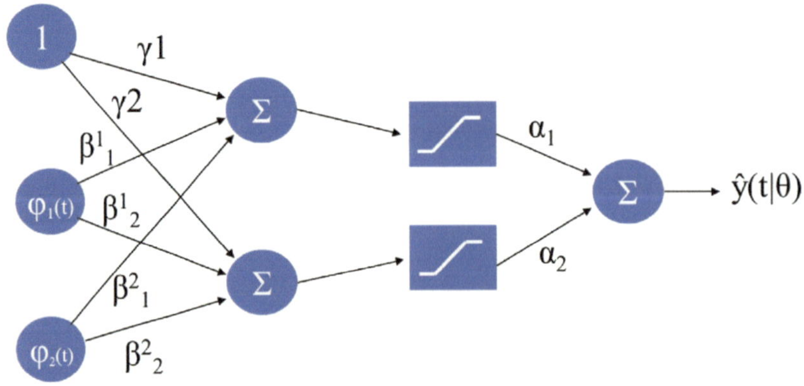

Figure 7.
A simple artificial neural net, ANN.

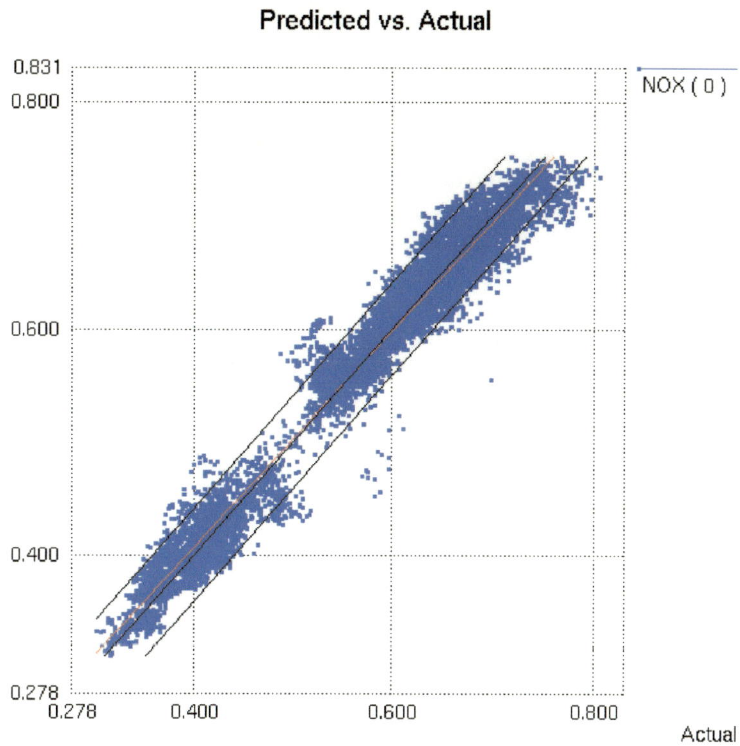

Figure 8.
A plot showing the correlation between prediction with an ANN and measurements of the actual NOx content in the exhaust gases from a power plant (coal fired boiler 4 at Malarenergi).

Experiment no	x_1	x_2	x_3
1	+	+	+
2	+	+	—
3	+	—	+
4	+	—	—
5	—	+	+
6	—	+	—
7	—	—	+
8	—	—	—
9	0	0	0
10	$\sqrt{3}$	0	0
11	0	$\sqrt{3}$	0
12	0	0	$0\sqrt{3}$

Table 2.
Factorial design of experiments with three important variables to predict a certain qualitative variable like paper property, lignin content, content of different plastics etc.

The first 8 experiments give the linear regression while the last four gives the non-linear components. As we vary all variables independently, we get the interaction between the variables directly. (+) means here a higher amount or concentration of the variable while (−) means a low. (0) is Origo and $\sqrt{3}$ is where a sphere is cutting the axis.

It is important to have an equal distribution in the whole sample volume of measurements. If a high concentration of samples around origo – the impact of the “real” samples will be too small. It is better to have a few good samples well distributed instead of many around origo or some other part of the space. By varying several variables at simultaneous also catches interactions between the variables. The reason while sometimes models built from only on-line data in a plant may have very little prediction power is if we have a number of important variables with controllers, and only get the white noise due to poor control. By really varying these variables in a systematic way as proposed by factorial design, we can build robust prediction models. If the models still are not that good, it may be because we are not varying or measuring all important variables. Then we should change the variables in the factorial design. If you do not know which variables are the most important you can start with the factorial design scheme in **Table 2** but add more variables and just vary them around origo and perhaps some other random point. From this first scan we can decide which variable to focus more experiments on.

The factorial design scheme can also be seen as values at the corners of a cube and where the axis crosses a sphere around the cube as seen in **Figure 9** below:

If it is expensive to run all experiments, you can make a reduced factorial design, where you principally pick some of the variants randomly and make a PLS model. You then add one or two experiments and see how much better it becomes and proceed until you feel satisfied. This can be illustrated as in **Figure 10**.

Principally the regression is made so that you start with a line through all data in the space and calculate the square of the distance between the point and the line. You add all values for all points. Then you change the direction and make a new try. This then proceeds until you have found a line that has least sum of square errors. You then make an axis perpendicular to this first line and proceed to find a plane.

One example can be seen in **Figure 11**.

$$\begin{aligned} \text{Strength} = {} & \text{A} + \text{B} * \text{concentration of filler} + \text{C} * \text{ration\_longfiber\_to\_shortfiber} \\ & + \text{D} * (\text{concentration\_of\_filler})2 + \text{E} * (\text{ration\_longfiber\_to\_shortfiber})2 \\ & + \text{F} * (\text{ration\_longfiber\_to\_shortfiber} * (\text{concentration\_of\_filler}) \end{aligned} \tag{6}$$

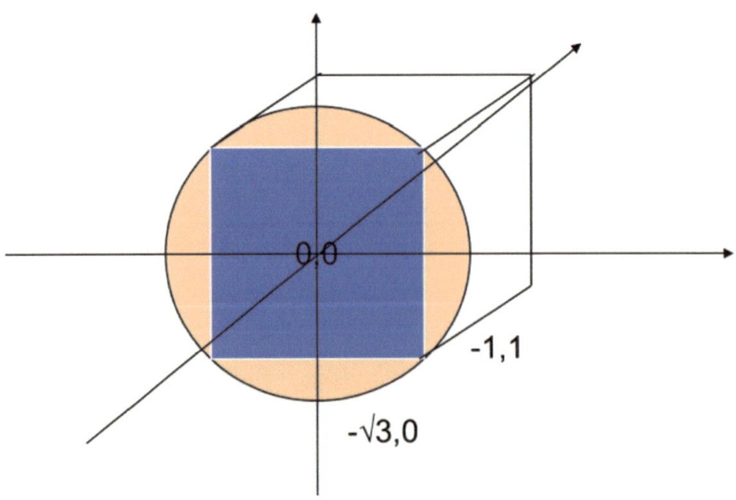

Figure 9.
Factorial design with values in all corners of the cube and where axis cross a sphere surrounding the cube.

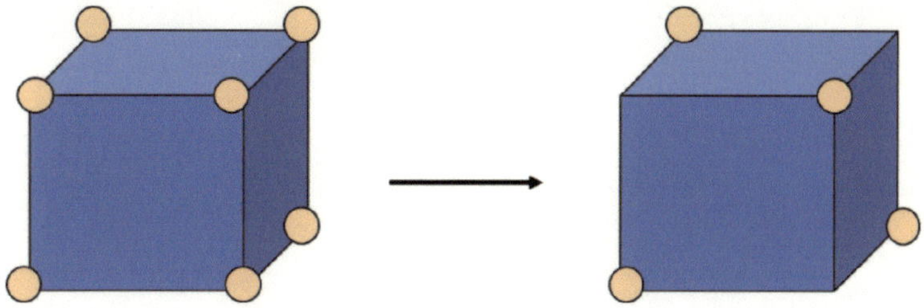

Figure 10.
Reduced factorial design.

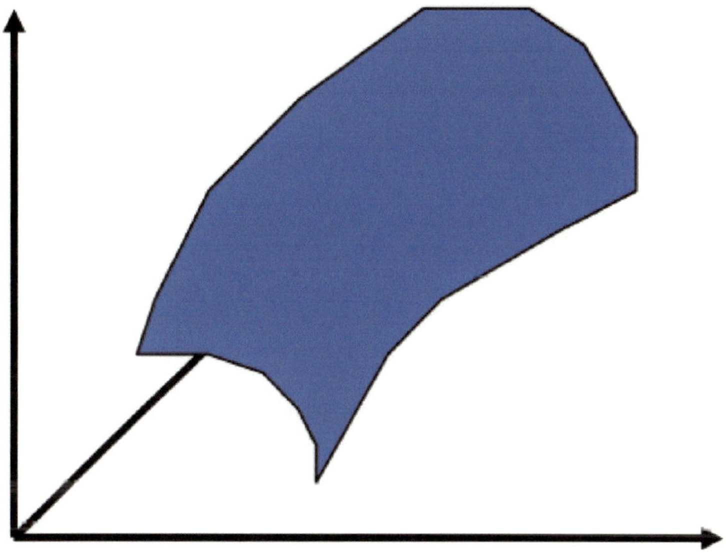

Figure 11.
The plane direction is corresponding to the line, the down wards bending the non-linearity and the cross bending of the surface shows interaction between the different variables x_1, x_2 and x_3.

In **Figure 12** we see what wavelengths have importance and to what degree for predicting the investigated property. At the top we have regression coefficients for AIL, Acid Insoluble Lignin, and at the bottom for ASL, Acid Soluble Lignin.

We can see from the regression coefficients in **Figure 12** that there is a significant difference between the spectra, indicating that the chemistry differs quite a lot. This as each wavelength corresponds to vibrations of a certain chemical bonding, like C-H, C-H2, C-O, C=O, etc. This example is taken from Skvaril Jan [16].

Confounding means that some effects cannot be studied independently of each other. This is very much the case in combustion processes, water treatment, process industries like pulp and paper etc.! This is why the factorial design of experiments make so much sense. In some cases, though there is no interaction between different variables, and then it might be OK to build linear models, but this is often more exceptions than the rule. There are a number of PLS methods. One popular version is PLS Regression which is presented by e.g. Svante et al. [17].

2.6 Fault diagnostics

It is interesting to determine both process and sensor faults. This can be performed in many different ways. You can listen to noise from an engine that

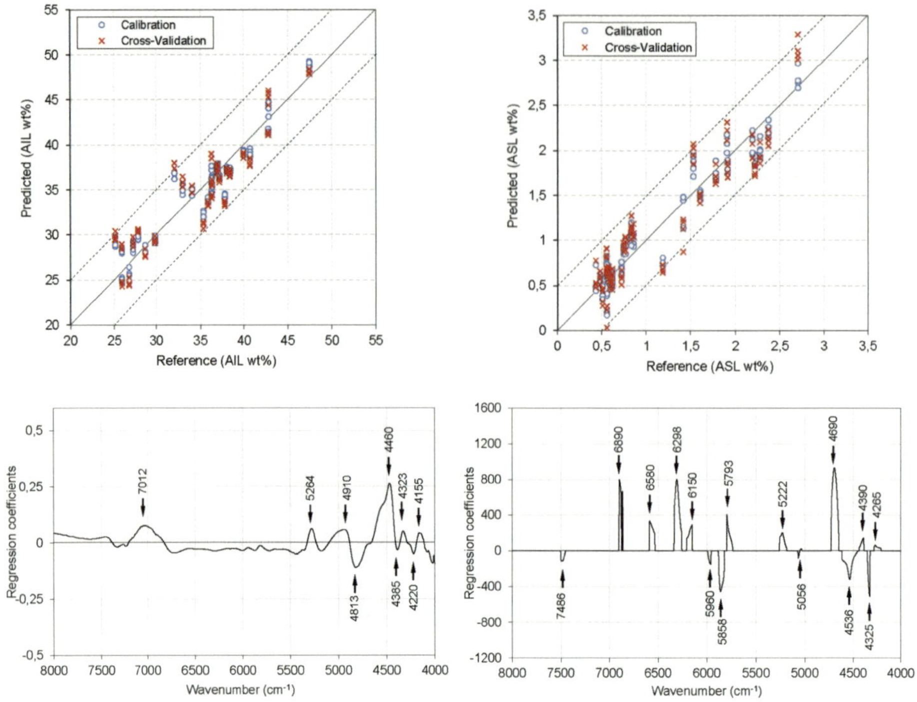

Figure 12.
Example of regression between wave lengths and lignin concentration in wood.

indicates some fault. Or you measure that the temperature has become too high somewhere. Fault detection can be systemized by using different tools and BN, Bayesian Networks, is a tool suitable for identifying causality relations and probability for different type of faults simultaneously.

2.6.1 Bayesian networks (BN)

Bayes was a priest in Scotland first discussing correlation versus causality. Correlation means that you can see how different variable are connected to each other, while causality means to take it a step further and also identify true dependence between a variable and a fault or similar. If we see that there is a correlation between homeopathic levels of a substance and effect on health, this can be a correlation but hardly that the homeopathic medicine is causing the good health. A lot of correlations are just random! With the Bayesian net you try to find the causality between different variables and a fault or similar and also quantify this. If we have a lot of experimental data we can use this to tune the BN, but if we do not have it but know from experience that there is a causality, we can make a reasonable guess of the importance in relation to other variables and use this for the BN. This gives an opportunity to make prediction models without "big data" and you can combine this input with real measurements in the plant.

Applications of BN for condition monitoring, root cause analysis (RCA) and decision support has been presented in e.g. Weidl G.,Madsen A L, Dahlquist E [18]; [19, 20] and adaptive RCA in Weidl et al. [21]. Weidl and Dahlquist [22] also has given a number of examples of RCA in pulp and paper industry applications like digesters and screens. In Weidl and Dahlquist [23] applications more generally for complex process operations are presented where object-oriented BN are utilized.

In Widarsson [24] Bayesian Network for Decision Support on Soot Blowing Superheaters in a Biomass Fuelled Boiler was presented and in.

If we have a number of BN variables U = {A_i} and parent variables pa(Ai) of Ai we can use the chain rule for Bayesian networks to give the probability for all variables Ai as the product of all conditional probability tables (CTP) $P(S_k I H_1, H_2, \ldots H_n)$. Here S_k is the child node which can be observed status, measured values by some meter, a trend or similar) and H_i is the parent node (assumed causes or conditions causing a change in the child node state). The CPT can be trained by real measurements with conditions and related failures or created by using experience by operators or process experts. This is of specific interest when you want to include possible faults occurring very seldom, but severe when actually happening. Data might also be created for training by running a simulator with physical models and with different faults.

The chain rule for all CTPs is as seen in Eq. 7.

$$P(\mathrm{U}) = \mathrm{P}(\mathrm{A}_1, \ldots, \mathrm{A}_n) = \prod_i \mathrm{P}(\mathrm{A}_i | \mathrm{pa}(\mathrm{A}_i)) \quad (7)$$

An example of a BN for a Root Cause Analysis function for a screen in e.g. pulp and paper industry can be seen in **Figures 13** and **14**.

2.6.2 Anomaly detection

If we have identified that a variable should be within certain limits or we have made a model using SVM or PCA or similar, we can see if the measured set of variables is within the boarders for a class or group. Both these types of measures can be used for anomaly detection. This can be very useful to identify if the process goes out of normal operations even if you have not passed the limits for a single variable.

2.7 Classification and clustering

2.7.1 Principal component analysis (PCA)

Svante et al. [25] have presented the tool PCA in an article already 1987. PCA is often in the same software package as PLS but has a different use. In the PCA we

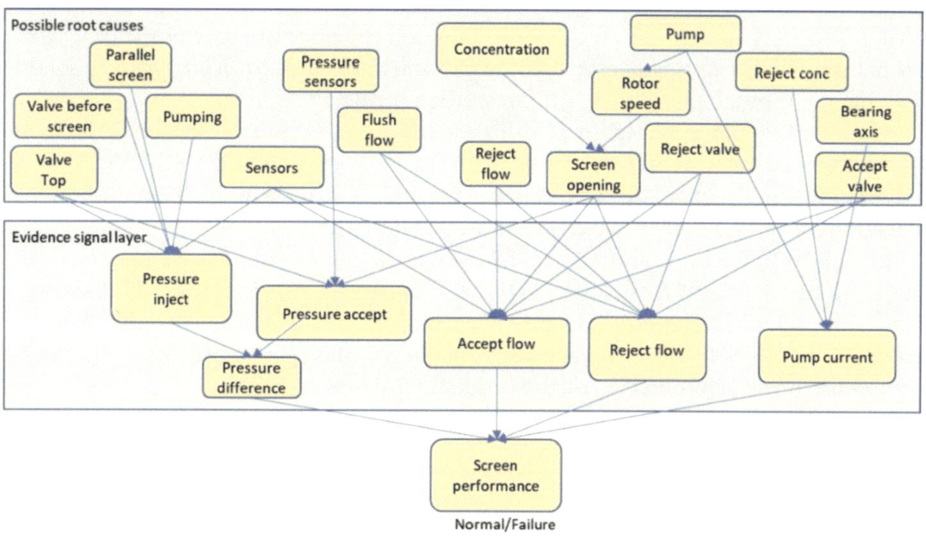

Figure 13.
A Bayesian model for RCA (root cause analysis) of a screen.

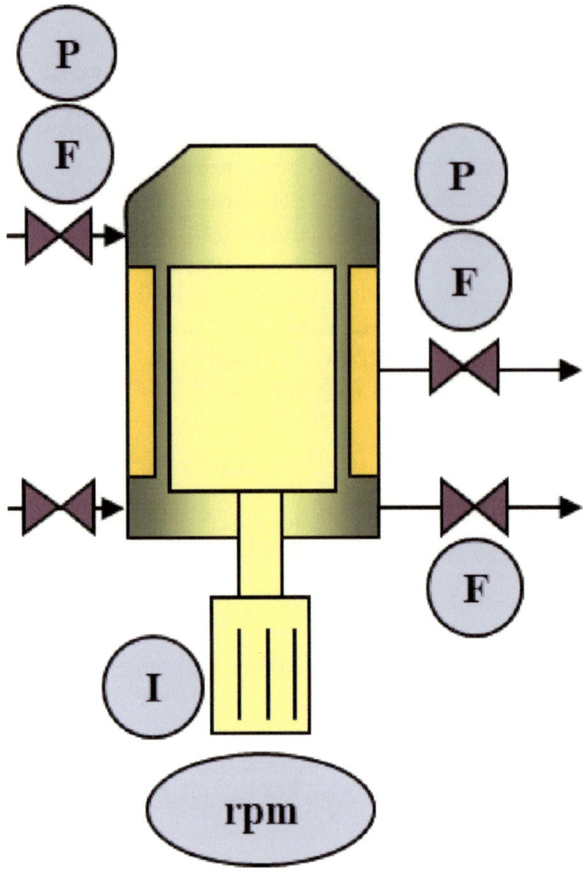

Figure 14.
A principal drawing of a screen with sensors.

plot all measured data onto different planes to see how the variables distribute in the plane. From this we can see that variables close to each other are affecting a certain property in the same way, while those on the opposite side of the diagram are having also the opposite effect. If the variables are close to Origo, we can believe they have not much effect at all on the studied property.

The score vector is a column of T. There will be one score vector for each single PC. Each experiment will have one value in the PC1 and one in the PC2. You plot all experiments in a coordinate system with PC1 and PC2. If we plot all experiments in a diagram with PC1 and PC2 we can get as in **Figure 15**.

In **Figure 15** we have plotted the time series of measurements and can see that there is a development from left to right as time passes by, along PC1. This shows that something is happening by time. We can also make a loading (p) plot. The loading plot shows how much each variable contributes to each PC. Each PC can be seen as the linear combination of the original variables

$$PC_i = \sum p_{ji} x_i \tag{8}$$

The loadings are the coefficients p_{ji}. Each variable can contribute to more than one PC. If we have more than two PCs, it can contribute to all PCs. In **Figure 16** we see the p-plot for a number of variables:

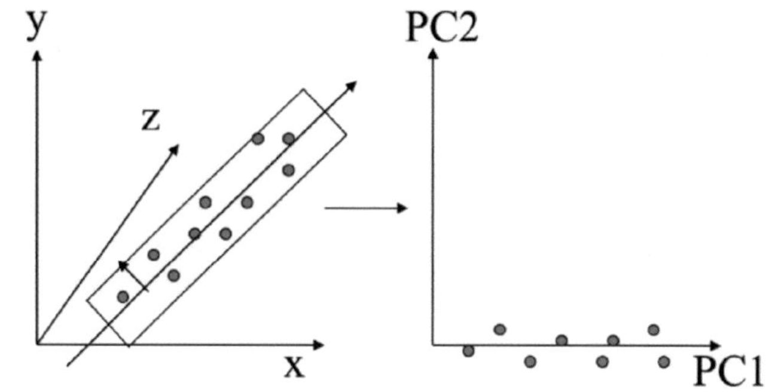

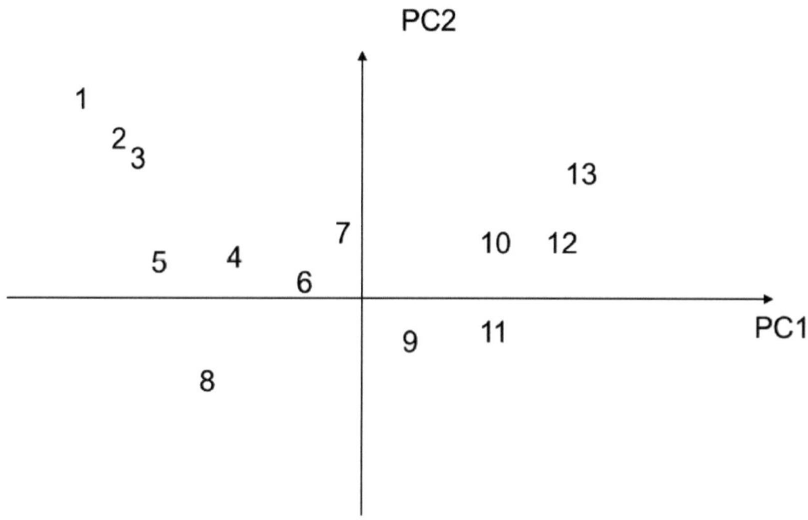

Figure 15.
Score plot (t). First sample no 1 at t = 0 and following no:s following time steps.

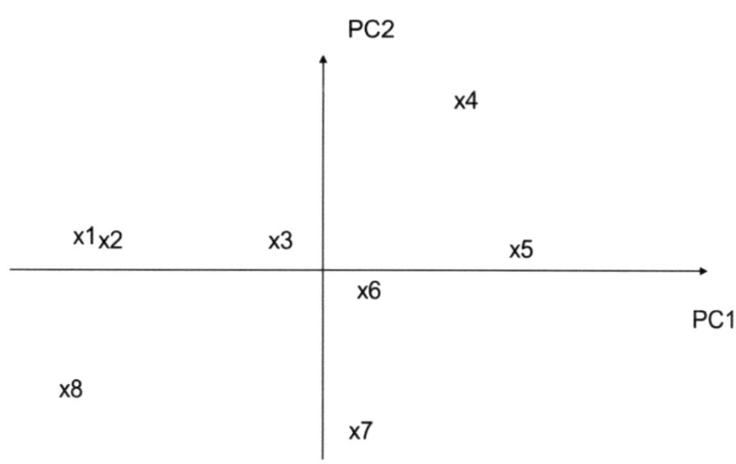

Figure 16.
P-plot for eight variables in the PC1 – PC2 coordinate system.

From **Figure 16** we can see that X3 and X6 have small impact while X4 and X8 have stronger impact but reverse to each other. X1 and X2 are following each other closely.

In **Figure 17** it can be seen that when the set of variables is within the circle the process is running OK, but when going outside you should take a look and try to get it inside the circle again. This is a bit towards diffuse control, like in the human body.

You can use the p-plot also to classify a number of faults. In **Figure 18** we see an example where vibrations, temperatures and electric power consumption was used to predict different type of faults. The faults were implemented at the lab and the variables measured. From this we could see that the variables were forming different patterns.

The PLS is principally partial least squares or projection to latent structures. Principally you do an interactive PCA for both X and Y matrices.

$$X = \sum T * PT + E \quad (9)$$

$$Y = \sum U * QT + F \quad (10)$$

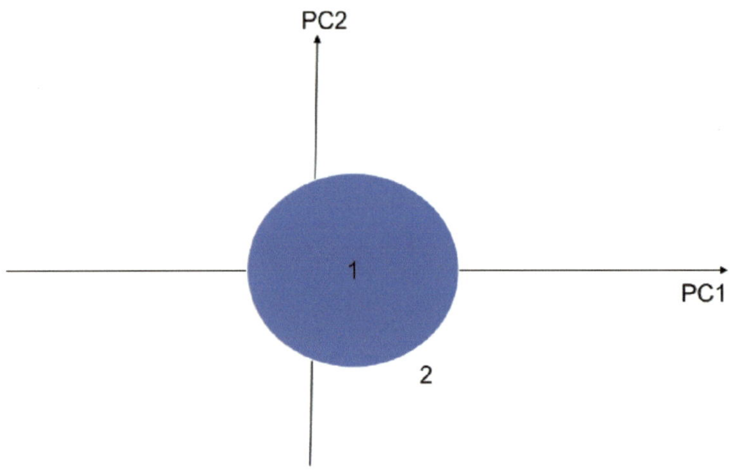

Figure 17.
Using the plot to control the process by keeping within a certain area of the PC1-PC2 space.

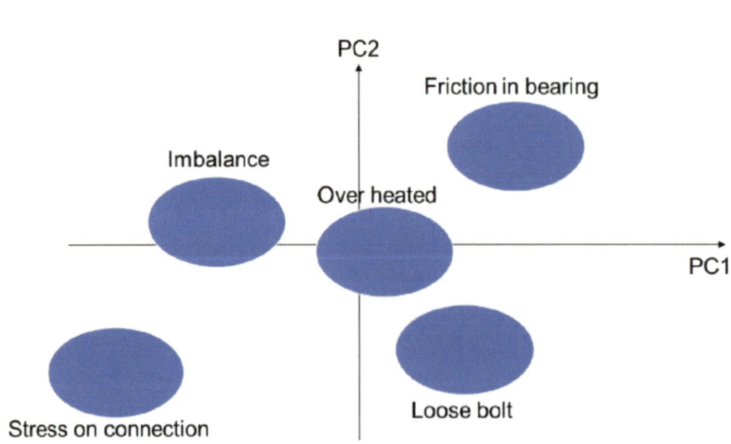

Figure 18.
Use of plots to classify different faults.

DOI: http://dx.doi.org/10.5772/intechopen.90741

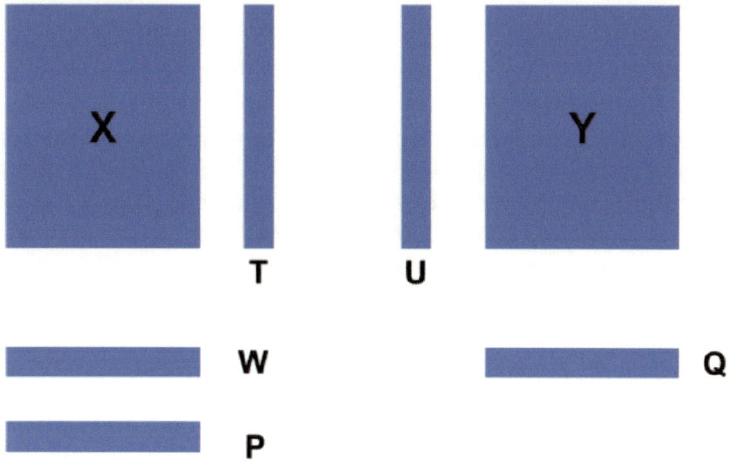

Figure 19.
The principles for PLS (partial least square) regression.

This can schematically be seen in **Figure 19** below.

U gives starting values for T, and T back to U iteratively. Interdependency. When the difference between two iterations is below a certain value, we take this as the result.

There are a number of versions of this. PLS2 general = all Y; PLS1 for each single Y; PCR also for each single Y, and no interactivity between Y and X (first X, then Y); PCR is often used by statisticians while PLS by application engineers normally.

The result from the PLS regression will be a ploynom. If only linear: $Y_1 = A + Bx_1 + Cx_2$. If also nonlinear: $Y_1 = A + B * X_1 + C * X_2 + D * X_1{}^2 + E * X_2{}^2$. If also interaction between variables: $Y_1 = A + BX_1 + CX_2 + DX_1{}^2 + EX_2{}^2 + FX_1X_2$. If we have more variables than two, we add X_3, $X_3{}^2$ and interaction between X_3 and the other variables, etc. These are used for prediction of Y_1. If you want to study several quality aspects using the same experiments, you add polynoms for Y_2, Y_3, Y_i in the same way, but with different constants of course.

2.7.2 Support vector machines (SVM)

In SVM we try to find the balancing point for different clusters and then try to distribute the different measured values as close as possible to one of these cluster balancing points. This is giving a similar type of clustering but is usually used for a big set of data where you want to find out how many clusters there might be. You can systematically test to have more or less clusters and see how the data fits from a statistical perspective into more or less clusters.

2.8 Adaptive control

In Narend S. Kumpati [26] Adaptive control using neural networks is presented. Since then MPC, both "fixed" and adaptive, have come to use in many applications in process industry. There is even a Journal of Adaptive Control and Signal Processing. In a recent number, April 2020, Merve et al. [27] discuss Improving transient performance of discrete-time model reference adaptive control architectures. This area is binding AI, modeling and control together.

3. Architectural structure

In **Figure 20** the structure implemented in the FUDIPO project (www.fudipo.eu) [28] with respect to different functions is outlined. In the chapter about the data structure Tieto has addressed different programs. These two are complementary. One is a set of commercial software linked into the Tieto HMI3 platform. Examples of the commercial tools are MatLab/Simulink for mathematical calculations and simulation, Hugin for Bayesian Network configuration and Dymola for Modelica implementation for simultaneous solver simulation.

In the second structure we have primarily open source programs like Red Node for configuration of the complete system, linking everything together. MatLab is replaced by Python and Simulink with OpenModelica, Dymola and these are then complemented by other simpler software for different functions. The idea is that you can test all functionalities together in the open source environment. If you have a smaller system you can configure and use this also for "the real case". If you have a bigger system demand you probably go for commercial software to also get support for the functions, and perhaps also make a service contract with someone who can support sustaining the system, and upgrading on a frequent basis, as the production plant is developing.

From this overview we can see that there are many possibilities with use of AI-tools, but it also takes some effort to understand which tools are useful to solve specific problems.

- The solutions must be **robust**. 100% of operational space must be covered in a reasonable way.

- Diagnostics must **detect** real faults, but **avoid** detect "false faults"

- **Autonomous** systems may be good, but you have to identify the boarders and **limits** and what are important functions to work with.

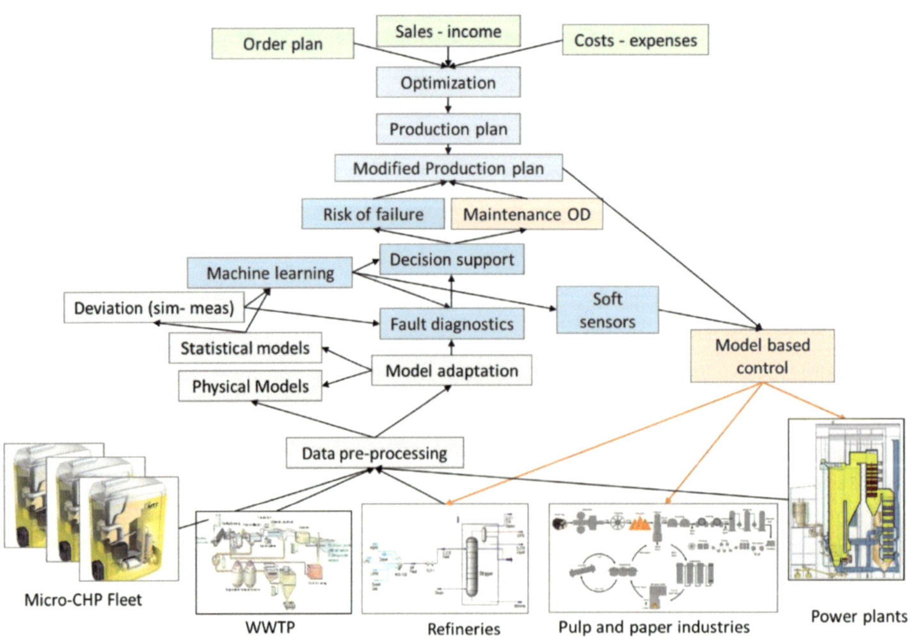

Figure 20.
Layout of a complete system where different level and functions are connected and integrated.

- Need to **define the problem** to solve!

- Optimization and adaptive systems and functions should include all important functions. To do so you also need to **vary the important variables**. You cannot train a system on constant values! Factorial design of "experiments" is then important.

- Many new tools are being accessible, but you **need to understand** how they work! Do not guess.

Acknowledgements

This project has received funding from the European Union's Horizon 2020 research and innovation programme under grant agreement No 723523.

Author details

Erik Dahlquist*, Moksadur Rahman, Jan Skvaril and Konstantinos Kyprianidis
Malardalen University, Vasteras, Sweden

*Address all correspondence to: erik.dahlquist@mdh.se

References

[1] Mosavi, A., Salimi, M., Faizollahzadeh Ardabili, S., Rabczuk, T., Shamshirband, S., & Varkonyi-Koczy, A. R. (2019). State of the art of machine learning models in energy systems, a systematic review. Energies, 12(7), 1301.

[2] Correia, F. M., d'Angelo, J. V. H., Almeida, G. M., & Mingoti, S. A. (2018). Predicting kappa number in a Kraft pulp continuous digester: A comparison of forecasting methods. Brazilian Journal of Chemical Engineering, 35(3), 1081–1094.

[3] Dahlquist E. Editor: Book “Use of process simulation in pulp and paper industry. Published by EU. Product of COST E 36. May 2008a. ISBN ISBN 978–91–977493-0-5

[4] Dahlquist, Erik (2008b) "Process simulation for pulp and paper industries: Current practice and future trend," Review Paper after Invitation. Chemical Product and Process Modeling: Vol. 3 : Iss. 1, Article 18. Available Open Source at: http://www.bepress.com/cppm/vol3/iss1/18

[5] Phatwong, A., & Koolpiruck, D. (2019, July). Kappa Number Prediction of Pulp Digester Using LSTM Neural Network. In 2019 16th International Conference on Electrical Engineering/Electronics, Computer, Telecommunications and Information Technology (ECTI-CON) (pp. 151–154). IEEE.

[6] Abbas, S., Khan, M. A., Falcon-Morales, L. E., Rehman, A., Saeed, Y., Zareei, M., & Mohamed, E. M. (2020). Modeling, simulation and optimization of power plant energy sustainability for IoT enabled smart cities empowered with deep extreme learning machine. IEEE Access, 8, 39982–39997.

[7] Karlsson, Christer P.; Avelin, Anders; and Dahlquist, Erik (2009) "New methods for adaptation to degeneration in process models for process industries," Chemical Product and Process Modeling: Vol. 4 : Iss. 1, Article 25. DOI: 10.2202/1934-2659.1127. Available Open Source at: http://www.bepress.com/cppm/vol4/iss1/25

[8] Lorencin, I., Andelic, N., Mrzljak, V., & Car, Z. (2019). Genetic algorithm approach to design of multi-layer perceptron for combined cycle power plant electrical power output estimation. Energies, 12(22), 4352.

[9] Karlsson Christer, Anders Avelin, Erik Dahlquist.:How to make better use of all the process data collected in process industry and power plants. 6th Eurosim congress on modeling and simulation, September 9-13, Ljubljana, Slovenia, 2007.

[10] Liu, Y., & Xie, M. (2020). Rebooting data-driven soft-sensors in process industries: A review of kernel methods. Journal of Process Control, 89, 58–73.

[11] Carlsson, A. Perspectives on the discovery of central monoaminergic neurotransmission. Annual Review of Neuroscience (Palo Alto, CA) 1987. 10. 19–40.

[12] Gattami Ather: Reinforcement learning for multi-objective and constrained Markov decision processes. 2019. Journal arXiv preprint arXiv: 1901.08978.

[13] Werbos Paul: A Menu of Design for reinforcement learning over time (p 67–95). In Miller Thomas, Sutton Richard and Werbos Paul (editors): Neural Networks for Control. 1990, Book ISBN 0–262–13261-3 MIT.

[14] Lindsten Fredrik, Thomas B. Schön, Andreas Svensson, Niklas Wahlström : Probabilistic modeling – Linear

regression & Gaussian processes February 23, 2017. Uppsala University Press.

[15] Ferreira S.L.C, R.E. Bruns, H.S. Ferreira, G.D. Matos, J.M. David, G.C. Brandao, E.G.P. da Silva, L.A. Portugal, P.S. dos Reis, A.S. Souza, W.N.L. dos Santos (2007) Box-Behnken design: An alternative for the optimization of analytical methods. Analytica Chimica Acta 597, 179–186.

[16] Skvaril Jan, Konstantinos G. Kyprianidis &Erik Dahlquist: Applications of near-infrared spectroscopy (NIRS) in biomass energy conversion processes: A review. Journal of Applied Spectroscopy Reviews, Volume 52, 2017 - Issue 8.

[17] Wold Svante, Michael Sjostrom, Lennart Eriksson: PLS-regression: A basic tool of chemometrics. Chemometrics and Intelligent Laboratory Systems 58, 2001, 109–130. www.elsevier.comrlocater chemometrics

[18] Weidl G.,Madsen A L, Dahlquist E. (2002) Bayesian networks for root cause analysis in process operation, European Journal of Operational Research, Special Issue on "Advances in Complex Systems Modeling".

[19] Weidl, G., Madsen, A.L. and Dahlquist, E. (2002a). "Condition Monitoring, Root Cause Analysis and Decision Support on Urgency of Actions". In Book Series FAIA (Frontiers in Artificial Intelligence and Applications), vol.87, A.Abraham et al. (Eds.), Soft Computing Systems - Design, Management and Applications, pp. 221–230. IOS Press, Amsterdam, the Netherlands

[20] Weidl G,Madsen A L, Dahlquist E (2002b) Condition Monitoring, Root Cause Analysis and Decision Support on Urgency of Actions, 2nd International conference on Hybrid Systems, Dec 1–4, Santiago,Chile

[21] Weidl G, Vollmar G and Dahlquist E (2003): Adaptive root cause analysis under uncertainties in industrial process operation, foundations of computer-aided process operations conference, USA, Florida, January 12–15, 2003

[22] Weidl G, Dahlquist E. (2002) Root cause analysis for pulp and paper applications, In Proceedings of 10th SPCI Control Conference, Pp 343–347, Stockholm, Sweden, June 3-5, 2002.

[23] Weidl G., Madsen A. And Dahlquist E.: Decision support on complex industrial process operations. Chapter no 18 p 313- 328 in the book Bayesian networks, a practical guide for applications. Editors Pourret O., Naim P. and Marcot B. John Wiley. ISBN 978–0–470-06030-8. 2008.

[24] Widarsson B, Karlsson C och Dahlquist.E: Bayesian Network for Decision Support on Soot Blowing Superheaters in a Biomass Fuelled Boiler, PMAPS, Sept 13–17,2004, Baltimore, USA.

[25] Wold Svante, Esbensen Kim and Geladi Paul: Principal Component Analysis. Chemometrics and Intelligent Laboratory Systems, 2 (1987) 37–52

[26] Narend S. Kumpati : Adaptive control using neural networks. (). In Miller Thomas, Sutton Richard and Werbos Paul (editors): Neural Networks for Control. 1990, Book ISBN 0–262–13261-3 MIT, p 115–142.

[27] Dogan K. Merve, Tansel Yucelen, Wassim M. Haddad, Jonathan A. Muse: Improving transient performance of discrete-time model reference adaptive control architectures. 27 April 2020. https://doi.org/10.1002/acs.3114. Journal of Adaptive Control and Signal Processing

[28] FUDIPO (2020) Description of an open platform based on Node Red for AI use in process industry, url: www.fud ipo.eu

Chapter 2

A Framework for Learning System for Complex Industrial Processes

Moksadur Rahman, Amare Desalegn Fentaye, Valentina Zaccaria, Ioanna Aslanidou, Erik Dahlquist and Konstantinos Kyprianidis

Abstract

Due to the intense price-based global competition, rising operating cost, rapidly changing economic conditions and stringent environmental regulations, modern process and energy industries are confronting unprecedented challenges to maintain profitability. Therefore, improving the product quality and process efficiency while reducing the production cost and plant downtime are matters of utmost importance. These objectives are somewhat counteracting, and to satisfy them, optimal operation and control of the plant components are essential. Use of optimization not only improves the control and monitoring of assets, but also offers better coordination among different assets. Thus, it can lead to extensive savings in the energy and resource consumption, and consequently offer reduction in operational costs, by offering better control, diagnostics and decision support. This is one of the main driving forces behind developing new methods, tools and frameworks. In this chapter, a generic learning system architecture is presented that can be retrofitted to existing automation platforms of different industrial plants. The architecture offers flexibility and modularity, so that relevant functionalities can be selected for a specific plant on an as-needed basis. Various functionalities such as soft-sensors, outputs prediction, model adaptation, control optimization, anomaly detection, diagnostics and decision supports are discussed in detail.

Keywords: learning system, soft-sensors, model predictive control, fault detection, isolation and identification, information fusion

1. Introduction

Despite recent economic growth, industrial plants are facing tremendous local and global competition. In order to maintain long-term competitiveness, industrial plants need to optimize their operation continuously for better quality, availability, flexibility and cost. As a consequence, industrial systems are becoming more and more complex due to the increasing coupling between highly nonlinear and stochastic subsystems or sub-processes. Often these systems include many control loops and operate under multiple operational constraints. Hence, the development of new methods and tools for optimal operation, monitoring and control of complex industrial systems is a matter of utmost importance. Rapid development of industrial automation, high-performance computing, artificial intelligence, machine learning, big data, cyber-physical systems, advance sensors, internet of things and industry 4.0, stimulated the industry-wide application of advanced methods and tools needed for optimal operation, monitoring and control. Although many advanced techniques for optimal operation, monitoring and control are already available and many more emerging day-by-day, the widespread use of these techniques within the industrial domain has been particularly limited [1–3]. There are numerous reasons identified to be accountable for the limited industry-wide application.

Although introduction of advanced automation could ensure better asset utilization, the enterprise must make sure that the newly available capacities are used effectively. Need for major infrastructure overhaul and resistance to change towards new systems that requires user's skill upgrade are two major issues that hindering the industrial application. The penetration barriers for technology niches are also quite high due to the fact that the industrial automation sector is occupied by only few multinational conglomerates. One can also blame the lack of pilot applications proving the robustness of these emerging techniques. Traditionally, advanced functionalities i.e. output prediction, optimal control, diagnostics and decision support, have been developed separately by utilizing different approaches and often with different model assumptions [4, 5]. Due to this segregated approach, the integration of different functionalities has been difficult and, consequently often neglected. However, each of these activities are closely related and cannot really be conducted individually on a isolated manner. For example, a fault in the system or a sensor failure can have a significant impact on the output prediction or control. Therefore, integration among different functionalities are essential. Due to their longevity, existing automation systems of large industrial plants mostly date from the past few decades. Often replacing these automation systems completely may not be economically viable. Hence, there is a need for an architecture that will allow easy integration of advanced functionalities with both existing and state of the art automation platforms of complex industrial systems. In order to get a structured view on industrial automation and how optimal operation, control and monitoring can leverage the benefit from such systems, a brief overview of the automation pyramid as presented in **Figure 1** can be helpful.

So what exactly is the automation pyramid? It is a graphical representation of the different technological levels of automation in a industrial plant that allow communication among different technologies within each level as well as between the different levels. The framework is defined by International society of automation (ISA) within ISA-95 that is the international standard for the integration of enterprise and control systems [6]. The first level of the pyramid, commonly referred as field level, consists of devices, sensors and actuators that are used to measure different process parameters such as flow, temperature, pressure or concentration and to manipulate different process variables via different mechanical, hydraulic, pneumatic, electrical or electronic devices. The next level, referred as control level, comprises distributed control or logical devices such as the programmable logic controller (PLC), distributed control system (DCS) or proportional–integral–derivative (PID) controller. The control level uses these control and logical devices to control or regulate the devices in the field level that actually perform the physical work. They receive inputs from all field level sensors to make decisions on what actions need to be taken by the filed level actuators to meet the predefined set-points.

An example of separation between field and control level is presented in **Figure 2**. Suppose the level of a tank need to be controlled to a predefined level in a industrial plant. A level sensor measures the level of the tank in real time and transfers this information to a PID controller. The controller adjust the position of a

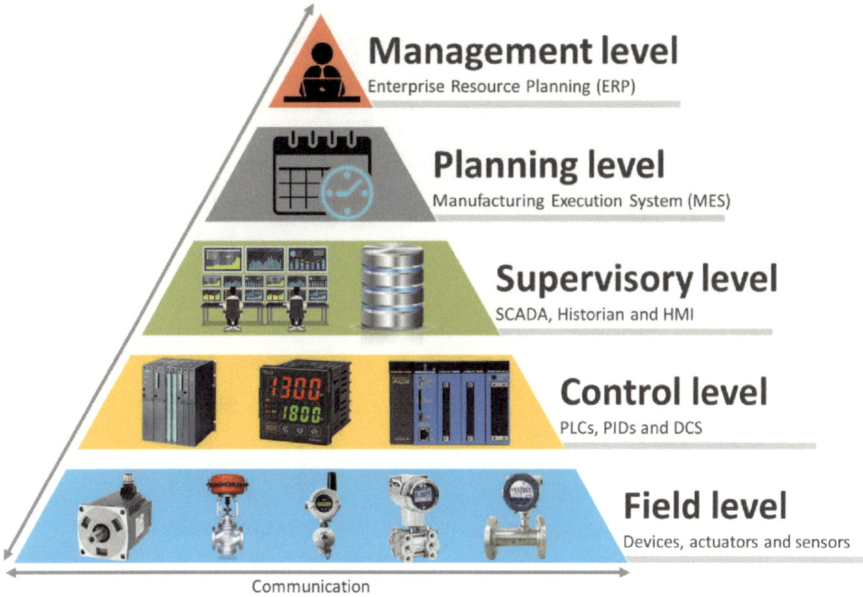

Figure 1.
The automation pyramid of a typical industrial plant.

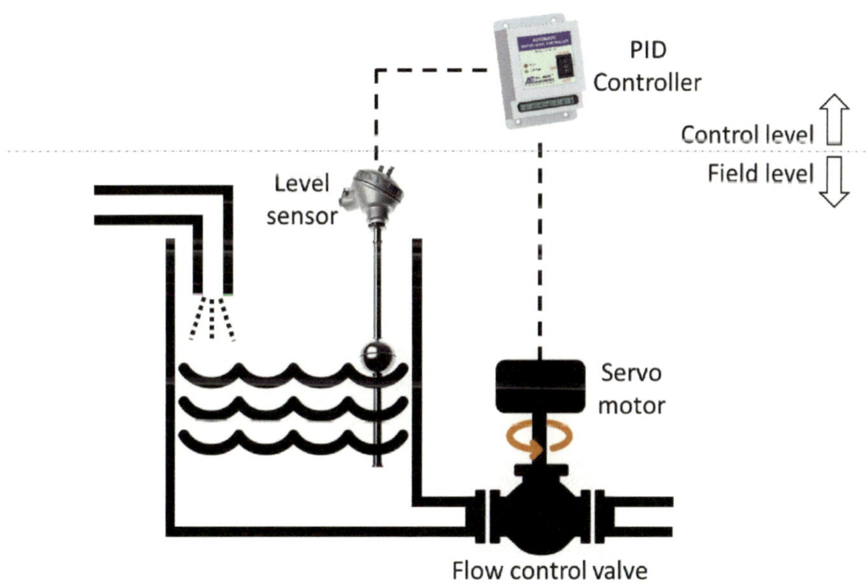

Figure 2.
Example of segregation between field and control level.

flow control valve by means of servo motor. In this scenario, the tank, level sensor, flow control valve amd servo motor belong to the field level and the PID controller belong to the control level. The supervisory control and data acquisition (SCADA) system correspond to the third or supervisory level that is used to access data and control multiple systems from a single location. The SCADA gathers information from all the subsystems and sub-processes of a industrial plant, carrying out necessary analysis and supervisory control and displaying the information in a logical and organized manner (**Figure 3**). For example, supervisory control algorithms calculate set-point values for the field level controllers (PIDs and PLCs).

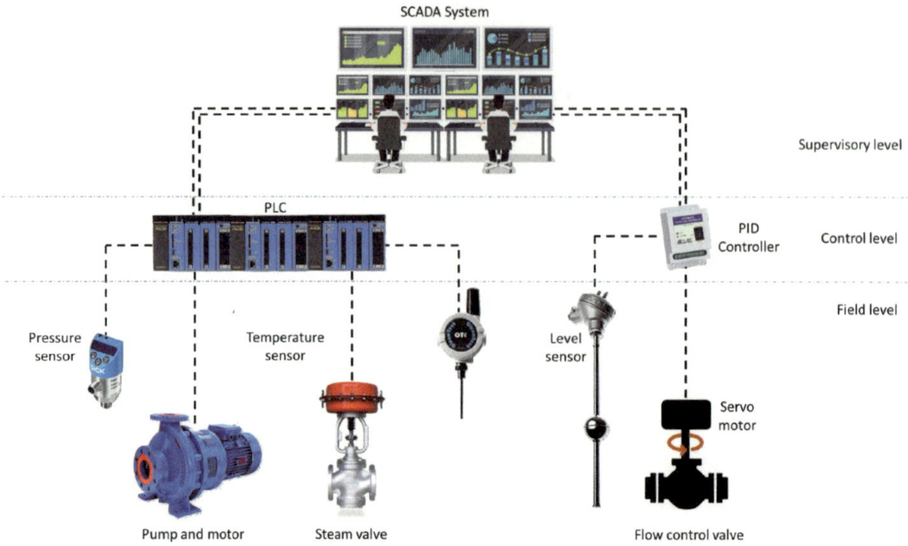

Figure 3.
Relation between supervisory and control level.

Human-machine interfaces (HMI) and workstations are also included in this level. Often this level uses process historians or databases, software programs that store the historical process data. Hence, it is possible to study the patterns and find abnormalities in the processes by the experts or automated programs.

The fourth or planning level includes the manufacturing execution system (MES). MES is used to monitor the entire production process in a industrial plant from the raw materials to the finished goods. A MES performs many activities including production scheduling, management of production equipment and labor, quality control, performance analysis and maintenance management. MES provides a holistic view on the production process and allow planners to make decisions based on the available information. At the top or management level, enterprise resource planning (ERP) systems are placed to establish plant scheduling methods and material management features. ERP is a integrated software that allows organizations to monitor day-to-day business activities from manufacturing, to sales, to procurement, to accounting, to project management, to risk management, and many more. A complete ERP package typically includes enterprise performance management tool that is used to plan, budget, predict, and report on an organization's financial results. To be inline with the fourth industrial revolution widely known as industry 4.0, the structure is becoming more of a pillar than a pyramid; this enables enhanced communication beyond existing layer boundaries as well as cloud computing functionality [7]. Irrespective of its structure, advanced methods and tools can bring benefits to all levels of the automation hierarchy by providing solutions for process monitoring, coordinated process control, integrated planning and scheduling of man, machine and materials through better decision support. However, a pyramid structure is chosen here due to its simplicity and relevance.

Typically, the process components are designed to meet the operational objectives that are essential for the optimal and economic operation of the plant. Nevertheless, in reality, the process variables encounter both arbitrary and sustained deviation from their targets due to external disturbances, inherent variability and uncertainties. This is where the control system comes into play, by actively manipulating the process to ensure stable operation of the plant while keeping the product quality and specification within the target. Due to their simplicity and robustness, more than 90% of all industrial control loops are based on PID controllers [8]. PIDs show superior performance as regulatory control of uni-variate problems, i.e. in regulating flow, temperature, pressure, level, and other variables. In principle, a PID evaluates the one-and-only process variable, decide if it is acceptable or not, and takes corrective measures if necessary. This scheme works well for control problems with only one variable or with several variables that can be manipulated independently. Despite their widespread usage, PIDs have multiple drawbacks when it comes to supervisory control of multivariate industrial processes with high level of non-linearity. Therefore, multivariate control techniques are particularly essential for supervisory control, whereas PIDs can still be used for uni-variate regulatory controls under a supervisory control loop. Different model-based and model-free multivariate process control techniques are widely studied by the research community. In particular, model-based control is widely used by the industry and has demonstrated an excellent track record [9]. However, advanced control concepts that depend on process models to maneuver the plant are prone to slow deterioration. Hence, model adaptation over time is essential to ensure optimal control of the plant.

Apart from a robust control scheme, fault diagnostics also have an important role in ensuring the optimal operation of a plant. In particular, soft faults and slow deterioration of process components over time reduce the nominal production capacity of a plant. It is often difficult to detect such faults just by looking at the process variables, and they frequently remain unnoticed until the problems become severe or lead to an unwanted plant shutdown due to component breakdown. These faults and deterioration can also affect the control system negatively and disturb the process stability.

A fault diagnostics system can be beneficial for a processing and energy plant in numerous ways. Early detection of process, equipment or component faults or deterioration can provide decision support for operators, engineers and managers at different levels, i.e. DCS, computerized maintenance management system (CMMS), MES and ERP. As a result, the operation of the plant, along with maintenance, production and inventory planning can be improved. For example, an early indication of a developing fault can provide decision support by initiating one or more suggested actions that the control system or plant operator can perform to prevent the fault development. If prevention is not possible, then early detection of such deterioration can provide an indication of the remaining useful life (RUL) of the affected component that, in turn, can provide an indication of when maintenance is needed. Once a maintenance action is planned, that can initiate procurement of the required spare parts and adjustment of the production plan based on necessity.

To achieve such cross-platform functionality, there is a need for an integrated framework for optimal control, diagnostics and decision support for the complex industrial systems. The framework needs to be generic enough to accommodate different systems with different levels of complexity. This is also necessary to cover the broad range of systems that can utilize such a framework, starting from single or multiple assets within a plant to a large fleet of assets spread over a large geographical area.

2. Framework for generic learning system

For better resource utilization, product quality and process efficiency, supervisory system of a modern industrial plant need to perform various activities including outputs prediction, model adaptation, control optimization, anomaly detection, diagnostics and decision support. In order to enable the supervisory system to perform all these activities efficiently, we propose a framework for the generic learning system that can be integrated to the supervisory system of the complex industrial plants. The architecture is very flexible and modular, so that relevant functionalities can be selected for a particular case study on a plug and play basis.

The framework is developed in a way so that it can be retrofitted to the existing automation platforms of a complex industrial plant. This provides solution to one of the major barrier that hindering the widespread use of modern techniques emerging for optimal operation, control and monitoring of complex industrial plants. Since the framework allows easy integration of the learning system to the existing automation platforms, the need for extensive infrastructural modification and skill development reduces drastically. The overall framework for the learning system is presented in **Figure 4**. The learning system is placed in the supervisory level of the automation pyramid. However, it actively supports decision making in both planning and management level. The learning system need process data as inputs to perform systematic computational analysis. The data are gathered from the process historian or the database. The first step before performing any analysis is data assurance that includes outlier removal and noise reduction by means of various data filtering techniques. Subsequently, different advanced analysis are performed on the data ans the results are written back to the process historian. Firstly, trend analyses are carried out on the data to identify any patterns in the process parameters. Important process outputs are predicted by using physics-based and data-driven process models. Advanced control optimisation techniques are applied to calculate most optimal set-points for the low level regulatory controllers. Different physics-based and data-driven anomaly detection and diagnostics algorithms are also applied in order to find process abnormalities and faults. As a final step, results from all these analysis are used to provide robust decision support with the help of information fusion techniques. Moreover, the architecture allows integration of state-of-the-art sensors for measuring feedstock properties, different process parameters that are needed to better operation and control of complex industrial processes. Human–machine interface (HMI) are also provided for the visualization and further analysis. This is a key part of the framework that the users i.e. operators, engineers and managers will directly interact with. Hence, the HMI need to be designed such a way that it is user friendly and useful for them. This will determine

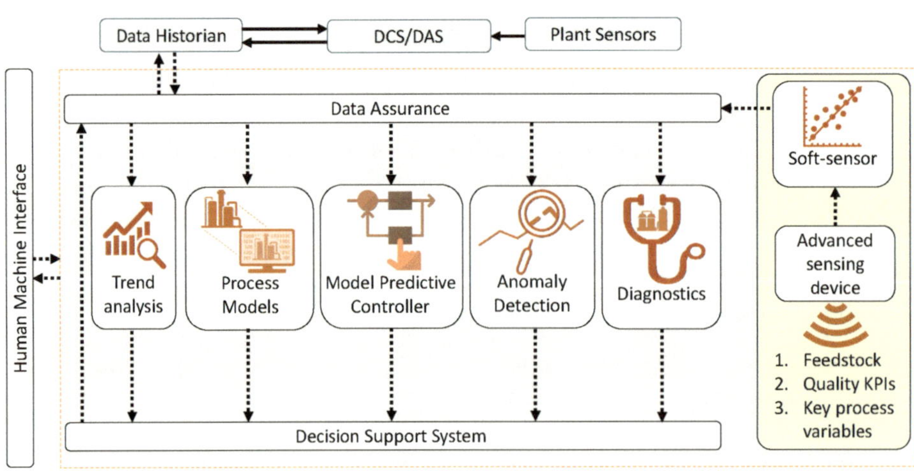

Figure 4.
Integrated framework for the learning system of complex industrial systems.

if the learning system will actually be used or not. Hence, the user need to be involved in the process of designing the HMI.

The different modules of the framework is discussed in the following sections.

2.1 Data assurance

Data assurance refers to different data preprocessing techniques that ensures accurate, reliable and meaningful analysis. The data preprocessing steps typically includes data cleaning, smoothing, scaling and grouping or binning [10]. Data cleaning particularly refers to detection and removal or replacement of outliers and missing data. Data smoothing on the other hand refers to removing noise from the data. Here in the data assurance layer, outliers in the data are detected and different noise reduction techniques are applied to refine the data. So what exactly meant by outliers in the data? An outlier is a measurement that differs significantly from other measurements in a dataset. The definition is quite broad in nature, allowing the analyst to decide on the boundaries that separate measurements to be considered as outliers from normal. Typically, outliers represent only a small fraction of the data and they do not follow the inner relationships present among different process variables. Very simple example of a outlier in a dataset is shown in **Figure 5**.

There are many readily available techniques that can be used for outlier detection. As each dataset is different, there are no common methods that can be applicable to every dataset. Rather, an analyst or domain expert, must examine the raw measurements and decide whether a value is an outlier or not and what methods can be used to detect it. Typically, statistical methods that are widely used for detecting outliers corresponding to significantly extreme values are mean and standard deviation, and median absolute deviation method. According to the mean and standard deviation method, a measurement is labeled as outlier if it more than three standard deviation away from the mean value. However, as both the mean and the standard deviation are sensitive to outliers, this method can be problematic in some cases. A rule of thumb is that for normally distributed dataset, mean and the standard deviation is a better choice. However, if dataset is not normal, the median absolute deviation can be used. In this case, absolute deviation from the median

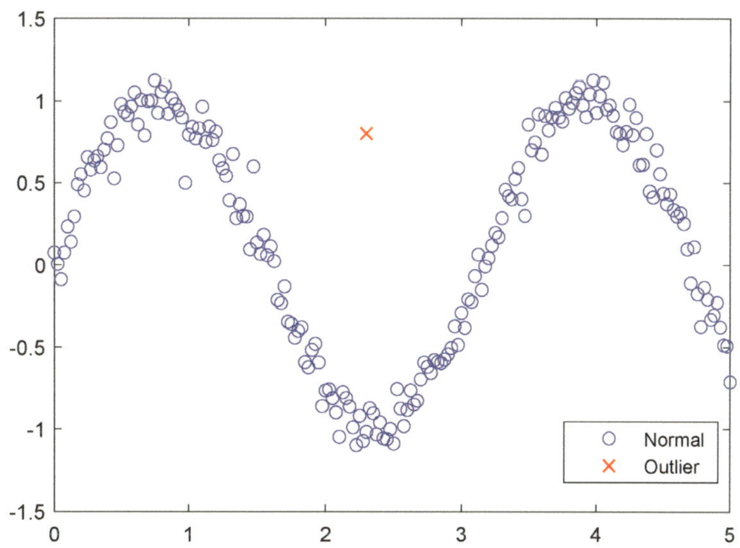

Figure 5.
Example of a outlier in dataset.

value is used instead. Normally, historical data or a window width is used to apply such techniques for time-series sensor readings.

Process data are subjected to noise. Hence, different noise reduction techniques are needed before performing different analysis on the data. A typical example of noisy sensor data and output data after smoothing is presented in **Figure 6**. However, one need to be careful when applying different noise reduction techniques. Too much data smoothing can filter out many useful information in data that can be important for different data analysis techniques. For noise reduction, time domain filters i.e. moving average filter, moving median filter, Savitzky–Golay filter, artificial neural network (ANN) and local regression smoothing, and frequency domain filters i.e. low pass, high pass and band pass filter are well known data smoothing techniques. Among these, moving median filter is simple but most powerful data smoothing technique. It particularly useful for eliminating unwanted noise from the time-series sensor data. Two of its main advantages are (a) median filtering preserves sharp edges and (b) it is very efficient for smoothing of spiky noise. However, presence of outlier in the data can effect the outcome of a moving average filter. Hence, such smoothing techniques need to be used in addition to the outlier removal step. The mathematical expression for moving average filter is presented in Eq. (1).

$$\overline{y_n} = median(x_{n-k}, \dots \dots, x_n, \dots \dots, x_{n+k}), \tag{1}$$

where, the window width is $(2k + 1)$ is one of the major tuning parameter for this filter. x_n and y_n are the nth sample of the input and output sequences. The filter is fast in terms of computational time and not really difficult to implement.

2.2 Trend analysis

Trend analysis, also known as temporal reasoning, is a very important tool for diagnostics and decision support in complex industrial processes. Typically humans are very good at detecting patterns and trends in historial data by visual inspection. This is the backbone of any manual supervision and monitoring strategy of a industrial plant. However, detecting pattern by a automated algorithm is a difficult

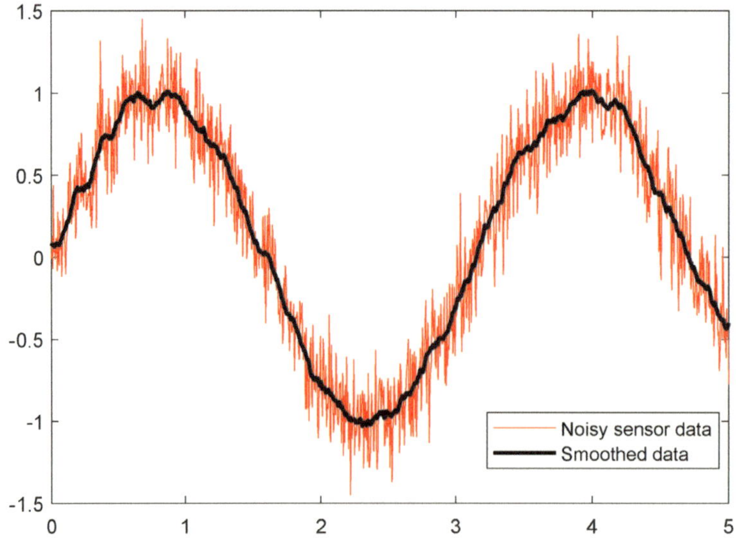

Figure 6.
Example of noise reduction.

problem. Generally, trends are difficult to quantify due to the non-deterministic artifacts and background noise that typically presents in measurements. With the fast evolution advanced data analytics, it is possible to identify meaningful trends in time-series data that can be used in automated process monitoring, diagnostics and decision support. Numerous methods exist for performing trend analysis, ranging from the relatively simple methods such as linear regression to more complex methods such as Mann-Kendall and Spearman's rho tests to identify nonlinear trends in time-series data.

In this work, the aim of trend analysis is to extract useful trends from the historical process data so that it can be used as a prior knowledge to the decision support system. Moreover, visualizing automated trend information to the operators can improve their reaction time to any unwanted process drifts and abnormalities. The trend extraction methods can be either qualitative or quantitative in nature. Qualitative methods has gained upper hand over quantitative methods on extracting high-level knowledge from the process data [11]. Hence, Qualitative methods are better suited as the input needed by the decision support system. As the name suggest, qualitative trend analysis attempts to provide qualitative patterns from the historical data by fetching the underlying short and long term trends.

The most common way of representing qualitative trends in data is the use of seven primitives (**Figure 7**) with constant signs of first and second derivatives, originally developed by Janusz and Venkatasubramanian [12]. However, Charbonnier and Portet [4] proposed a self-adaptive qualitative trend analysis method by utilizing the first three primitives: steady (A), increasing (B) and decreasing (C). The method is further developed and applied to many industrial applications [13, 14]. The method divides online process data into linear segments to extract underlying trends. Real-time self-adaptation of the tuning parameters are performed to detect the variations and artifacts presents in the data. An example of trend fitting by using self-adaptive qualitative trend analysis approach is shown in **Figure 8**.

2.3 Process models

Process models, also known as mathematical models or simply models, are abstractions of real processes or systems that are used to characterize behavior of the processes or systems, given that the inputs are known [15]. Typically, such models can be used for prediction, control, fault detection, etc. Depending on the the modeling approach, models can be widely classified as first-principle, empirical

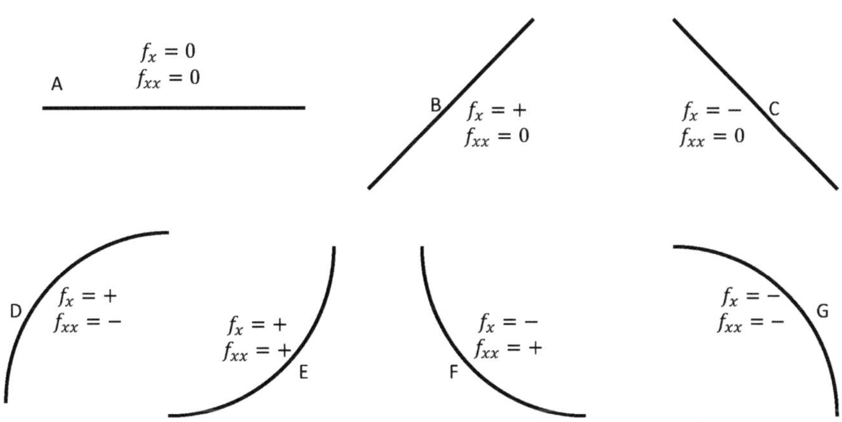

Figure 7.
Most common primitives.

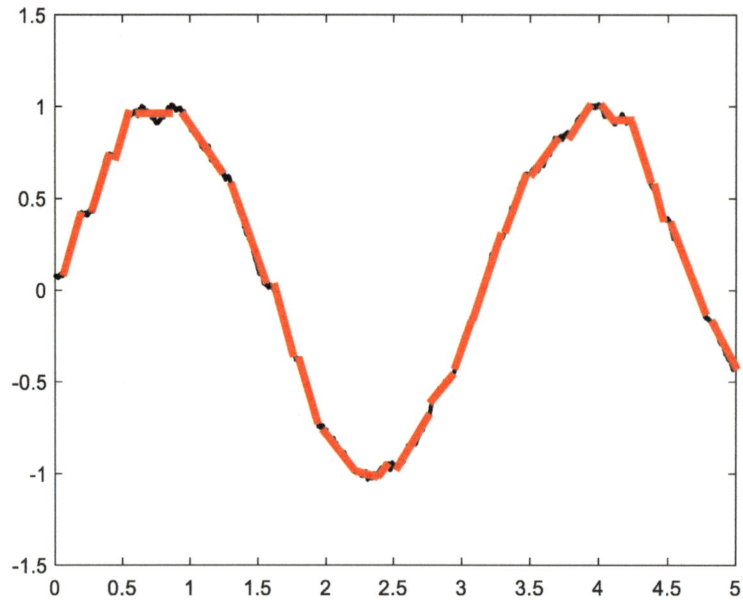

Figure 8.
Trend fitting example by self-adaptive approach.

and hybrid models. First-principle models, also known as White-box models, are based on mathematical equations that explain the physical, chemical or other basic principles. On the other hand, empirical models are based on data or observations that occurred in the past. These kind of models are also known as black–box models, as they relate inputs to outputs without revealing any knowledge of the internal working principles. Hybrid models are obtained by combining both first-principle and and empirical approaches. Process models can be further categorized into steady-state and dynamic models. A steady state-model is based on the assumption that the system is in equilibrium, and is thus time-invariant. This type of model is useful for system design but not for control applications. On the other hand, a dynamic model accounts for the time-dependent changes in a system and can therefore capture the transient behavior of the system. At the end the selection of the modeling approach and model types entirely depends on the purpose of the models. In this work, process models are used for output prediction, control and diagnostics purposes. Both first-principle and and empirical models are considered in order to take advantage of the benefits and avoid drawbacks associated with them.

The complexity of process models can vary widely, from simple conceptual models or linear models to high-fidelity computational fluid dynamics (CFD) models, depending on the purpose of the modeling work. Added model complexity almost always comes with a cost of high computational time that may impede the online application. There is no common modeling approach that fits the needs of all applications. Rather, the modeling approach for each application needs to be selected on the basis of the relevant purpose.

Typically, all theoretical process models are based on general conservation principles i.e. mass, energy and momentum balances, chemical kinetics, physical phenomenon such as friction, diffusion, compaction, and/or component specification. Most of the modeling work start with the assumption that some property is conserved within the system boundary. The general conservation principle can be formulated as Eq. (2).

$$\begin{matrix}\text{Rate of} \\ \text{accumulation}\end{matrix} = \begin{matrix}\text{Rate of} \\ \text{inflow}\end{matrix} + \begin{matrix}\text{Rate of} \\ \text{generation}\end{matrix} - \begin{matrix}\text{Rate of} \\ \text{outflow}\end{matrix} - \begin{matrix}\text{Rate of} \\ \text{consumption,}\end{matrix} \tag{2}$$

Assuming the physical property under consideration is $X(t)$ where t is the independent variable for representing time. If the rate of change in inflows and outflows are denoted by $\dot{x}_{in}$ and $\dot{x}_{out}$; and rate of change in generation and consumption are denoted by $\dot{g}(t)$ and $\dot{c}(t)$ within the system boundary shown in **Figure 9**, a general balance law can be written in following form as in Eq. (3).

$$\frac{dX(t)}{dt} = \dot{x}_{in} + \dot{g}(t) - \dot{x}_{out} - \dot{c}(t), \tag{3}$$

This general balance law can be adapted for all three fundamental quantities: mass, energy and momentum, in order to model different industrial processes.

Reaction kinetics modeling another important aspect of process model development. For simplification let us consider a chemical reaction (Eq. (4)), where product C is formed by the reaction between reactants A and B.

$$A + B \rightarrow C, \tag{4}$$

Typically, the rate of reaction for a chemical reaction depends on principal quantities like temperature, pressure, and composition. For the sake of simplicity, let us assume that the effect of pressure is negligible in this case. Hence the rate of reaction r_c can be expressed as Eq. (5).

$$r_c = k * C_A{}^{\alpha} C_B{}^{\beta}, \tag{5}$$

where C_A and C_B are the concentration of reactants A and B, and k is the reaction rate constant. α and β are the exponents of concentration corresponding to each reactant. The rate constant k and the exponents α and β must be determined experimentally by monitoring how the rate of a reaction changes as the concentrations of the reactants are changed. The reaction rate constant k is temperature dependent and generally expressed according to the Arrhenius equation (Eq. (6)),

$$k = A_i * e^{-\frac{E}{RT}}, \tag{6}$$

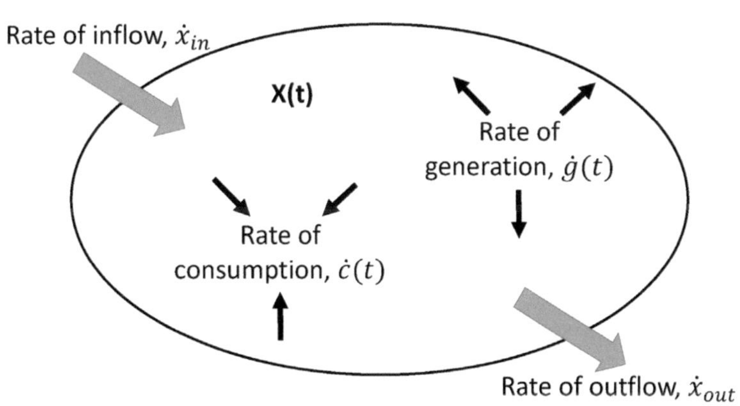

Figure 9.
An example of a system boundary within which physical properties are considered to be conserved.

Where, A_i, E, R and T pre-exponential factor, activation energy, universal gas constant and temperature of the reaction, respectively.

Depending modeling purpose the process models can also include more complex physical phenomenon such as diffusion, friction, compaction, porosity, velocity etc. A first-principle model usually consists of the three types of equations i.e. algebraic equations (AEs), ordinary differential equations (ODEs), and partial differential equations (PDEs).

Typically, dynamical systems are described by differential equations. Often a set of AEs are solved to find numerical solution of a set of ODEs. Generally, PDEs are used to describe processes with distributed parameters [16]. Partial derivatives with respect to both time and space resulted in models that computationally expensive to solve. Often lumped approximation is considered by assuming infinitesimally small continuous stirred tank reactors (CSTRs) in series. By assuming ideal mixing, it is possible to avoid changes of parameters in space inside a infinitesimal CSTR. Consequently, it is possible to model the process by using differential-algebraic equations (DAEs). DAEs are commonly solved by using various numerical methods. Many dynamic system modeling tools use their own solvers for this purpose. For example, One of the popular solver used by OpenModelica and Dymola is DASSL. The basic principle of DASSL is not unique, it replaces the derivative part with a difference approximation and solve the resulting system of equations with a Newton method [17]. However, great care in parameter initialization is necessary to ensure numerical convergence or fast convergence.

2.4 Model predictive control

Model predictive control (MPC) refers to a range of control algorithms for feedback and feed-forward control based on the receding horizon philosophy, where a set of optimal control moves are calculated according to the prediction of future behavior of the plant based on a process model. Using a process model, the MPC optimizer is able to estimate the consequence of past inputs on future outputs. As presented in **Figure 10**, at every control step, the MPC attempts to optimize future behavior of the plant by evaluating future sequential control moves over the prediction horizon. The controller then only executes the first step of the previously evaluated optimal control moves. The entire process is repeated again before the next control move.

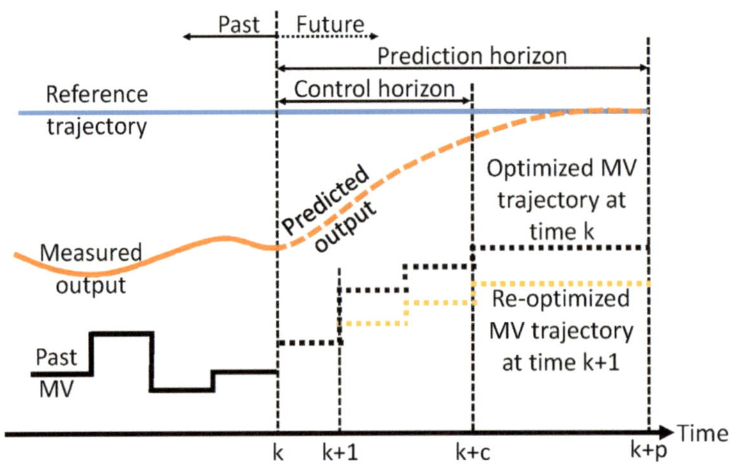

Figure 10.
Schematic representation of model predictive control. MV: Manipulated variable.

MPC provides superior performance particularly for processes that have multivariate interactions between inputs and outputs, which is a common traits of complex industrial systems. We argue that for highly complex processes MPC alone might not be the solution; particularly for processes where feed-stock properties varies unpredictably due to natural variation. In such cases, feed-forwarding the feed-stock variation to MPC will provide tighter control of the process. The scheme for a feed-forward MPC concept is depicted in **Figure 11**, where the feed-stock properties are feed-forwarded along with plant measurement to the MPC. A process model utilizes these information to make better prediction about the future outputs. The MPC optimizer computes the optimal control moves by solving a constrained finite-horizon optimization problem in which the cost functions make use of model predictions. The operational constraints are incorporated in the optimizer to ensure compliance. Additionally, the MPC also uses feedback to compensate for inaccuracies in the model and ensure convergence.

In reality, the cost function is a mathematical expression that is either minimized or maximized to find a best solution among all possible feasible solutions. Here, the cost function is expressed to find a sequence of incremental manipulated variables (MVs) over a control horizon of c samples, as presented in Eq. (7). The cost function minimizes a weighted sum of future squared errors of the outputs $y(k+i)$ and a weighted sum of increments in the sequence of MVs $\Delta u(k+i)$, while limits for Mvs and limits for predicted process variables are considered as a form of constraints in Eq. (8).

$$f(k) = \sum_{i=1}^{p} \|\Gamma e(k+i)\|_2^2 + \sum_{i=0}^{c-1} \|\wedge \Delta u(k+i)\|_2^2, \tag{7}$$

subject to constraints,

$$\begin{aligned} &y_{min} < y(k+i) < y_{max} && \forall i\ [1, p] \\ &u_{min} < u(k+i) < u_{max} && \forall i\ [0, c-1] \\ &\Delta u_{min} < \Delta u(k+i) < \Delta u_{max} && \forall i\ [0, c-1] \end{aligned} \tag{8}$$

In this minimization, the future errors $e(k)$ are calculated over a prediction horizon of p samples according to Eq. (9),

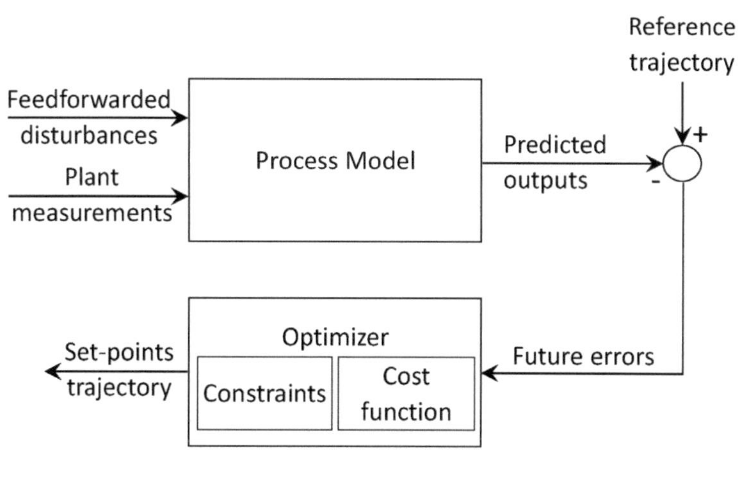

Figure 11.
Scheme for feed-forward MPC.

$$e(k) = s(k) - y(k), \tag{9}$$

where, $s(k)$ denotes the reference set-point trajectories. For outputs prediction a dynamic model can be used, or if an observer based formulation is utilized then a reduced order state-space model identified from the dynamic model, or linearised model from the step tests of a real plant can also be used.

2.5 Anomaly detection

Anomalies are the unusual, unexpected, abnormal patterns in a signal or a process variable. The term anomaly comes from the Greek word "anomolia" that means uneven or irregular. So how anomalies differ from faults? Faults are unexpected malfunctions in one or more components of a process that are not a failure or breakdown. However, faults may result in failures or catastrophic breakdowns if not resolved in time. On the other hand, anomalies only tell us there might be something abnormal with the system or a signal but it not necessarily means there is a fault in the system. Anomalies can occur due to many reasons other than faults. Maybe it is hard to accept but real systems are continually anomalous in many ways. Interestingly, anomalies can be positive or negative in nature depending on the context and interpretation [18]. Due to its nature, anomaly detection creates significant noise. However, detection of such abnormal conditions in the process can assists the operators on decision making so that they can react in time to avoid or correct the situations associated with them. Here, for the decision support system, anomaly detection is an additional source of information that will assist in robust decision making. We will also take the opportunity to distinguish between outlier detection and anomaly detection functions as we use both of these techniques in our framework. In outlier detection, we detect and remove or replace data that are either missing or illegitimate (e.g. a negative flow-rate) or very far away from rest of the data. In anomaly detection, we detect abnormal pattern in data and forward this information to the decision support system.

According to the general failure mode curve (**Figure 12**), a new machine runs with good health condition for some period of time. Then it reaches a point H where

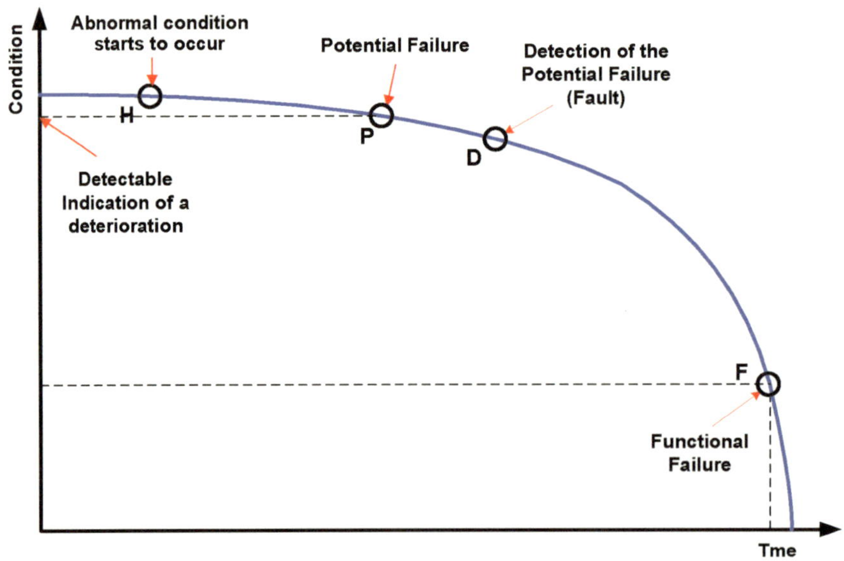

Figure 12.
Equipment failure mode diagram (adapted from [19]).

degradation starts to occur due to some damage-causing conditions. Point P represents the time where potential failure is recognized. The degradation progresses and then reaches a point where it can be detected. In general, the abnormal condition between P and F falls within the detectable range. The range between P and D refers to anomaly whereas between D and F refers to a fault [19]. Anomaly can also be a discrete event causing a rapid shift in measurement changes [20]. The goal of anomaly detection is, therefore, to detect the potential failure as early as possible.

Anomaly detection is extensively studied within many different application areas including credit card fraud detection, finance, cyber-intrusion, network monitoring, and many industrial plant monitoring [21, 22]. The simplest form of industrial anomaly detection technique can be as simple as logging an alarm if a sensor reading drifts away from a predefined upper and lower boundary. However, there are quite many anomaly detection techniques explored by researchers; which can be broadly categorized in three groups: (a) statistical techniques i.e. principal component analysis (PCA), histogram, Gaussian mixture models, Gaussian Kernels, etc., (b) cognitive techniques i.e. expert systems, finite state machine, etc., and (c) machine learning techniques i.e. clustering, classification, etc.

Anomaly detection is an important step in the process of fault diagnostics, and can be performed using measurement deviations or residuals as illustrated in **Figure 13**. A threshold-based detection or a binary logic can be applied. According to the threshold-based anomaly detection, the residuals should be very close to zero or lie within the threshold when the system is running in a normal operating condition and at least one residual should deviate noticeably from zero when an anomaly occurs. As a threshold, a Gaussian distribution of the residuals is often assumed in order to take into account variations due to measurement uncertainties. In the case of the binary logic, the residual is considered as a signal which is zero when the system is functioning properly and different to zero when some abnormal behavior is observed.

There are a variety of methods available for anomaly detection starting from the conventional model-based or statistical approaches to the more sophisticated machine learning techniques. Model-based methods rely on system models combining the theoretical knowledge with the test or actual performance data. When an abnormal condition (or a discreet fault event) occurs somewhere in the system, it produces deviations in measurements from their expected reference values. An accurate system modeling followed by a robust residual generation and proper threshold selection is critical. Machine learning techniques usually treat the anomaly detection task as a pattern recognition problem. The algorithm tries to learn a decision boundary from

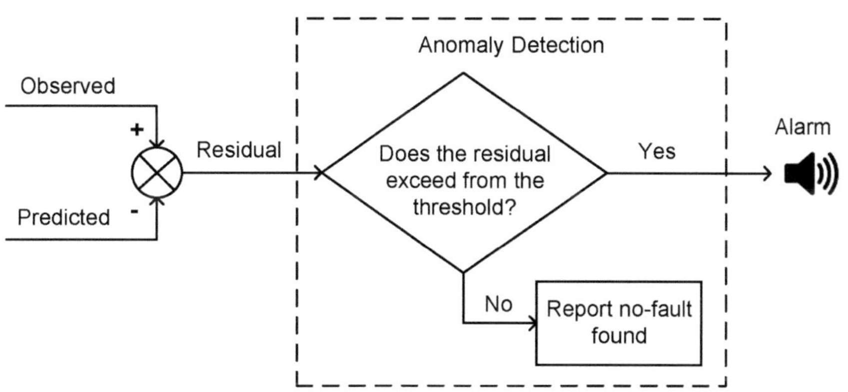

Figure 13.
Anomaly detection schematics.

the training data (i.e. the normal data). The detection accuracy can be evaluated using the standard detection decision matrix as presented in **Table 1**.

In machine learning algorithms anomalies can be detected in a supervised or unsupervised way. In the former case, labeled data is used for training. The labels can be binary, e.g. yes/no, one/zero, normal/abnormal, and fault/no-fault. ANN, support vector machine (SVM), and K nearest neighbor (KNN) are examples of the widely used supervised classification algorithms. For the unsupervised case, the normal and abnormal classes are distinguished based on their similarity using distance or density functions. Hierarchical clustering (HC), self-organizing map (SOM), K-means and K-medoids are some of the common unsupervised clustering algorithms.

In ANNs, a fault detection task is considered as pattern recognition. During training sample patterns of the two classes are feed into the network and the network tries to recognize the patterns based on their corresponding output labels. Among ANNs, an autoassociative neural network (AANN) is more suitable for an anomaly detection [23]. First, the model is trained on a normal data as input and output (**Figure 14**). For the normal input data, the difference between the model output and the target output will be close to zero, while for abnormal input patterns, at least one of the output residuals will deviate noticeably from zero.

Actual	**Predicted**				
	Abnormal	**Normal**	**Total**		
Abnormal	True abnormal	False normal	K1 + K2	Detection rate	Missed detection rate
	K1	K2		K1/(K1 + K2)	K2/(K1 + K2)
Normal	False abnormal	True normal	K3 + K4	True normal rate	False alarm rate
	K3	K4		K4/(k3 + K3)	K3/(K3 + K4)
Total	K1 + K3	K2 + K4	K1 + K2 + K3 + K4	Detection accuracy	
				(K1 + K4)/ (K1 + K2 + K3 + K4)	

Table 1.
Detection decision matrix.

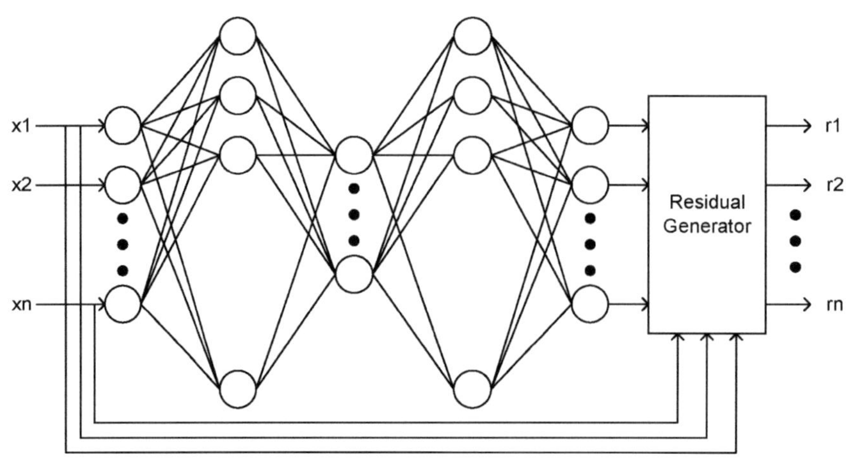

Figure 14.
Architecture of an AANN for anomaly detection.

According to KNN algorithms, anomalies are data points located farthest away from the normal data points or in low-density regions if weighted distances are considered (see **Figure 15**). After estimating all distance values, they need to be sorted in descending order. Anomalies are data points with the largest distance values. Then, the test data points that fall in the top n% distance range are considered as anomalies, where n is user defined value. The Euclidean function is the most convenient distance function in KNN.

A support vector machine is another type of supervised learning classifier. It is a binary classifier in its nature that separates two different classes by maximizing the margin between them. If one of the classes to be distinguished is taken as positive the rest of the class will be considered as negative. The classifier will, therefore, learn a boundary to separate the positive and negative classes as illustrated in **Figure 16**. The purpose of the support vector machine is to maximize the separation distance (margin) between the two classes. The type of SVM used for anomaly detection is called a one-class SVM. In this case, the model is trained only on the normal data class, and anything deviated from the normal class is considered an anomaly. The one-class SMV maps training data patterns into a high-dimensional

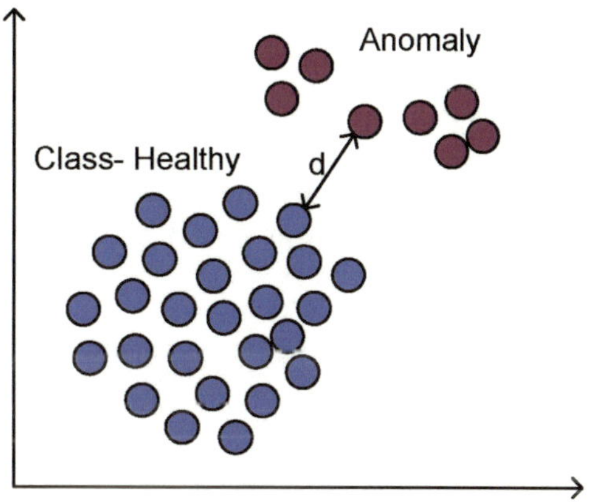

Figure 15.
Anomaly detection using a KNN.

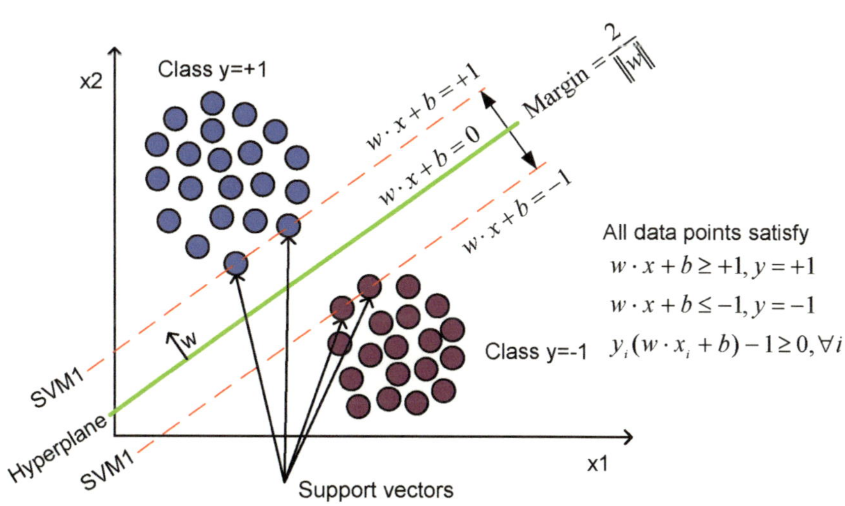

Figure 16.
The SVM classifier for linearly separable classes.

feature space using the kernel function and finds the maximum margin that separates the training sample and the origin. **Figure 17** shows a linear one-class SVM for data points in a 2-dimensional space.

The function to be minimized in order to maximize the margin between the origin and the training class is

$$\min_{w,\xi,\lambda} \frac{1}{2}\|w\| + \frac{1}{vm}\sum_{i}^{m}\xi_i - \lambda \quad \text{s.t.} \quad (w, \psi(x)) \geq \lambda - \xi_i, \xi_i \geq 0 \quad and \quad 0 < v < 1, \tag{10}$$

The decision function is given as

$$f(x) = sgn\,(<w, \Phi(x)> - \lambda), \tag{11}$$

Applying Lagrange multiplication yields the following quadratic programming to be optimized

$$\min \frac{1}{2}\sum_{i,j}\gamma_i\gamma_j k(X_i.X_j) \quad \text{s.t.} \quad 0 \leq \gamma_i \leq \frac{1}{vm} \quad and \quad \sum_{i}^{m}\gamma_i = 1, \tag{12}$$

where γ is the Lagrange multipliers, k is the kernel function used to project the input feature into the feature space, λ is an offset parameterizing a hyper-plane in the feature space, and m is the number of training data points. There are different types of kernel functions for instance, linear kernel, polynomial kernel, radial basis function (RBF) kernel, and Sigmoid kernel.

2.6 Fault diagnostics

After detecting an anomalous condition, fault diagnostics, also known as process diagnostics, aims to determine and provide specific information about the possible cause. Often this process can be also quite independent from the anomaly detection

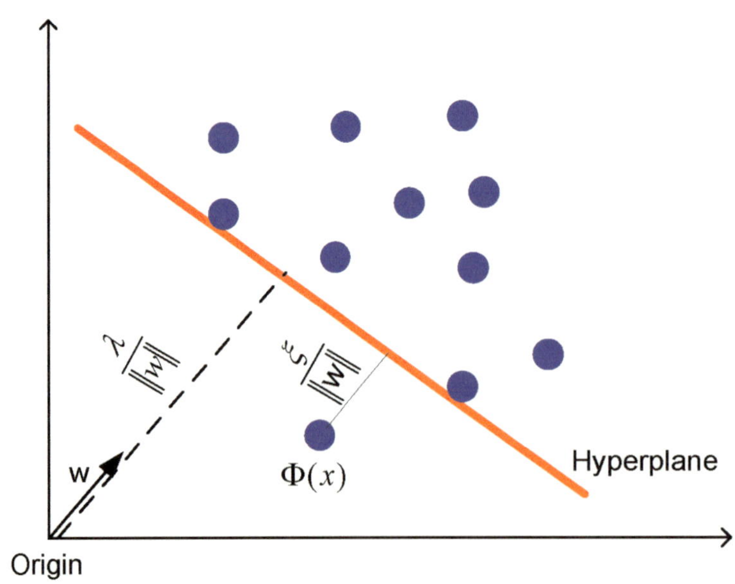

Figure 17.
One-class SVM.

layer. Typically, process diagnostics consists of three steps known as fault detection, isolation, and identification. Fault detection and isolation (FDI) are often performed simultaneously by use of physical models or data-driven techniques, as in the case of anomaly detection. Fault detection is the step to determine if a fault is present or under development in the process and the time of fault occurrence. Fault isolation refers to the technique of pinpointing the location of the faulty component(s) of a process, such as devices, sensors, actuators, controllers etc.

Typically, FDI methods are widely classified in three categories: model-based methods (typically first-principles, state-space or input–output models), model-free (also known as data-driven) methods, and knowledge-based (or rule-based) methods. All these methods have their own advantages and disadvantages. Hence, in the realm of process diagnostics, there is no silver bullet to address every single case. For applications where the process is difficult or expensive to model, or sufficient information is not available to model the effects of all possible anomalies and faults accurately, data-driven and machine learning methods have been developed in the years. Similarly to the techniques used for anomaly detection, classification techniques such as ANN, KNN and SVM are often used to assign the measured data points to the cluster indicating the faulty component. However, simultaneous faults or malfunctioning in more than one component require more complex methods. All the methods have various degrees of sensitivity to measurement noise.

On the other hand, fault identification refers to the way of estimating the severity or magnitude of the fault, and providing information on whether the process can continue to operate as usual or if a corrective action (in extreme case, shut-down) needs to occur. Typically, the extent of deviation in measured parameters can give an indication on the severity of the fault. However, simultaneous faults in different components may have opposite effect on measured parameters, hiding the real problem magnitude and rendering this step quite challenging.

A combination of model-based and data-driven approaches for fault isolation and identification is often preferred when an accurate numerical model of the process is available. A common approach is to include health indicator factors, or state variables, in the model. Health indicator variables can represent e.g. a fouling coefficient in a heat exchanger or a flow capacity deviation in a pump, compressor, or turbine. Such variables can be varied when simulating the process until the model outputs match the observed measurements from the real system. Various optimization techniques such as genetic algorithm (GA) have been used for this purpose. This method can often perform isolation and identification together, when the health indicator factors are allowed to take multiple values. One drawback commonly experienced is the so-called smearing effect, mostly induced by noise and model uncertainties, where the effect of anomalous measurements tends to be "spread" over multiple health factors even when only one single component is actually faulty. To overcome this, preprocessing of the data to reduce noise is usually necessary; downstream processing of the obtained health indicators through machine learning techniques is also a solution to improve the isolation and identification accuracy.

2.7 Advanced sensors

The learning system is designed to work with the data that is collected in the database. In order to best utilize this, it is important to understand the inherent properties and qualities of the data gathered about the process. Data is gathered from multiple sensors located in different parts of the process. Sensors are devices that provide output signals based on a certain input that represents a physical quantity. These devices can be more or less complex, ranging from straightforward ones that measure pressure and temperature to more complex ones that determine other physical and chemical properties. The most measured parameters in the process industry are temperature, followed by pressure and flow rate. However, this does not mean that these parameters can be measured accurately.

Sensors are based on different principles to provide useful information about the system. Most physical properties of interest cannot be measured directly but are rather obtained by utilizing different principles, converting a property that can be easily measured to the one of interest. Temperature is often measured with a thermocouple, which utilizes the thermoelectric effect where temperature differences are converted to electric voltage. Two dissimilar metal wires are connected at one end in an electric junction. Once the temperature changes at the junction, it creates a voltage that can be measured with a voltmeter and is a function of temperature. This is the case for the majority of sensors used in the process industry, requiring some sort of model to convert the measured parameter to useful output.

Measurements of a process parameter carry with them a certain uncertainty. This can arise from many sources and will propagate to the final result. The more the sources of uncertainty and the more complex the process of getting to that result, the higher the final uncertainty in the measured data. This difference in the measured value from the actual one can be either random or systematic. Random differences contribute to random signal noise, whereas systematic differences can be due to bias or deterioration of the sensor. An unsteady process, where the values of the parameters fluctuate, will make it even harder to obtain accurate measurements.

For the development of a learning system, measurement data from the process can be used to monitor the operation of the different modules. This can provide information on the performance of the different components. Another very useful piece of information, particularly for the process industry, is information on the properties of the feedstock that is coming into the process. This can provide a feed-forward signal, enable the prediction of the properties of the final product, and allow the optimal control of the process. This can be done with a more advanced sensor, which in essence requires a more complex model to convert the measurement to the property of interest. Such sensors are often referred to as soft sensors. As long as the sensor and the property of interest are in the same location in the process, the model is part of the sensor, regardless of its complexity. If the property of interest is in a different part of the process than the one where the sensor is installed, the model behind the sensor becomes a model of the process rather than a conversion of the sensor input to useful output. These advanced sensors are subject to uncertainty and noise in the same way as the simpler ones. However, uncertainty and noise in the measurement can increase when there are more components and models in a measurement chain, and this affects how the data can be used.

With regards to the measurement of feedstock properties, a particularly promising technology is near infrared (NIR) spectroscopy. This technique is based on the difference in absorbance of light in the near infrared field by different chemical bonds in the molecules that are illuminated. This can in turn provide detailed information about the chemical properties of the material that is measured. The measurement head itself will provide a spectrum of absorbance in a range of wavelengths and this information can be calibrated against the desired physical properties of the material that are of interest for the specific application. This is typically done at first in a laboratory environment, and the models obtained from the laboratory experiments are then transferred or adapted to the real environment. NIR has been shown to be very capable of predicting key properties of the incoming feed for a range of different processes [24–26] in a much faster way than time consuming lab analyses and as such can form the base of an advanced sensor in the process industry.

An example of a lab setup for the analysis of fuel samples with a Fourier transform NIR spectrometer is shown in **Figure 18** for refuse-derived fuel (left) and woodchips (right). NIR spectroscopy can be used for both solid and liquid fuels, and a spectrum of a hydrocarbon mixture is shown in **Figure 19**. The spectra obtained from the NIR instrument are matched to the property of interest for every sample analyzed and a calibration model is built using a statistical analysis of the data (typically referred to as chemometrics for spectroscopy applications). This results in a calibration curve like the one shown in the right hand side of **Figure 19** where the quantity of interest predicted by the model from the spectrum of the sample is compared to the quantity of interest measured in the lab. The 45° line in the figure represents a perfectly accurate prediction (the predicted value is the same as the measured one), but small deviations from the measurements always occur and the accuracy of the model is depicted by the width of the area between the dashed lines.

In order for the NIR-based soft sensor to be used in a real process environment, the head needs optical access to the feedstock. An example of an installation of a NIR sensor in a pulp and paper mill is shown graphically in **Figure 20**. In this case the optical access is provided through an observation hole. The NIR spectra obtained by the instrument are converted to the desired property through a model,

Figure 18.
Lab setup for the analysis of fuel samples with a Fourier transform NIR spectrometer for refuse-derived fuel (left) and woodchips (right).

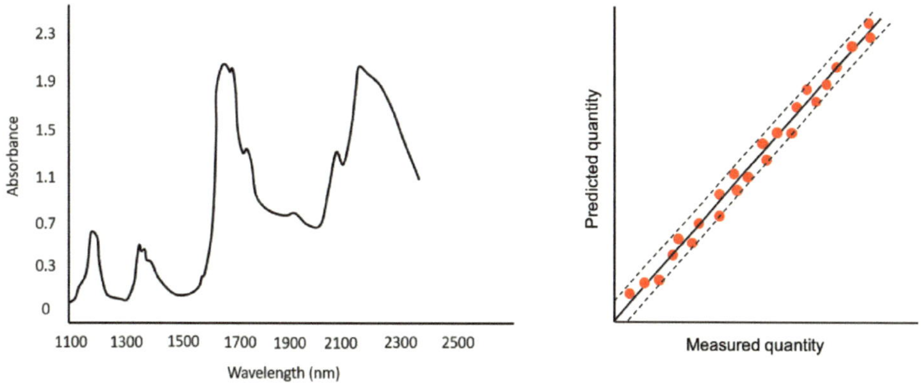

Figure 19.
NIR spectrum of hydrocarbon mixture (left) and soft sensor model calibration (right).

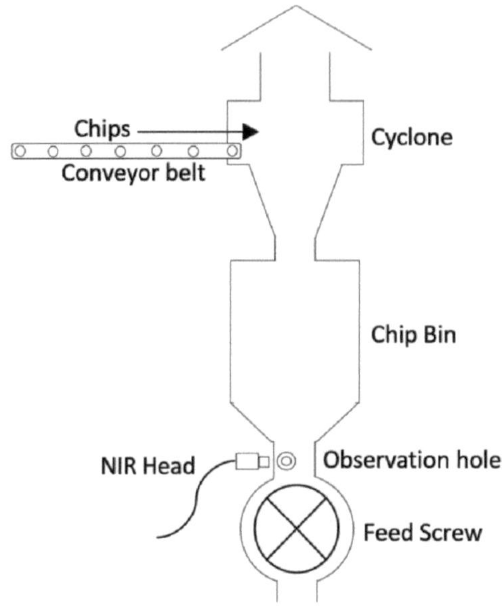

Figure 20.
Schematic of the NIR soft-sensor location in a pulp digester.

with the entire setup of the measurement head, the analyzer and the model behind it constituting the soft sensor. The information from the sensor can be delivered to the database of the factory and from there it can be fetched and used as a feed-forward input signal for the control of the process.

The main advantage of using a NIR sensor is that it can provide fast and non-intrusive measurements of the properties of the feedstock material, which can then be used to optimize the operation of the process itself. Variations in the properties of the feedstock will inherently affect the process and the optimal operating conditions of the downstream components will change depending on these properties. Real-time information on the variation of the key properties of the feedstock can therefore be very useful when combined with model predictive control in order to determine the optimal operating conditions in real-time, with the information from the sensor used as a feed-forward signal for the control of the process.

2.8 Decision support system

A decision support system (DSS) is a computer-based program that supports decision-making activities, for example at operation, planning or management levels. Through the analysis of large amounts of data, a DSS provides decisions for uncertain, unstructured, or rapidly changing problems, which either complement or replace human reasoning. The system can be either fully automated, fully dependent on human actions, or a hybrid. However, the hybrid approach is widely acceptable where the DSS incorporate a human–computer interaction and the cyber part usually provides a range of information that operators or managers use to decide on an action [27]. A DSS can for example have access to a database of historical events and corresponding decisions, and retrieve cases similar to a current event to suggest a possible action. If we linked it back to the automation pyramid, typically a DSS will directly interact with the supervisory level and influence the decision making in planning and management level. Within the learning system architecture, the DSS is used to support decision making at all three levels. Based on the outcome from individual sub-components of the learning system, the DSS will assists the operators by providing information about the health status of the process equipment. The system can also suggest possible actions under a particular process fault. This can also trigger series of actions that will assist decision making at planning and management level. Due to a fault severity if a equipment requires maintenance, the DSS will assist the maintenance planner by providing RUL of the affected equipment. Additionally, if a process equipment becomes unavailable the production planning need to be adjusted accordingly. This will effect the decision making in ERP level since both production material and spare-parts inventory need to be adjusted accordingly.

According to literature, DSS can be model-driven, data-driven, knowledge-driven, document-driven or communication driven [28]. A model-driven DSS employs statistical, financial, mathematical, analytical, simulation or optimization models for decision support. In model-based DSS, possible scenarios are simulated with the aid of models to take optimal decisions; for example, the optimal maintenance interval can be calculated with the aim of minimizing the total costs. Developing a model-driven DSS is a complex, time consuming and expensive process that requires a considerable level of expertise. A data-driven DSS requires access to and manipulation of time series of internal and external data. It is the most common of the five types of DSS. The success of a data-driven DSS always depends on the access to accurate, well-structured and organized data. A typical knowledge-driven DSS contains a rule-based algorithm such as decision tree or similar [29]. The knowledge from the expert is stored in form or rules such as "if sensor A is faulty and control system is functioning, schedule maintenance in XX hours". Thus, automated decisions can be taken by analyzing massive amount of data and applying predefined rules. In this work, we will only focus on knowledge-driven DSS particularly highlighting an example of probabilistic approach.

More sophisticated algorithms aiming at simulating human reasoning in a probabilistic manner are built from Bayesian belief networks (BBN). Bayesian networks represent a culmination of Bayesian theory of probability, which can be summarized as in Eq. (6). The equation represents a casual statement of the kind, where X causes Y and Y takes the role of an observable effect of X. $P(Y)$ is called the prior probability, while $P(Y|X)$ is called the posterior probability. The factor that relates the two, $P(X|Y)/P(X)$, is called the likelihood ratio.

$$P\left(\frac{Y}{X}\right) = \frac{P\left(\frac{X}{Y}\right)}{P(X)} * P(X), \tag{13}$$

A BBN is a probabilistic graphical model that represents factorization of joint probability distribution [12]. It provides a comprehensive way to handle uncertainty in mathematical computation, consequently widely used for representing uncertain knowledge. Bayesian probability differs from classical probability by the fact that classical probability does not put any weightage to the evidence while Bayesian probability always comprises of a certain degree of belief in the evidence [13]. The most beneficial aspect of a BBN is that it can be constructed either by training with historical data, and with limited data set or even in the absence of data only by integrating expert knowledge. A BBN has two major parts: a qualitative or structural part, consisting of nodes and connections, and a quantitative part that is a set of conditional probability distributions. Typically, each node corresponds to a unique random variable (e.g. occurrence X), while each edge or connection corresponds to a conditional dependency. This qualitative structure is referred to as directed acyclic graph (DAG). The term "acyclic" refers to the fact that the direct connections are static causal probabilistic dependence and cycles are not allowed (e.g. if X causes Y, Y cannot cause X). Constructing a BBN involves building the structural part of the BBN or DAG and specifying the conditional probabilities also known as parameters. A BBN can be constructed completely manually from expert knowledge, completely automatically from data, or through a combination of a manual and automatic technique, where partial knowledge about the structure or the parameters are learnt from the data.

For maintenance planning purposes, the DSS would combine and fuse together information coming e.g. from different diagnostics approaches, maintenance history, operator observations, etc.

2.9 Applicability of the framework for fleet level monitoring

The presented framework can be applied both to single large units (ie a complex industrial plant) and to multiple smaller units, in a fleet management approach. The requirements for fleet management shape the framework in a multitude of ways.

In order to manage multiple assets at the same time, the level of detail of the simulation may be reduced. This may result in less complex models and different requirements of the level of control and management. This can further increase the modularity of the framework. Different levels of control and management may not be desired and approaches used may be less complex. In essence, a framework that focuses on fleet management will not focus on the optimization of a single system with a multitude of sensors at the first instance. This requires an approach that allows the removal or deactivation of functions as desired. This is in line with the development of a framework that can be generic and applicable to different cases, but further highlights the need for modularity.

The increase of the number of assets being monitored through the framework requires more models that simulate the operation of the assets. One option is to have a different model for each system, which can result in thousands of models being employed in the platform. However, the units of the fleet being managed are inherently very similar to each other; they are copies of each other with minor differences which arise from manufacturing uncertainties. The units will be operating at different conditions, which will affect the degradation of components and sensors in a unique way for each system. However, the different assets can be represented by a single model that simulates their performance and a different set of tuning parameters for each one. This can reduce the load in the framework platform.

The management of the fleet will result in large amounts of data being collected, a set of data for each asset in the fleet. Since the different assets belong to the same family, and their operation is similar, the data from the entire fleet is useful for the management of all assets independently and collectively. A system that learns from the operation of different assets can gain more knowledge than one that is focusing on a single asset. This creates more challenges for data management, but also provides more knowledge for possible faults, and can allow the prognosis of remaining useful life and other parameters of interest with greater confidence. From the framework perspective, this requires more instances for visualization of data from both the single unit and the fleet, and an analysis of the different trends.

3. Conclusions

Over the last few decades there has been an significant exploration of new techniques and tools to improve product quality and process efficiency of complex industrial processes. There is a need for a framework that will allow integration of different tools for optimal operation, control and diagnostics to enable robust decision support. As a stepping stone, a generic learning system architecture has been developed that allows easy integration with existing supervisory system of industrial plants. The architecture enables inclusion of different functionalities as individual modules. The system can therefore be easily adapted according to the different requirements of different cases. The architecture is flexible enough to be implemented in a remote server with a web-based interface or run locally in a isolated server. As a final reflection, utilization of such a learning system in addition to the existing supervisory systems can only be justified by demonstrating quantifiable economic benefits. Only then will all stakeholders be on board for adoption of such a system. Another aspect that is often neglected is that the system users, i.e. plant operators, engineers and managers, need to be involved from the very beginning of the process from development to implementation of such a system.

Acknowledgements

The authors gratefully acknowledge the financial support from European Union's Horizon 2020 research and innovation programme under grant agreement No 723523 through FUDIPO project (http://fudipo.eu/). The authors also thank entire FUDIPO team for their inputs and ideas without which the framework would not come this far.

Conflict of interest

The authors declare no conflict of interest.

Nomenclature

$X(t)$	Physical property under consideration
t	Independent variable for representing time
$\dot{x}_{in}$	Rate of change in inflow
$\dot{x}_{out}$	Rate of change in outflow
$\dot{g}(t)$	Rate of change in generation
$\dot{c}(t)$	Rate of change in consumption
A, B	Reactants
C	Product
r_c	Rate of reaction
α, β	Exponents of concentration
k	Reaction rate constant
A_i	Pre-exponential factor
E	Activation energy
R	Universal gas constant
T	Temperature of the reaction
c	Control horizon
$y(k+i)$	Future plant output

$\Delta u(k+i)$	Increments in manipulated variable
$e(k)$	Future errors
p	Prediction horizon
$s(k)$	Reference set-point trajectory
γ	Lagrange multiplier
k	Kernel function
λ	Offset parameterizing a hyper-plane in the feature space
X	Cause of Y
Y	Observable effect of X

Abbreviations

ISA	International society of automation
PLC	Programmable logic controller
DCS	Distributed control system
PID	Proportional–integral–derivative
SCADA	Supervisory control and data acquisition
HMI	Human-machine interface
MES	Manufacturing Execution System
ERP	Enterprise resource planning
CMMS	Computerized maintenance management system
RUL	Remaining useful life
HMI	Human–machine interface
CFD	Computational fluid dynamics
AE	Algebraic equation
ODE	Ordinary differential equation
PDE	Partial differential equation
CSTR	Continuous stirred tank reactor
MPC	Model predictive control
MV	Manipulated Variable
PCA	Principal component analysis
ANN	Artificial neural network
SVM	Support vector machine
KNN	K nearest neighbor
HC	Hierarchical clustering
SOM	Self-organizing map
AANN	Autoassociative neural network
RBF	Radial basis function
FDI	Fault detection, isolation and identification
NIR	Near infrared
DSS	Decision support system
BNN	Bayesian belief networks
DAG	Directed acyclic graph

Author details

Moksadur Rahman*, Amare Desalegn Fentaye, Valentina Zaccaria, Ioanna Aslanidou, Erik Dahlquist and Konstantinos Kyprianidis
Malardalen University, Vasteras, Sweden

*Address all correspondence to: moksadur.rahman@mdh.se

References

[1] Rahman M. Towards a learning system for process and energy industry: Enabling optimal control, diagnostics and decision support. Licentiate dissertation, Mälardalen University, Västerås, Sweden; 2019.

[2] Lee J, Davari H, Singh J, Pandhare V. Industrial Artificial Intelligence for industry 4.0-based manufacturing systems. Manufacturing Letters. 2018; 18:20–23.

[3] Diez-Olivan A, Del Ser J, Galar D, Sierra B. Data fusion and machine learning for industrial prognosis: Trends and perspectives towards Industry 4.0. Information Fusion. 2019; 50:92–111.

[4] Leitch R, Quek C. Architecture for integrated process supervision. In: IEEE Control Theory and Applications. vol. 139. IET; 1992. p. 317–327.

[5] Baldea M, Harjunkoski I. Integrated production scheduling and process control: A systematic review. Computers & Chemical Engineering. 2014;71:377–390.

[6] ANSI/ISA-95. Enterprise Control System Integration Part 3: Activity Models of Manufacturing Operations Management; 2005.

[7] Hummen R. What is TSN? A Look at Its Role in Future Ethernet Networks [Internet]; 2017. Available from: https://www.belden.com.

[8] Miskowicz M. Event-Based Control and Signal Processing. Boca Raton, FL, USA: CRC Press; 2018.

[9] Mehta BR, Reddy YJ. Chapter 19 - advanced process control systems. In: Industrial Process Automation Systems. Oxford: Butterworth-Heinemann; 2015. p. 547–557.

[10] García S, Luengo J, Herrera F. Data preprocessing in data mining. 1st ed. Cham, Switzerland: Springer International Publishing; 2015.

[11] Zhou B, Ye H. A study of polynomial fit-based methods for qualitative trend analysis. Journal of Process Control. 2016;37:21–33.

[12] Janusz ME, Venkatasubramanian V. Automatic generation of qualitative descriptions of process trends for fault detection and diagnosis. Engineering Applications of Artificial Intelligence. 1991;4(5):329–339.

[13] Charbonnier S, Portet F. A self-tuning adaptive trend extraction method for process monitoring and diagnosis. Journal of Process Control. 2012;22(6):1127–1138.

[14] Zhang L, Cai S, Hu J. An adaptive pre-warning method based on trend monitoring: Application to an oil refining process. Measurement. 2019; 139:163–176.

[15] Mikles J, Fikar M. Process modelling, identification, and control. Springer Berlin Heidelberg; 2007.

[16] Ghasem N. Modeling and simulation of chemical process systems. Boca Raton, FL; 2019.

[17] Frisk E. Simulation of differential-algebraic equations [Course material]. Linkoping, Sweden: Department of Electrical Engineering, Linkoping University; 2020. Available from: https://www.fs.isy.liu.se/Edu/ Courses/Simulation/OH/dae3.pdf.

[18] Martí L, Sanchez-Pi N, Molina J, Garcia A. Anomaly Detection Based on Sensor Data in Petroleum Industry Applications. Sensors. 2015;15(2): 2774–2797.

[19] Jaw LC. Recent advancements in aircraft engine health management (EHM) technologies and recommendations for the next step. In: Turbo Expo: Power for Land, Sea, and Air. vol. 46997; 2005. p. 683–695.

[20] Simon DL, Bird J, Davison C, Volponi A, Iverson RE. Benchmarking Gas Path Diagnostic Methods: A Public Approach; 2008.

[21] Sebestyen G, Hangan A, Czako Z, Kovacs G. A Taxonomy and Platform for Anomaly Detection. In: 2018 IEEE International Conference on Automation, Quality and Testing, Robotics (AQTR). New York, USA: IEEE; 2018.

[22] Agrawal S, Agrawal J. Survey on anomaly detection using data mining techniques. In: 19th International Conference on Knowledge Based and Intelligent Information and Engineering Systems. vol. 60. Amsterdam, Netherlands: Elsevier B.V.; 2015. p. 708–713.

[23] Sadough Vanini ZN, Meskin N, Khorasani K. Multiple-Model Sensor and Components Fault Diagnosis in Gas Turbine Engines Using Autoassociative Neural Networks. Journal of Engineering for Gas Turbines and Power. 2014 5;136(9).

[24] Andersson N, Wilson DI, Germgård U, Germgard U. Validating continuous kraft digester kinetic models with online NIR measurements. Proceedings of the American Control Conference. 2002;5:3783–3787.

[25] Santos VO, Oliveira FCC, Lima DG, Petry AC, Garcia E, Suarez PAZ, et al. A comparative study of diesel analysis by FTIR, FTNIR and FT-Raman spectroscopy using PLS and artificial neural network analysis. Analytica Chimica Acta. 2005;547(2): 188–196.

[26] Skvaril J, Kyprianidis KG, Dahlquist E. Applications of near-infrared spectroscopy (NIRS) in biomass energy conversion processes: A review. Applied Spectroscopy Reviews. 2017 9;52(8):675–728.

[27] Kallestrup KB, Lynge LH, Akkerman R, Oddsdottir TA. Decision support in hierarchical planning systems: The case of procurement planning in oil refining industries. Decision Support Systems. 2014;68: 49–63.

[28] Felsberger A, Oberegger B, Reiner G. A Review of Decision Support Systems for Manufacturing Systems. In: SamI40 workshop at i-KNOW '16 October 18–19, 2016, Graz, Austria; 2016.

[29] Prasad D, Ratna S. Decision support systems in the metal casting industry: An academic review of research articles. Materials Today: Proceedings. 2018;5(1, Part 1):1298–1312.

Chapter 3

AI & Digital Platforms: The Market [Part 1]

Örjan Larsson

Abstract

This essay aims to describe the dynamics at play in the field of industrial AI, where the significant efficiency potential is driving demand. There are rapid technological development and increasing use of AI technology within the industry. Meanwhile, practical applications rather than technical development itself are creating value. The primary purpose of the article is to spread knowledge to industry. It is also intended to form the basis of the Swedish innovation program PiiAs ongoing work around open calls and targeted strategic innovation projects. The basic approach taken is to investigate both industry demand for AI and how the supply of technology is developing. AI takes in a broad and dynamic range of concepts, but it should also be considered in an even broader context of industrial digitalisation. The article has been divided into two sections: *The Market*, in which we assess the development and the consequences on the factory floor; and *The Technology*, which provides a more in-depth understanding of the structures of industrial IT and machine-learning technology. The article concludes with four practical examples from the industry.

Keywords: AI, artificial, intelligence, PiiA, blue, institute, automation, algorithmisation, platform, data, process, industry, IndTech, digitalisation, digital, twin, ecosystem

1. Introduction

This article has two primary purposes: the first is to provide the industry with an evaluation of the importance of AI development as a force for change and the second to create an internal basis for the Swedish Innovation program PiiA's future development efforts, within which AI can be described as the next phase of industry's digitalisation. Both these objectives are naturally compatible with the overall ambition of the report: to reach our target group of industry leaders and to serve as a source of knowledge for ongoing activities within relevant companies.

Technological, industrial development is awash with grand ambitions that have turned into mere passing fads and costly dead ends. With this in mind, throughout our work in assessing the development of AI, we have endeavoured to take into account the magnitude and direction of different vectors of change. On the one hand, we have attempted to understand the power of demand for AI by assessing the economic impacts at a macro level. We have focused on productivity and qualitative values at various stages of industry value systems. On the other hand, we have attempted to assess the range of available technologies by analysing initiatives taken on a global scale and through focused academic research. We have also put considerable effort into understanding the major commercial—or applied—forces that are crucial to development, both in the short and medium term.

We have also strived to place AI development in the context of current systemic developments, as characterised by the 'platformisation' of company IT resources. By this we mean the transfer of automation and IT support to the cloud—a trend that is creating new competitive dynamics. Finally, we have attempted to translate this big picture into real impacts on the factory floor and to revisit well-known concepts such as organisational development which—with the help of the raw power of AI technology—have the potential to make the previously impossible, possible.

The project was a collaboration between PiiA and Blue Institute, with valuable input from Blue Institute's network of CEOs and industry leaders on all levels. A big thank you is extended to everyone who contributed to this study.

2. The market

The computerisation of global industry began in earnest in the 1980s. The use of microprocessors made it possible to automate in new and efficient ways, and the process, automotive and electronics industry made significant productivity gains. Now, the world is entering a new technological and economic paradigm in the form of digitalisation, the first wave of which has already transformed the media and communications industries. In a second wave of change, the financial sector and trade will be transformed, and, under the third wave, the wider industry will be transformed. At the same time, artificial intelligence is emerging as the next—and probably most significant—stage of digitalisation.

For the manufacturing industry, this will mean that many companies in subcontracting arrangements will have access to cost-effective technology that allows for further automation and productivity increases. The impacts of AI for the process industry will not be yet another emptying of factory floors, but rather an opening of the way to achieve levels of process development that were previously unobtainable. AI is also expected to bring new levels of integration to the entire value system, on the way to achieving the ultimate vision of self-organisation. It may also change the structure of the industrial landscape; concepts such as 'ecosystems' and 'platforms' are fast becoming commonplace descriptions, even in traditional industries.

Our current stage of development could be described as 'increasing algorithmisation' (see **Figure 1**). From a developmental context, it represents a megatrend that is both supporting human beings—and enabling us to be replaced by computers. The trend began with the mainframes of the 1960s and continued with the microprocessor revolution of the 1980s and 1990s. Then in the 2000s, came the scalability, mobility and cost-effectiveness of digital platforms. Now, with AI and machine learning emerging as the next phase, the pace of development is set to increase further.

Demand for industrial AI is growing as the understanding of the value that the technology can potentially release grows. Various technological developments, which have been taken place over the past decade unbeknownst to the general public, are now coming to fruition and can directly be viewed within the context of analyses of potential economic effects and, increasingly, real-life business cases and investments.

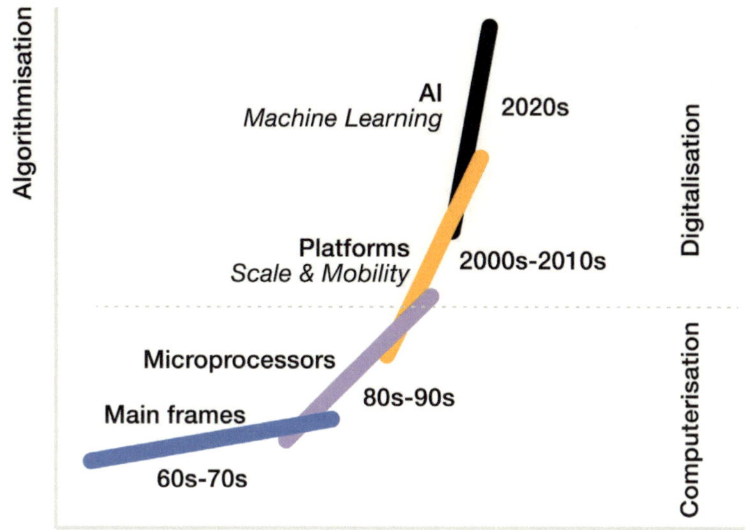

Figure 1.
AI and machine learning effectively add an extra 'gear' that will allow for increasingly advanced algorithms that increase efficiency and create new customer value within the industry. Source: Blue Institute (2019).

These changes have the potential to produce significant economic impacts for the global industry, and they may be particularly marked in the context of societal challenges related to population growth, climate and the environment.

The million-dollar question is: Will we see an explosion of AI, and its disruptive establishment across the global industry? Or, will this, still somewhat unwieldy, technology lead to more sporadic changes in the short term? Either way, our considered assessment is that AI is here to stay, that AI truly has the potential to change the world and industry, and that AI will be looked back on as a real revolution for the production economy.

In this section, we will look at the forces underlying supply-and-demand development trends for AI within the industry. A functioning market dynamic is a crucial prerequisite for ongoing industrial transformation pressure, something that we aim to highlight in this study by addressing three issues:

1. What are the expected value-creating effects of AI?
2. Will AI development and the supplier system be able to meet the demand for AI technology that is arising from this potential value creation?
3. Is this development sustainable, or are we seeing a 'hype' phase which will eventually fade, with the actual market breakthrough set to occur several years into the future?

In addition to these issues, this report illustrates, from various perspectives, how the impacts of AI will benefit the industry at the system/platform and operational development levels. The second part of the report will provide an in-depth study of the possibilities and challenges of AI technology.

2.1 Demand

We start this section by looking at the demand side of things (see **Figure 2**), as well as discussing the stance companies might take. This includes examining

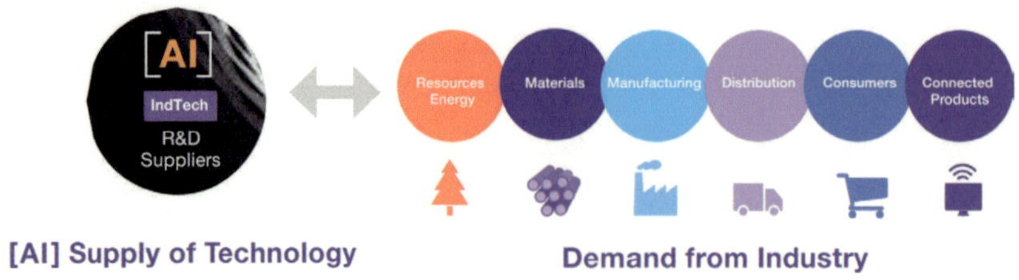

Figure 2.
The market for AI technical solutions and machine learning is expected to grow by 40% per year, while demand within the industry is expected to be driven by the significant potential value gains that can be created using the technology.

more significant developments that will lead to a future digital economy based on business ecosystems and digital platforms. We also introduce the concept of 'best practice' and provide an orientation model for individual companies wanting to assess their position and preparedness for change.

2.1.1 Substantial value effects within production systems will drive demand for AI

Our fundamental hypothesis is that demand for AI within the industry will correlate with the value that can be extracted from production via more effective analytical tools. We assume that the expected growth effects within the sector will lead to activities at the company level, which in turn will drive demand for AI technology.

According to the Vinnova study Artificial Intelligence in Swedish Business and Society, there is evidence to suggest that the general growth potential within value creation might be realised twice as fast in an economy with extensive AI utilisation, compared to one with limited utilisation.

The many dynamic effects of AI development and the changing regulations around it, also come into play, and these are expected to produce growth effects for the world economy. AI will also contribute to systemic effects, as business ecosystems and digital platforms are developed that transform the manufacturing industry into an information industry. These virtual value systems are decoupled from physical systems and so allow for new organisational models that echo the transformation that media, finance and commerce sectors are already undergoing.

Several studies have attempted to estimate the economic effects of AI at the macro-level; in this work, we have incorporated insights from three reports by Accenture [1], McKinsey Global Institute [2] and PwC [3]. According to consulting company PwC, AI's contribution to the global economy in 2030 will amount to an estimated USD 15.7 trillion. This means that in 2030, with the impact of AI, global GDP will be 14% higher than it would be without AI or the equivalent of China and India's combined GDPs.

The productivity impact corresponds to USD 6.6 trillion, while USD 9.1 trillion is expected to be produced from impacts on the consumer side. PwC's analysis also includes areas such as trade, transport, finance and health care.

Consulting company Accenture believes AI's global economic impact will be equivalent to USD 4.8 trillion in increased profitability during the period up to 2022, which does not contradict McKinsey's or PwC's analyses that have other timeframes.

In its report, Notes from the AI frontier: Modelling the impact of AI on the world economy, consulting firm McKinsey Global Institute (MGI) calculates that the effects of AI in all of the report's sectors will generate an impact of between USD

3.5–5.8 trillion or when expanded to include all available advanced analysis methods, on top of machine learning, USD 9.5–15.4 trillion.

When limited to the resource, process and manufacturing industry's value system, it is in the range of USD 1.7–2.3 trillion according to estimates in this study, or 3–6% per year of the global industrial sector's total assets.

Industrial productivity improvements are estimated to amount to 1.2% per year until 2030 or in the order of USD 1 trillion. Comparisons can also be made with other major technological shifts. During the nineteenth century, the steam engine increased labour productivity by an estimated 0.3% per year (although the disruptive effect eventually became quite considerable). The robotisation of the industry in the 1990s produced a 0.4% increase, and the consequences of IT development during the 2000s are expected to deliver a 0.6% increase. AI has at least twice the inherent potential.

'Added value of between SEK 22 and 45 billion per year could be unlocked for PiiAs industries'.

Placed in a Swedish context and related to PiiAs target industries, we estimate that added value of between SEK 22 and 45 billion could be unlocked per year. For PiiAs sectors this represents an average increase of between SEK 3 billion and SEK 7 billion per industry, of which approximately half would be productivity-related, with value also unlocked at the consumer level, through factors such as quality, time savings and better-targeted offerings.

The purpose of the comparisons above is more to illustrate the order of magnitude involved than to present precise figures. Even if the cited studies were to be greatly exaggerated, the effects would certainly be significant. We conclude that the movement that has now been set in motion has few parallels in history in terms of change potential. For companies and businesses, it means there will be few, if any, players who can afford to pass up the competitive improvements that AI will eventually deliver.

And there is a good reason to prepare in advance for the coming changes. Looking from a broader perspective, lopsided distribution of AI development is likely, with an ever-increasing gap between the performance of various countries, companies and workers. In terms of countries, China and the United States are the two nations that currently account for the majority of all AI-related activities, and they are thus the best positioned. Developed industrial economies such as Germany, Japan and Canada and smaller commercial economies such as Sweden and Finland are well placed. They should also be motivated by low productivity-development gains in recent years.

Economies with more modest foundations, such as India, Italy and countries in Southeast Asia, generally have less favourable conditions. Still, they could use their particular strengths within specific categories to build specialised AI capabilities. However, developing economies with low investment capacity, weak skills and weak digital infrastructures run the risk of falling behind.

In this section, we will work out how the concepts of AI, ecosystems, networks and platforms are interconnected. And how they contribute to value-creating market dynamics.

2.1.2 Systemic effects from digital platforms and business ecosystems

Among the most considerable value-creating effects are expected to come from changes at the system level. As previously mentioned, we foresee development in a form that can best be described as the transformation of the manufacturing industry towards becoming an information industry. This should not be read as a prediction of the demise of the production economy—preferably one that business leaders will have to manage two logical frameworks.

This development has also been called the 'platform economy'. In this section, we will work out the differences and connections between the concepts of AI, ecosystems, networks and platforms. We will also outline how they contribute to value creation and, as a result, to the demand for AI and market dynamics.

The connections between the concepts of 'networks' and 'platforms' lend themselves to being described with metaphors from biological ecosystems. These 'ecosystems' can be thought of as robust, scalable architectures that can automatically solve complex, dynamic problems, including self-organisation, self-governance, sustainability and scalability.

In the business ecosystem, there is a network logic between the companies involved which, in turn, is supported by a digital ecosystem characterised by a distributed peer-to-peer network model. The latter can also be described as a digital platform that makes relationships between companies and other organisations in the business network possible through transactions and technical support. A curated ecosystem reflects the balance between competition and collaboration in an open, dynamic and free market.

The term 'business ecosystem' was first mentioned in a 1993 article in Harvard Business Review [4]. The article presents the idea that companies not only belong to industries but are parts of business ecosystems that extend across different industrial and knowledge sectors. The term digital business ecosystem originated when the word 'digital' was added to the business ecosystem concept as a reference to the socio-economic development made possible through information and communication technology [5].

The classic effects of network logic affect how the number of users in the network influences the value development for each user, i.e. the so-called 'positive-network effect'. Adverse network effects, on the other hand, occur in poorly managed networks that reduce value development for each user. The positive network effect is, of course, the foremost and most sought-after competitive advantage within network logic. Consequently, the critical prerequisite for effective networks is to use digital platforms and other features to increase in size, thus increasing the value generated via network effects.

2.1.3 Platform: a transformative concept

The concept of 'platform' is thus a transformative one with the potential to bring about significant changes within business logic, economics, and society at large. Any company for whom information on factors such as supply-and-demand status, customer needs, trends, and willingness to pay is an essential asset is very likely to participate in the platform revolution.

The concept of 'platformisation', then, is used as a strategy for operating multi-faceted platforms and connecting buyers, sellers and other stakeholders, without necessarily owning the products or services being sold.

In textbooks, traditional linear value systems are likened to value chains or pipes [6]. Platforms represent a transformation from linear structures to a matrix complex of relationships between connected producers and customers. They collaborate through the resources, properties and services provided by the platform's technology.

This development has, in its first wave, affected sectors where the product itself is information, such as the media, entertainment and financial sectors. The concept of 'scale without mass' [7] is vital. Unlike physical products, which have high fixed costs plus substantial marginal costs that are reduced per unit should production

DOI: http://dx.doi.org/10.5772/intechopen.93098

be scaled up, digital products, for the most part, have near-zero fixed costs and marginal costs. With the internet as a distribution network, it is, therefore, possible for companies with small fixed assets and a low number of employees to quickly scale up to become international businesses.

The empirical evidence suggests that the platform model vastly outperforms the linear value system when the right conditions are in place. Examples of such successes can be found among today's major tech companies, including Google, Amazon, Microsoft and Apple—all of which are also known as mega-platforms [8].

2.1.4 How digital platforms are changing traditional industry

The uniting factor in both network and platform logics is the need to match and facilitate connections between producers and buyers, regardless of the type of goods being exchanged (see **Figure 3**). Industry organisations will change as a consequence of the competitive advantages that platforms can provide within meeting places.

Platforms make it possible to bring new value for customers with low marginal costs to existing physical products—to achieve scale without mass—and we are already getting an early indication of how the industry will separate physical production logic from virtual data-driven logic. The automotive industry is experiencing shrinking margins in vehicle manufacturing and is developing business models that address mobility while being based on AI platforms.

The industrial technology suppliers of tomorrow will not just sell hardware but will also develop into connected suppliers of efficiency and quality within production systems based on analysis, with machine learning delivered in collaboration between human and artificial intelligence. The process industry will not just sell materials, but also data on these materials based on AI analyses that increase the quality and efficiency of the manufacturing industry.

Uncoupling physical assets from the value they create also means that certain products can be marketed as services in the market for best possible use through greatest value creation, rather than being linked to a specific owner. The result is that both efficiency and value can increase—dramatically, in some cases.

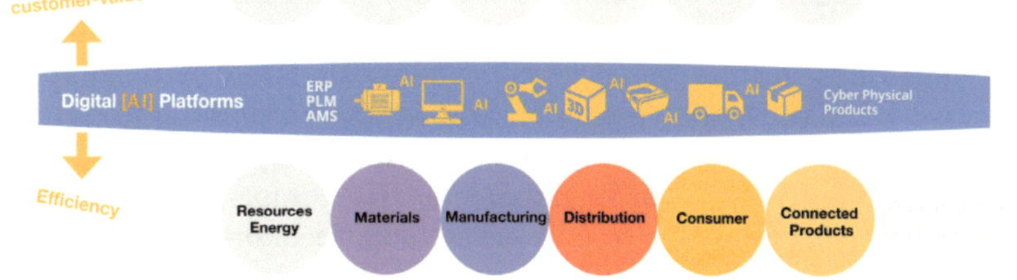

Figure 3.
Physical value chains and the flow of raw materials, materials and products are complemented by equally elementary data streams that enable digital twins at different levels, including entire processes and value systems. The technology and connectivity that makes physical production, as well as the digital twin, describes the digital platform (or ecosystem). Data streams and twins make it possible to create digital business ecosystems, i.e. meeting places where markets are created in new ways. Source: Blue Institute, 2019.

Platforms also have the potential to change the cost structures and pricing in physical production. Once someone launches a digital AI platform that allows for trade and provides free marginal production capacity on a larger scale, purchasing prices for semi-manufactured products will theoretically fall at the same rate at which the available capacity is filled. Such a day is probably not too far away. There are also estimates that digital platforms that match labour to needs (once again with the help of AI) have the potential to increase global GDP by 2% by 2025 and create 72 million full-time jobs [9]. It is not surprising that a new word is being increasingly used: algorithm economics. In the same way that apps have changed people's communication with machines, AI algorithms will revolutionise the development between the machines.

We conclude that while the business economics doctrine will undoubtedly continue to exist once the resources, process and manufacturing industry becomes an information industry, how it is followed will be revolutionised.

We are in the process of leaving an industrial era in which scalability in supply-side economics has been the single biggest driver; as more units are efficiently produced, the cost per unit diminishes. This has driven corporate mergers, globalised supply chains, oligopolies and monopolies. The largest companies have the most massive volumes and cost advantages that are difficult for smaller competitors to achieve. In the transformation into a digital and AI-driven platform economy in which physical products are paired with digital, scalable services, similar constellations will also be created through large-scale demand economics.

The demand economy is driven by aggregated and visible demand, social networks, app-development and other phenomena that make networks bigger and more valuable to all users. The impacts will be just as difficult to absorb as within the large-scale production economy. Scale within the demand economy is the foundation for positive network effects and therefore a future driver of the global economy.

The advantages of platforms over linear value systems will lead to the disruption and dissolution of many industrial businesses. The continuous improvement of physical value chains will be complemented by developments through which data streams will become equally important for competitiveness. These data streams will pave the way for digital twins to be created of objects, machines, processes and, ultimately, the whole value system—all physical production and logistics. Advances are being made towards achieving the vision of self-organising value systems, one of the core concepts of Industry 4.0. The physical world and the computer world will become two sides of the same coin.

One problem that needs to be solved in this context is the ownership of data. Who owns the data that companies generate? Today, there is no real regulatory framework, and the various industries collecting data are uncertain about how much, and what kind of data they should share with other companies. In answer to this, initiatives are underway in several computer labs, and we see examples of public, open laboratories, in areas such as forestry and traffic data. This is one approach to systematising data collection. But better-defined structures and agreed-upon standards are needed to define, describe and share data safely.

Ultimately, the transformation for classic industrial companies involves managing two different logics; the massive scale of the supply economy is not going away, and at the same time the ability to create demand with economies of scale is becoming a significant differentiating competitive factor. The skill lies in being able to handle both.

The use of the term 'platform' as found in 'digital platform' can be traced right back to the very early days of computerisation and the concept of 'computing platforms'. From the first mainframe computers, via the client-server model with its personal computers and networks, and into the era of digitisation, the word has been used to define hardware platforms and software platforms, or, to put it another way, general operating systems. The three development paradigms mentioned above are in turn called the first, second and third platforms. We are now in the era of the third platform, more complex and more intertwined than ever before, and characterised by the fact that computing power is found almost everywhere. Ready for use by people and objects, through the Internet of Things.

2.1.4.1 Industry case study: Mälarenergi Smart Flows: optimisation of the district heating network

Through its Smarta Flöden (Smart Flows) project, Swedish company Mälarenergi aims to use AI to optimise the production of district heating based on streaming data. Its goal is to avoid overproduction while continuing to provide reliable district heating to customers. The project receives funding from PiiA and is a collaboration between Mälarenergi, RISE Västerås, Mälardalen University, ABB, Sigholm and Evothings Labs. The Smart Flows project combines learning systems with Industrial IoT and cloud services to enable fully automatic optimisation of industrial process flows. The project is also part of the larger-scale work to create a 'City Control Room'.

Mälarenergi AB is a commercial company that supplies electricity, district heating, water, district cooling and fast communication solutions, primarily within the Mälardals region. The company also sells electricity to private and corporate customers throughout Sweden. The Group is owned by the City of Västerås and has a turnover of approximately SEK 3 billion.

2.1.4.1.1 The challenge

Measuring, understanding and predicting flows of materials, gases and liquids are central to many process industries, and, as a result, these flows are often subject to continuous optimisation. Air flows in mine ventilation and distribution flows for wastewater are good examples of process flows. Optimising and automating these flows has the potential to produce considerable savings in energy and total costs, which in turn can create positive environmental effects.

Process flows are rarely in a constant state, and instead are continuously developing as demand changes, new infrastructure is expanded, or as customers come and go. As a result, there is a clear need to make the industrial systems adaptable and teachable. District heating systems are an example of a system in which changes take time.

"The hope is to eventually be able to create a hybrid solution between the learning system and the physical model."

From the time that production is increased at a plant, it can take several hours before consumers' scores of kilometres away can feel the benefits. But by using real-time data and a connected distribution network, plants will be able to anticipate needs and quickly make decisions about increased or decreased production. The Smart Flows project uses the Internet of Things and cloud services from Microsoft Azure to manage historical and close-to-real-time data dynamically.

The project also has an operational development dimension through which the goal is for customers to be able to buy services in the form of comfortable indoor temperatures which be individualised.

2.1.4.1.2 The experience

The system takes in more than 15,000 properties, ranging from private homes to commercial properties and industries. During the project's first year, significant time has been spent collecting data. Measurement data from all Mälarenergi's district heating plants over the past three years have been collected. The majority of values are hourly, but where possible, 15-minute values have also been sampled. Mälarenergi appointed an internal analysis group to analyse the more than one billion data points gathered.

Meanwhile, in parallel, a project team has conducted several user studies, creating profiles for different user categories to understand the operation's visualisation and analysis needs.

The first predictions made with the AI system were based on the factors of distribution time, weather, and social behaviour. The most successfully generated predictions have been made concerning the heating needs of the building itself. Here, the weather is a substantial contributing factor, and as long as there are good weather forecasts, predictions can be created that adequately reflect reality. The social behaviour of customers is the least reliable factor. If a single customer showers or runs a lot of hot water for a specific period, there will be a massive potential impact on district heating. A variety of methods are being tested to improve the accuracy of social behaviour modelling.

The project has also tested physical models of the district heating network. The results show that the model can predict the dynamic/moving behaviour of the district heating system in terms of heat dissipation. The hope is to eventually be able to create a hybrid solution between the learning system and the physical model, a model with validation capability for the real-time learning system.

One example is where the learning system might want to send a certain amount of heat/water. The physical model can then say whether or not it is physically possible to do so and calculate whether this would result in the water reaching the customers on time. If it proves to be possible according to the laws of physics, then the sending of the heat can proceed. If it is not physically possible, then the exercise serves as a valuable input to the learning system. The algorithm will learn that this exact procedure is not feasible for the next time.

During the period 2019–2021, the goal is to have a complete learning system in place covering all Mälarenergi's district heating customers, and that is compliant with industrial standards.

Source: PiiA, Mälarenergi, RISE.

2.1.5 In search of best practice

The old business wisdom to 'follow the money' takes on new resonance for companies using AI as a transformation tool. In sectors where added value has traditionally been created through marketing and sales, it may be prudent to focus AI effort in the same areas. If operational excellence is crucial—as is the case within the process and manufacturing industry—then there are good reasons to invest heavily in AI for the supply chain and production processes, but also to develop new products and to add new customer value. And in terms of primary resources and materials, there is the potential to increase value for customers who can contribute to top-line growth.

Returning to the corporate and economic side of things, it is time to pose the question of whether the business case for AI is settled and whether now is the right time for significant investment. There may be little question of the way things are heading when we look at the bigger picture, but as we all know, the devil is in the details. In general, machine learning is a powerful technology that so far requires specialised knowledge and incredibly careful preparations, tests and validations before it can deliver.

Against this background, the primary purpose of this paper is to engage and contribute to Swedish industry's practical knowledge and preparedness for action, and to seek out best practice. While we are well aware of the current advantages and disadvantages of the technology, future developments may progress very quickly.

One model that is widely used within the Blue Institute and PiiA are the S-curve (see **Figure 4**) [10]. In the context of the digitally-driven industrial shift that we currently find ourselves in, we are preparing to leave the S-curve's initial innovation phase with its lab experiments and industry pilots, to move into the next stage with trailblazers leading the way in seeking a best practice that delivers results; we call this the 'best practice' phase.

Best practice, in turn, lays the foundation for an accelerated transformation of an industry. Experience from previous technological shifts has shown the power of good role models. For example, over just a few years in the 1980s, the Swedish pulp and paper industry became the world leader in computerised automation. One explanation for this is that company leaders were inspired by their Swedish colleagues in the sector and shared their experiences. When industry leaders dare to take the lead, rewards await in the form of competitive advantages. If you can get others to follow, industrial benefits can be created on a large scale.

Examining the development of applied industrial AI and using the empirical evidence we have through, among other things, PiiAs project base, we can identify three types of companies in different stages of the curve (see **Figure 5**):

- The majority of companies—an estimated 70%—belong in the 'aspiring for insights' category. They realise that change is coming, but still lack readiness and ability, which must therefore be developed.

- We are now seeing the rise of the 'innovation pilot' category to which an estimated 20% of businesses belong. They are engaged and have dared to take the first steps down the path to applied industrial AI or are receiving help in making preparations for applications on a larger scale.

- The 'accelerator' category includes a small group of pioneers, estimated to be less than 10% of companies, who have found their own best practice solutions and are ready to scale up and transform their businesses using AI as a tool.

In this report, we return to three prerequisites for succeeding with applied AI in industry, examining them from different perspectives:

Leadership and adaptability involve creating appropriate change teams with the skills needed for the task ahead. Still, it also involves taking into account the job changes that AI will eventually lead to. This includes having the ability to collaborate between humans and machines—collaborative intelligence—and understanding the consequences this has on the organisation and working models. To put the question of jobs into perspective, an estimated 14% of the global workforce will need to change their job duties as a consequence of AI [1].

Also crucial is data, both from an ownership perspective and a quality perspective. Information is the raw material of AI technology, which is then converted into money with the help of algorithms. The final fundamental prerequisite is security and risk management. These challenges also feature in Vinnova's 2018 study [11], and we will look more closely into these aspects in the second part of the report: The Technology.

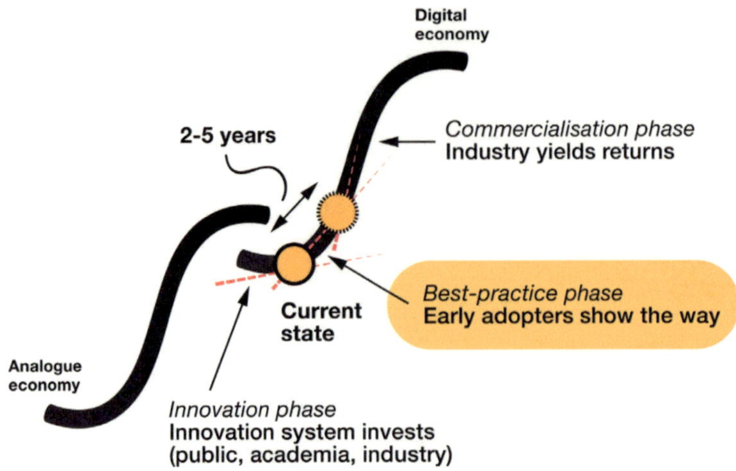

Figure 4.
Digital development with AI as an essential component leaves the pure innovation phase. After that comes the 'Best Practice' phase in which trailblazers dare to experiment and inspire others. Source: Blue Institute, 2019.

In the race to the top of the S-curve, it's crucial to address the challenges that crop up along the way. This starts with the 'aspiring for insights' category, gaining the insights they need to understand the opportunities AI presents and to know the conditions within their own companies. Such companies may need to analyse their data management, organisational data strategy and the value of their data. It may also be a good idea to lay the groundwork for rules and relevant policies for data security management within the company. This might include minimising the risk of data breaches, as well as security measures for people and assets. It is increasingly common for policies for managing data in connection with AI applications to address ethics and the risks of skewed, biased data sets when learning specific analysis models.

Those in the 'innovation pilots' category, meanwhile, have gained insights and probably also received help with organising their data and testing machine learning on one or more suitable processes. Within PiiA's empirical data, we see companies at this stage that are trying out different methods and suppliers to gain knowledge and decision-making expertise for the next step, which we have referred to as 'accelerators' in this model.

Those in the 'accelerator' group now need to increase the pace of implementation and, therefore, transfer the responsibility for transformation to their line organisations, along with appropriate expert support. These development steps also come with increasing demands on the ability of companies to manage job transformation, as well as data as a strategic asset, as well as the security and ethical issues around data usage.

A study by McKinsey examined 400 AI applications in 19 different sectors. It found that in 69% of cases, AI was a means to improve existing, more straightforward, analytical methods. Entirely new applications accounted for only 16% of cases. In comparison, the remaining 15% of cases were unable to benefit from deep learning technology for reasons, including a lack of data.

McKinsey Global Institute, Notes from the AI frontier: Applications and value of deep learning, 2018.

2.1.6 Organisational development

Applying AI is essentially a matter of organisational development. It's a skill for which different corporate leaders will have different aptitude levels. A driven

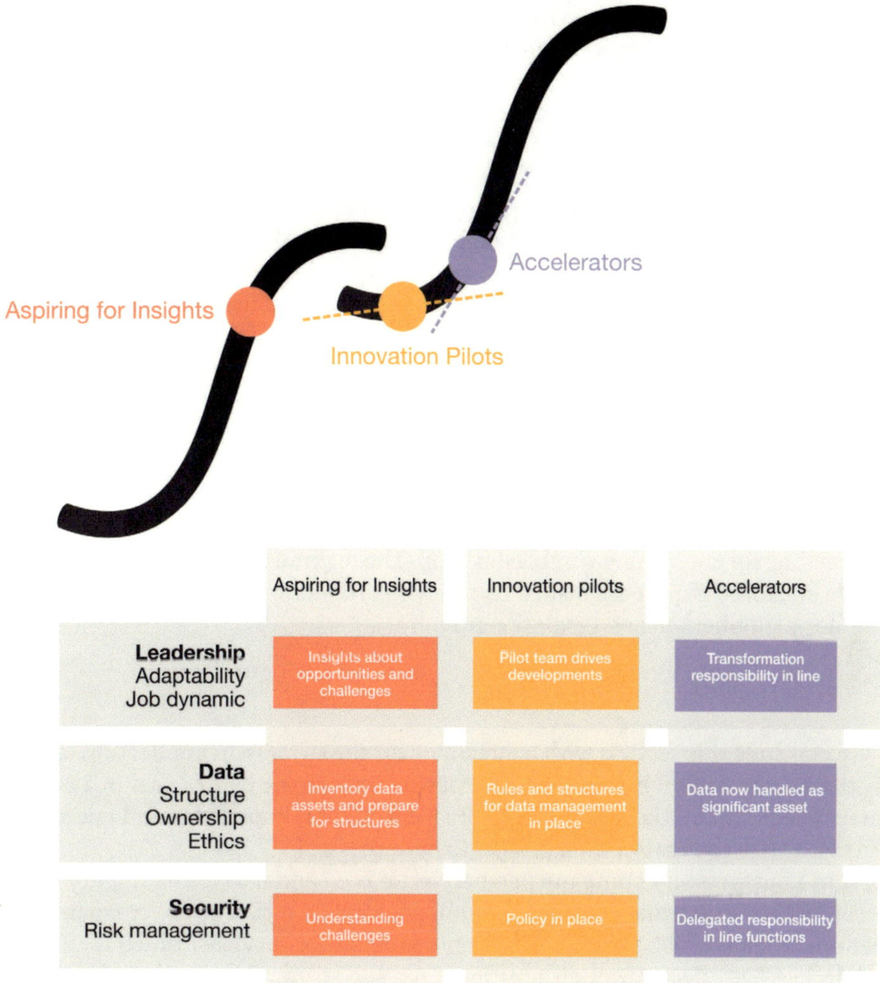

Figure 5.
Three typical development steps for implementing AI in an industrial context. From creating insight into opportunities and challenges, to the more full-scale transformation of a company's processes. Source: Blue Institute, 2019.

individual is sure to see AI for the powerful tool that it stands to become and will also have the ability to create teams in which creativity, process knowledge, and a knowledge of tools, methods and good leadership all make a difference.

Most complex processes in the supply chain stand to benefit from artificial intelligence and machine learning. In simplified terms, the methodology can be divided into four parts:

1. Data collection, preparation, and training of the model.

2. Using the trained model for analysis and prediction.

3. Using the analysis for augmentation, i.e. enhancing or increasing human abilities.

4. Using the analysis for automation, which can now be developed to new levels. It additionally becomes possible to introduce automation into areas where its implementation was previously seen as too complicated or expensive.

AI and automation are often used as interchangeable concepts, but the underlying technology differs. Automation describes systems that are programmed to perform specific repetitive tasks, such as an industrial robot which repeats the same step over and over again or a word-processing program which can repetitively perform what previously manual tasks were. AI systems, on the other hand, are designed to find patterns, to learn from experiences and to make consistent decisions. AI does not need specifically programmed paths to determine how it should behave in different situations (see **Figure 6**). Together, AI and automation may represent the next step in streamlining various processes within industry, whether these be in production or administration. Automated machines use data; AI understands data; so, they complement each other.

Augmented reality (AR), for example, has advantages in situations where people tend to perform poorly at consistently monitoring processes. AR can provide support when such monotonous tasks transition into critical business situations. AI-supported AR also helps warehouse workers and truck drivers to keep track of goods and products. AR can help process operators carry out routine checks on machines and processes, as well as providing support to service personnel, and speeding up emergency troubleshooting. The technology makes it possible to provide enhanced expert assistance remotely to production facilities far from the technology supplier's nearest expert centre.

2.1.7 The AI flywheel

We conclude this section with a metaphor for the successful application of AI—the data or AI flywheel. It is a concept that nods to the fact AI that in a business context needs an 'inertia mass'—a combination of data, knowledge and energy that all interact with each other for a project to be successful.

From a business and company perspective, it is essential to create mighty inertia masses, within which machine learning innovations (as a part of the company) fuel other operations, which in turn can become products or services, which in turn can provide leverage effects through additional AI, and so on. Those companies that are first to succeed in this area will be the few surprising 'platform' companies capable of growing with information – scale without mass (see also the section 'How digital platforms are changing traditional industry').

The AI Flywheel concept is often associated with Amazon Web Services (AWS). Amazon likes to use the word flywheel to describe how different parts of its business function as a perpetual motion machine, within which more data yields better products, more customers, and even more data and so on. The company's machine learning platforms create momentum throughout the organisation. Offering

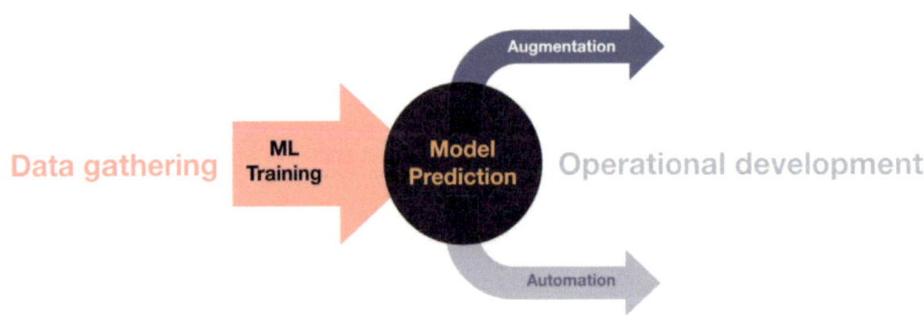

Figure 6.
The streamlining of the industry's supply chains is enabled using machine learning (ML) which makes predictions of process behaviour based on collated data. These predictions can then be used to enhance people's abilities or to automate processes fully. Source: Blue Institute, 2019.

machine learning to outsiders as a paid service is in itself profitable. The fact that such projects also generate data provides even more leverage.

AWS grew by 40% in the first quarter of 2019 with an operating margin of 29%. Somehow, Amazon has cracked one of the business world's riddles; how to create small innovative teams within a much larger (bureaucratic) business. Agile teams that learn quickly develop competence in many AI-related areas and then spread the knowledge to the rest of the organisation in useful, coherent and collaborative ways that create value throughout the organisation. It is impressive. But how can traditional industry adopt the logic of the AI flywheel? In this context, we will content ourselves with exploring the concept of the basic organisation—the team—which makes more prominent strategies possible.

The most common mistake when companies adopt AI is that they start focused on the technology and not the business needs. Hiring data scientists and giving them access to data to build 'something interesting' is very likely to lead to a dead end.

As discussed, teams with different competencies are required, of which the four basic ones are: product or production managers (who can describe, in detail, which problem to solve), systems engineers (who know what data can be used), computer scientists (who know how to build useful models), and cross-discipline specialists. This latter group can be called DevOps engineers or translators (Development-Operations is a term borrowed from the IT industry and agile development, while translators is a term borrowed from McKinsey). In this context, this invaluable category of people (usually made up of engineers) create commitment and knowledge and can move relatively freely between production, process and operational development, as well as customer's needs and preferences.

The members of the team use different tools and work together to solve the group's challenges. But each is ideally a person with the ability to scale up their operations and fill the flywheel metaphor with power and torque.

Of course, more or less similar competence combinations have always been used to develop products and operations. What is new is machine learning technology and a requirement for expertise in this area, at least for the time being. For companies that are advancing in the AI world, it is necessary to take serious steps to create such pilot teams. For companies that are scaling up, the accelerators, such teams are still needed but should push up against the boundaries and away from regular business operations. In such circumstance, there is a potential to achieve flywheel effects at the company level.

Mining Magazine asked 115 mining companies where they saw the most significant opportunities for AI. The answers are interesting, and the priorities identified would resonate with those in a range of sectors outside mining.

1. Better decision management and error minimisation.

2. Understanding market trends and customer behaviour.

3. Discovering mineral deposits.

4. Autonomous vehicles and drilling units.

5. Automated monitoring of health and safety risks.

6. Increased productivity.

7. Production planning.

8. Maintenance planning.

9. Automation and support for regulatory and team compliance.

10. Rescheduling after unforeseen events.

2.2 Supply

We will now move on to analyse the supply side of things by assessing technology development and the supplier system. We will do this firstly through an overview of the area of AI development with a brief conceptual summary, followed by describing the structure and strategic challenges of the supplier industry. The chapter ends with our conclusions about the effects on the market system.

Behind the applied development of digital technology for the industry lie significant investments. In simple terms, they can be described as three development hubs, each of which, according to Blue Institute estimates, accounts for about one-third of an impressive SEK 1.5 trillion invested globally in R&D each year (see **Figure** 7):

- The driving forces for the first hub—private-public innovation collaborations—consist of national ambitions along with industry insights on the values at stake in the fourth industrial revolution. These have generated significant investments through which private and public national capital unites in various programs around the world. Industrie 4.0 in Germany is among the most renowned. In Sweden, the Strategic Innovation Programs have been established focusing on selected growth areas. By and large, these private, public investments amount to 500 billion annually.

Figure 7.
The applied development of digital technology for the industry is being driven by historical investments in what can schematically be described as three fields of influence or developmental hubs. Each of these accounts for about one-third of an impressive 1.5 trillion invested in technology development each year. Source: Blue Institute.

- The second development hub consists of large tech companies' annual R&D investments in the construction of cloud services and investments in AI. The Big Five—Apple, Alphabet, Microsoft, Facebook, Amazon, plus IBM—are estimated to invest nearly 500 billion a year.

- The third hub consists of the traditional ICT industry plus automation providers.

- A significantly more fragmented industry, but a further estimated 500 billion is invested in research and development.

Development projects are now starting to leave the laboratories on all fronts and to arrive on the market, first as innovation projects, then as best practice, and then, eventually, as robust commercial offerings.

The driving forces behind each hub are essential: a return is required on those large investments; standardisation work is about to yield results, and the world's industrial leaders are beginning to understand the vast sums and value at stake with the impending transformation of the industry. Last but not least, a dynamic is arising as the three developmental hubs start to propel each other, as development results are released onto the market. This means that the momentum for the whole system increases further.

2.2.1 Significant breakthroughs in AI technology

Significant breakthroughs in AI are coming to the public's attention more and more frequently. This is occurring across all application areas. Initiatives within foundational research and product development are producing visible results, and the development curve is growing steeper.

1. The most mundane and yet revolutionary example is personal assistants such as Alexa, Siri, and Google Assistant, which are continually learning more, and making themselves known via our phones and calendars.

2. Estonia wants to make its government and judiciary as efficient as possible and so is developing an AI model to act as a judge in minor legal cases within where the value of the dispute is less than EUR 7000.

3. The OpenAI development institute recently unveiled a pre-developed language model (GPT-2) that can generate realistic texts in different kinds of style and prose. The text robot is so powerful that the Institute is refraining from releasing the fully trained model due to the risk that it may contribute to the spread of so-called 'fake text'.

4. In March 2019, the Google company DeepMind presented a model capable of diagnosing complex eye diseases in real-time. In thirty seconds, Google cloud algorithms can provide a detailed prognosis with the same precision as world-leading eye specialists.

5. In January 2019, a research group at Columbia University announced that they had made significant progress by creating a robot that can imagine itself. After a day of intensive training, it was able to adapt to different situations, manage new tasks and detect and repair injuries in its own body.

2.2.2 Basic AI concepts

The concept of Artificial Intelligence (AI) lacks established unambiguous definitions and demarcations. The nature of the field allows for broad philosophical, social and mathematical discussions. AI Research in itself is both specialised and dispersed across subfields that often lack contact with each other. This makes the area in its entirety challenging to comprehend fully. However, for this analysis, we have chosen to use the same definition that Vinnova used in its study of artificial intelligence for the Confederation of Swedish Enterprise and Society, 2018, namely:

'The ability of a machine to mimic intelligent human behaviour. Artificial intelligence is also the designation of the science and technology field that aims to study, understand and develop computers and software with intelligent behaviour.'

AI can thus be defined as the ability of machines to perform cognitive functions that we associate with human minds, such as perception, reasoning, learning, interacting with the environment, problem-solving and, ultimately, even creativity.

A dominant theory for assessing the characteristics of AI systems is the so-called Turing-test [12]: a computer passes the test if a person, after having asked several written questions, cannot discern whether the answers have come from a human or a machine.

The set of abilities [13] which are considered essential to enable artificial intelligence to be experienced as humanly intelligent include:

- Natural language management (NPL).
- The ability to store knowledge.
- An automated ability to reason.
- Machine learning to make discoveries tailored to given conditions.
- Vision technology to see.
- Robotics for moving or manipulating objects.

Another central figure of thought in AI is the rational, intelligent agent. The agent is a piece of software, an algorithm, which is expected to operate autonomously and be able to sense its environment, endure for a long time and adapt to changing conditions, as well as setting up and reaching goals. All the properties of the Turing test are also valid for the rational agent to function.

2.2.3 Narrow and broad AI

AI can be classified in many different ways, but a standard description is narrow versus broad or general AI. All the AI that exists today is narrow or rather; specialised. Our intelligence, however, is general. If at some point in the future, AI becomes general, it will probably change society fundamentally. When, and if, this will occur is debated and time spans from ten to several hundred years from now—or never—have been suggested. A marginal part of AI and machine learning development today touches on general artificial intelligence. The majority of development resources are focused on making narrow/specialised AI more effective.

DOI: http://dx.doi.org/10.5772/intechopen.93098

2.2.4 Machine learning

Machine learning is an area of computer science that explores methods of getting computers to learn from data without having been programmed for the task. The area is related to statistics and pattern recognition (see **Figure 8**).

Machine learning has been the prevailing developmental track for practical applications of AI for a few decades. Progress has been made through the application of machine learning to increasingly larger sets of data. In a relatively short time, different machine learning subtypes have been developed, within which algorithms are continually being improved and adapted for various applications.

Those in the sector talk of 'supervised learning', which means that an AI algorithm uses sets of data to 'train' while receiving feedback from people to learn when the relationship between given inputs and outputs meets the requirements. Unsupervised learning means that the network works without prior knowledge. The computer must teach itself the underlying structures only using the input provided and not through any pre-given response.

Deep learning is a type of machine learning that can process a wide range of data points, may involve more straightforward data processing, and can provide more accurate results than traditional machine learning approaches—although it requires a greater amount of data to do so.

Deep learning connects software-based 'neurons' in a neural network. The network can receive large amounts of input and process it through multiple layers that learn more complex functions for each layer. Once the network has learned, for example, what an object looks like, it can recognise the same item in a new image.

Reinforcement learning means that the algorithm is rewarded when it is successful, for example, through the accumulation of points in a game using a step-by-step approach to reach the maximum score. The algorithm remembers the successful features and outcomes and corrects itself for ever better results. It learns by discovering. This method, which is inspired by the brain's dopamine system, is used when there is insufficient training data available, when the ideal, ultimate goal cannot be defined explicitly or when the only way to learn about the environment is to get started and interact with it.

Reinforcement learning is the latest breakthrough in machine learning and received widespread publicity when the AlphaGo computer program from Google-owned DeepMind in 2015 defeated one of the Chinese champions in the board game Go. Since then, the technology has gained several commercial breakthroughs and is used, among other things, to streamline the operation of gas turbines, wind

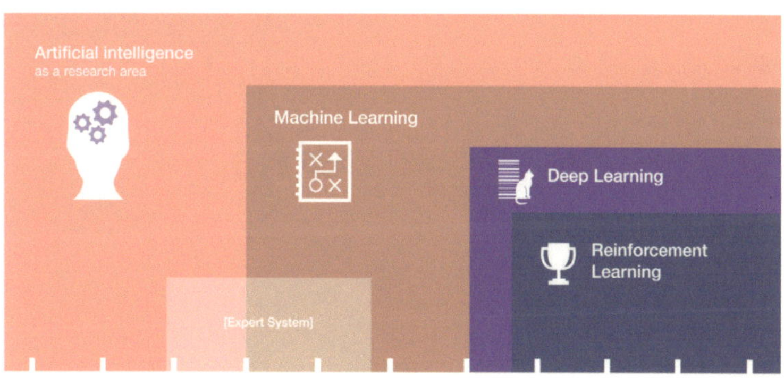

Figure 8.
AI and different learning methods put in a development perspective.

turbines and energy use in computer halls. The method has good potential for future industrial applications.

For an in-depth look at machine learning, we refer to Part 2 of the report. In this section, we will continue with an analysis of the supplier system.

2.2.5 The availability of AI is dominated by significant platform suppliers

The range of applied AI technology is increasing at a rapid pace: we are seeing infrastructures, tools, algorithms, data and pre-trained AI models for various purposes, all offered as standard products by all major platform providers (see **Figure 9**). The development of automation suppliers means that industrial control systems will also get built-in machine learning capabilities. The telecommunication industry is beginning to offer distributed and cloud-integrated edge technology that shares IoT concepts. There is also an increase in specialised AI providers for different applications.

Technology providers of all categories contribute to the quick commercial distribution of machine learning technology, and several market studies show strong anticipated growth in the coming years. According to the analysis company Markets & Markets [14], the market for machine learning, language management and vision systems will grow from about USD 22 billion in 2018 to more than USD 190 billion in 2025.

This corresponds to a growth rate of almost 40%. IT consultants and system integrators are also seeing business opportunities and are gaining knowledge around the new tools. According to various studies, AI development within the IT consultancy industry is seen as among the most pronounced technological breakthroughs of all time [15].

But it is the big tech companies that are driving the lion's share of commercial AI development. The platform companies Apple, Alphabet, Microsoft, Facebook, Amazon and IBM together have an estimated value of over USD 4 trillion. They account for 55% of the value of the Nasdaq 100 Index (see **Figure 10**).

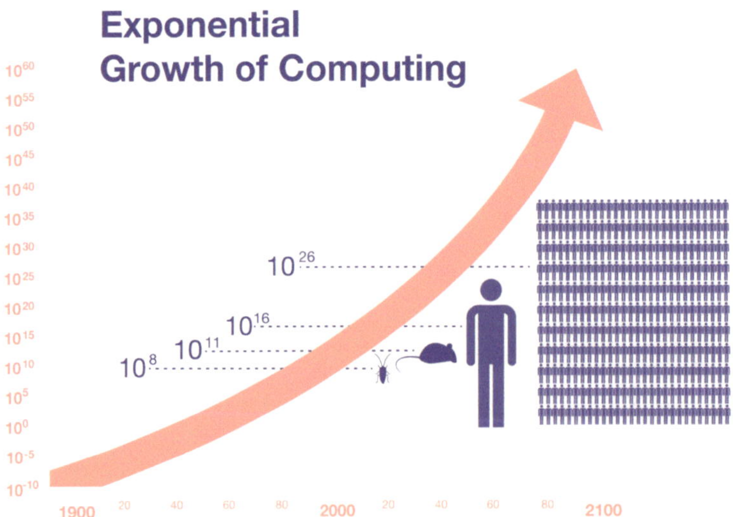

Figure 9.
Computing power is a prerequisite for the development of artificial intelligence. It took 90 years to reach the first million instructions per second (MIPS) per $1000—now 1.2 MIPS/$1000 are added every hour. Source: Ray Kurzweil and KurzwelAI.net.

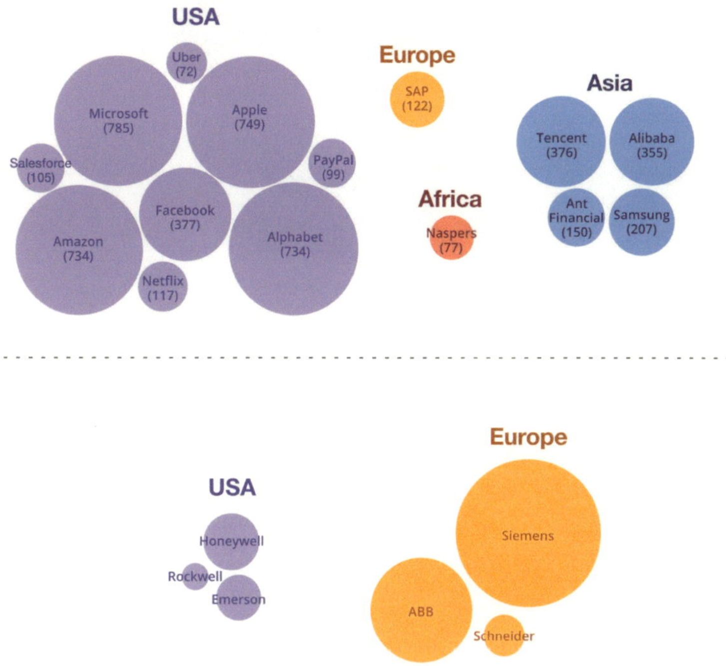

Figure 10.
Top: Platform companies in the US have a definite lead in the development of AI platforms. Market values December 2018. Below: Comparison between the three largest IndTech companies in Europe vs. the USA. Source: EC, EU industrial policy after Siemens-Alstom, Blue Institute.

It is these companies that are behind the commercialisation of AI in the West. In Asia, Tencent, Alibaba and Samsung are dominating, while Europe lacks corresponding strengths. In comparison, however, Europe has advantages in industrial technology companies (IndTech). The European companies ABB and Siemens are significantly larger than their US counterparts.

The long-term AI strategies of the platform companies include a large R&D component reinforced by acquisitions. For example, Google's purchase of DeepMind for USD 400 million, Twitter's acquisition of Magic Pony for USD 150 million and Microsoft's purchase of Github for USD 7.5 billion. To ensure the availability of top academics from universities, employees are being offered high salaries, unlimited computer and computing resources and minimal bureaucracy [16].

The concentration of resources, expertise and access to data is therefore currently focused on a few global commercial players. This is a part of the platform war, the battle for market domination over cloud services within which the mightiest battle is between Microsoft, Amazon Web Services and IBM.

Underlying this growing market landscape is the quiet market dominance of the platform companies. Generic cloud products reach end-users directly or via domain providers. Within the industrial context, automation, process and machine suppliers can add industry-specific value.

Automation suppliers operate in this way, serving as targeted market channels that increase the value of the platform companies' large-scale production of computing power and machine learning (see **Figure 11**). Two groups of more independent initiatives flank these platform alliances and centres. One consists of companies that sell predictive analysis solutions and build individual platforms. According to a qualitative evaluation by analysis company Forrester [17], this segment is led by SAS, IBM and SAP, with a long tail of smaller players.

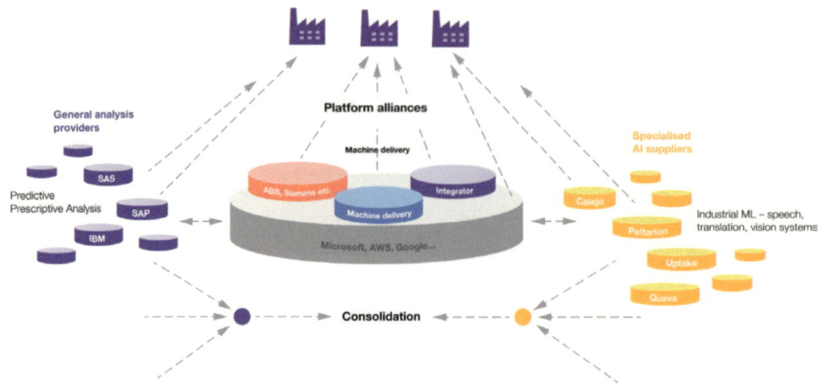

Figure 11.
For industrially applied AI, three groups of suppliers can be distinguished. In the middle are the large general platform providers, which are creating more and more alliances with companies that can serve as value-adding specialised channels of the platform suppliers' large-scale AI offering. These are flanked by general analysis players, within which there are several large companies, as well as specialised industrial suppliers of various sizes. The dynamics of the industry are expected to give rise to significant consolidation. Source: Blue Institute, 2019.

The second flank is made up of specialised companies that supply systems for speech, language, vision and generally applicable machine learning platforms for industry and others. There is a similarly long tail of small and medium-sized players. The sector is immature, heavy with development and likely to undergo further consolidation.

2.2.6 Automation, industrial IT and digitalisation lead to IndTech

As demand for digital platforms increases and the boundaries between industrial IT, automation and other domains become blurred, more and more players are becoming interested in industrial technology. Cloud service providers Microsoft, IBM and Amazon are building alliances and challenging traditional automation providers such as ABB, Siemens, Emerson and Rockwell.

A second challenge for automation suppliers comes in the form of ICT companies. Ericsson, Cisco, Huawei, Nokia, Samsung and other industry operators are looking for applications for 5G technology, and they consider the industry's Internet of Things an opportunity. The goal of 5G is to make wireless technology available for applications that have significantly higher bandwidth, speed and reliability requirements than personal use applications. According to Ericsson, operators stand to increase revenues by 34% if the process industry and electricity industry increase the use of wireless communication [18]. Ericsson is supporting this development through its IoT Accelerator Platform Initiative. This is a one-stop-shop that will make it easy and safe to connect IoT modules and that will also assist in translating the technology into a business setting.

Suppliers of industrial IT and automation now need strategies to deal with platform companies as well as IoT infrastructure.

The dominance of the platform suppliers makes it impossible for automation companies to avoid dependency on their resources, and the challenge for them will be to create relationships that develop the industry's strengths (industrial, process and customer relationships) and increase customer value without becoming marginalised in the platform war. The platform and ICT companies can, by extension, be expected to contribute to making automation solutions more uncomplicated and more cost-effective and also to add new value. Intelligent apps in intelligent ecosystems are a development trend that has the potential for a significant impact, thanks to the fact that platform companies are opening up their APIs.

Platforms provide process and machine suppliers with additional automation and the potential for advanced in-house analysis. Machine suppliers and the automation industry share an ambition to build connected competence centres for optimisation and fault remediation in customer facilities. By extension, this strategy is also about competition for the valuable data that can be mined from industrial manufacturing.

A new image for the industry's suppliers is emerging, where the ability to create real customer value will distinguish winners from losers. If IndTech suppliers succeed in this, they will have a much more developed role in future industrial value systems as highly specialised vertical suppliers of efficiency and quality. At the same time, the process flows will be held together using collaborative logistic systems.

2.2.7 Industrial case study: focus on mining companies: the ENSAF project: energy and safety diagnostics.

There is currently a significant interest in the early diagnosis of problems in underground mining facilities. There is a trend towards achieving fully automated mining, meaning that should hazard arise underground, increasingly there are few or no personnel around to address them. This makes it crucial to have a capacity for early detection of risks from fires caused by factors such as the overheating of vehicles, equipment, cables and so on.

It is possible to detect the risks of overheating early by placing sensors in facilities and on mining vehicles, which then continuously transmit information to a central diagnostic system. With this approach comes the need to continually train staff in different fire scenarios and in managing different situations.

Project ENSAF (a PiiA funded project) is a collaboration between ABB, RISE, Mälardalen University and Epiroc, which owns one-third of Mobilaris. With the assistance of the Swedish mining companies, who are involved in helping set project requirements and who act as sounding boards for the work, the project is aiming to find a solution to the significant challenges that fires pose in mines. The proof of concept, which involves the fitting of suitable sensors to one of Epiroc's vehicles in one of Boliden's mines, is planned to start at the end of 2019. Data collection will continue into 2020.

Boliden is a high-tech metals company with its mines and smelters, and it is working over the long-term to guarantee society's access to the base and precious metals; from the mining of ores (minerals) to the production and delivery of high-quality metals to the industry. Its production capacity is high due to experience, innovation and advanced technology, developed in collaboration with various Nordic technology and engineering companies. Approximately 5800 people work at Boliden, and its operations are conducted in Sweden, Finland, Norway and Ireland.

'In the event of a fire, the smoke, in particular, poses a serious threat to both people and appliances. It is, therefore, important to be able to detect if a fire is about to start'.

2.2.7.1 The challenge

The destructive impacts of mining fires can be significant, both in terms of human suffering and in terms of costs and lost revenues. On an annual basis, about one fire per week occurs in a Swedish mine, with the majority started by vehicles moving about the mine. Sweden has been spared from major mining fires in modern times, but in the global sector, it happens all too often. Take the well-documented case of the Pike River coal mine in New Zealand, wherein 2010 some 29 people died following several gas explosions. In addition to all the human suffering, the accident put the plant out of service for 45 days due to fire extinguishing and remediation work. The production loss corresponded to half a billion Swedish Crowns, in addition to all the restoration costs and elevated insurance premiums. Additionally, large penalties can be imposed if the root cause of a mining accident can be attributed to safety deficiencies.

In the event of a fire, the smoke, in particular, poses a severe threat to both people and appliances. It is, therefore, essential to be able to detect if a fire is about to start. Other types of risks that can arise include leaks on hydraulic lines, which may cause oil under high pressure to produce intense sprays or fog formations.

2.2.7.2 The experience

The ENSAF project is creating a system that leverages all existing fixed measuring sensors in mines and on vehicles and links various measurements (such as temperature, hydrocarbons, CO_2 and CO concentrations, relative humidity and flow) to each other via simulation models. The aim is to identify problems at an early stage and nail down as precise a location for the problem as possible. The information collected is used as input to a decision-tree model to assess the risk of fire and also to determine the content of any toxic gases that may be hazardous to humans or machinery through corrosion. It will be possible to follow real-time developments in the mine and compare the measurement data collected with the simulations.

Development of the system is primarily conducted by Epiroc and Mobilaris, ABB and MDH, with Boliden acting as a sounding board and contributing with experience on mining conditions. RISE contributes, among other things, with knowledge around fire and protection, as well as conducting fire tests in its premises in Borås. ABB sees the potential, through the conceptualisation of development, for a complete solution that could be offered commercially, incorporating its automation system. Epiroc provides the measurements from the mining vehicles, with the data collected in its Certiq system. It then communicates the safety information to Mobilaris.

In terms of sensors, the project has been able to detect, among other things, gas formations caused by cables loaded with currents higher than they are rated for. Smoke detectors are used, to detect not only the shape of smoke but also oil mist that can occur through leakage. Thermal cameras can be used at longer distances to detect temperature increases on, for instance, cables.

Today, Certiq collects object data (e.g. hydra-like oil level, engine power, etc.) 24/7 from several thousand mining vehicles across the world. Since the establishment of ENSAF, the system has gone from communicating and gathering data from a few hundred vehicles to some 3000. The sensors tested under ENSAF are now implanted in Certiq, which transmits sensor data to Mobilaris. Here, gas values from the vehicles in the mine shaft, for example, can be monitored and may trigger alarms. In a fire situation caused by a mining vehicle, it is possible to correlate all available data and perform root cause analyses, and, with the help of deep learning, provide answers to the cause of the fire. In the future, artificial applications could anticipate possible fire situations and suggest appropriate maintenance activities to avoid fires.

Using measurements and analyses, ventilation can be adapted to suit real needs. This can save energy without risking functionality and provide a good working environment for both people and machines. Close to 50% of the energy consumed in an underground mine goes to ventilation. In the case of a fire hazard, the system is set to minimise the risk to underground personnel. The system is based on sensors that can communicate with each other to increase communication security locally and also with central systems that can give an overall picture of the situation.

Sources: ABB, PiiA, MDH.

This chapter continues in part 2.

Author details

Örjan Larsson
Blue Institute and Strategic Innovation Program PiiA, Västerås, Sweden

*Address all correspondence to: orjan.larsson@blueinst.com

References

[1] Accenture, Reworking the Revolution. Dublin, Ireland: Accenture Strategy; 2018

[2] McKinsey Global Institute. Notes from the AI Frontier: Modeling the Impact of AI on the World Economy. New York City, New York, United States: McKinsey; 2018

[3] Rao AS, Verweij G. Sizing the Prize. London, UK: PWC; 2017

[4] Moore JF. Predators and prey: A new ecology of competition. HBR. 1993

[5] Nachira F. Towards a network of digital business. Ecosystems. 2002

[6] Parker G et al. Platform. Revolution. 2016

[7] Brynjolfsson et al. Scale without Mass: Business Process Replication and Industry Dynamics. Boston, US: Harvard Business school; 2008

[8] Schwarz A, Larsson S. Plattformssamhället.Plattformssamhället: Den digitala utvecklingens politik, innovation och reglering. Lund, Sweden: Fores; 2019

[9] Microsoft, The Future Computed. Redmond, Washington, US: Microsoft Corporation; 2019

[10] Larsson Ö. Blue Institute. PiiA Insight. Västerås, Sweden: Swedish IndTech; 2018

[11] VINNOVA. Artificiell intelligens i svenskt näringsliv och samhälle. Stockholm, Sweden: Vinnova; 2018

[12] Turing AM. Computing Machinery and Intelligence. Oxford: Mind Oxford University Press; 1950

[13] Russel N. Artificial Intelligence, A Modern Approach. London, UK: Pearson; 2016

[14] Markets & Markets, Artificial Intelligence Market. Pune, India: Markets and Markets; 2017

[15] Konsultkompaniet, Så blir it-branschen 2019. Stockholm Sweden: Konsultkompaniet; 2019

[16] Financial Times AI academics under pressure to do commercial research. 2019

[17] The Forrester Wave: Predictive Analytics and Machine Learning Solutions, Q1. 2017

[18] Dagbladet S. Ericsson ikapp och förbi Huawei I 5G-racet. 2017

Chapter 4

AI & Digital Platforms: The Technology [Part 2]

Örjan Larsson

Abstract

This essay aims to describe the dynamics at play in the field of industrial AI, where the significant efficiency potential is driving demand. There are rapid technological development and increasing use of AI technology within the industry. Meanwhile, practical applications rather than technical development itself are creating value. The primary purpose of the article is to spread knowledge to industry. It is also intended to form the basis of the Swedish innovation program PiiAs ongoing work around open calls and targeted strategic innovation projects. The basic approach taken is to investigate both industry demand for AI and how the supply of technology is developing. AI takes in a broad and dynamic range of concepts, but it should also be considered in the even broader context of industrial digitalisation. It is not just a question of technology development, but equally about application knowledge. Realising the full potential of AI requires the ability for change within individual companies, but also to handle exchanges and interactions in changing ecosystems. The article has been divided into two sections: *The Market*, in which we assess the development and the consequences on the factory floor; and *The Technology*, which provides a more in-depth understanding of the structures of industrial IT and machine-learning technology. The article concludes with four practical examples from the industry.

Keywords: AI, artificial, intelligence, PiiA, blue, institute, automation, algorithmization, platform, data, process, industry, IndTech, digitalization, digital, twin, ecosystem

1. Introduction

This article has two primary purposes: the first to provide the industry with an evaluation of the importance of AI development as a force for change. Second to create an internal basis for the Swedish Innovation program PiiA's future development efforts, within which AI can be described as the next phase of industry's digitalisation. Both these objectives are naturally compatible with the overall ambition of the report: to reach our target group of industry leaders and to serve as a source of knowledge for ongoing activities within relevant companies.

Technological, industrial development is awash with grand ambitions that have turned into mere passing fads and costly dead ends. With this in mind, throughout our work in assessing the development of AI, we have endeavoured to take into account the magnitude and direction of different vectors of change. On the one hand, we have attempted to understand the power of demand for AI by assessing the economic impacts at a macro level. We have focused on productivity and qualitative values at various stages of industry value systems. On the other hand, we have attempted to assess the range of available technologies by analysing initiatives taken on a global scale and through focused academic research. We have also put considerable effort into understanding the major commercial – or applied – forces that are crucial to development, both in the short and medium-term.

We have also strived to place AI development in the context of current systemic developments, as characterised by the "platformisation" of company IT resources. By this we mean the transfer of automation and IT support to the cloud – a trend that is creating new competitive dynamics. Finally, we have attempted to translate this big picture into real impacts on the factory floor and to revisit well-known concepts such as organisational development which – with the help of the raw power of AI technology – have the potential to make the previously impossible, possible.

The project was a collaboration between PiiA and Blue Institute, with valuable input from Blue Institute's network of CEOs and industry leaders on all levels. A big thank you is extended to everyone who contributed to this study.

2. From market to technology

2.1 One platform to rule them all...

"One Ring to rule them all" was the theme of Tolkien's Lord of the Rings. The parallel with the power of algorithms – and thus the importance of platforms – is not so far remote from that idea. In the platform war, there is currently a battle for dominance of the market, with resource concentration an ongoing feature. At the same time, most companies have unique needs for which the general services that the big tech companies offer in their public clouds are not enough. Most companies, therefore, now have some form of a privately-owned platform (private cloud). One trend in the market is for hybridisation between one or more public clouds and a company's private environment.

The issue of cloud complexity is becoming increasingly topical as more and more specialised domain-specific clouds/platforms are launched to the market (**Figure 1**). The concept of vertical clouds has taken hold and complements the original "general" clouds, which are now also called horizontal clouds. Vertical clouds represent industrial verticals and subprocesses within each vertical. They typically take the form of a PLM cloud, an MES cloud, an automation cloud, and so on. Here we can see, as described above, how automation vendors are now adding vertical domain-specific clouds or platforms to their offerings.

Another contemporary trend is locally distributed clouds known as edge computing or sometimes fog, which we touched upon earlier in the report. Edge computing is expected to have an increasing significance as AI technology increasingly requires a local capacity to complement the core resources of server halls.

One practical, short-term solution to get all these clouds working together through hybridisation solutions. The big hope for a long-term solution for industry in operational applications lies in standardisation which would make it possible for different environments to be combined in the same physical facilities. Standardisation work is going on within ISO, among other areas. The idea that we, like Windows, would become a de-facto standard is not likely, even though Microsoft is the provider that is currently the most successful in production-related applications.

DOI: http://dx.doi.org/10.5772/intechopen.93579

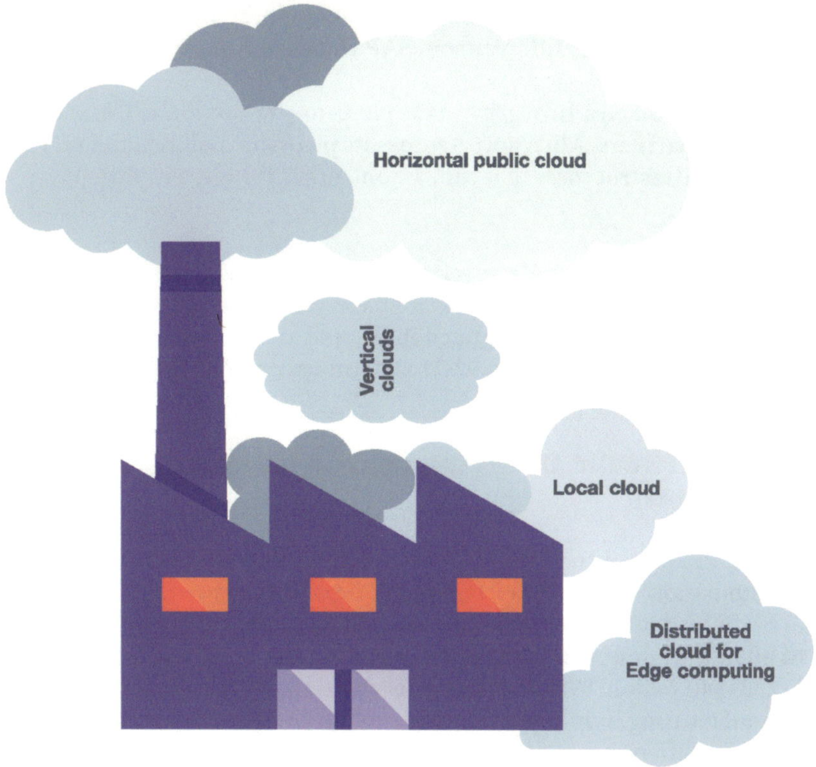

Figure 1.
Cloud structures are becoming increasingly complex. Public, vertical, local and distributed clouds are four varieties of the cloud with related services and infrastructures. These platforms are suitable for coordinating through hybridisation and, in the longer term, common standards.

2.2 Automation suppliers

There is currently a developmental trend that has the highly consolidated automation industry and leading companies such as ABB, Siemens, Emerson and Rockwell all moving in the same direction. Their common goal is to achieve market platforms that are specifically tailored to the industry. While these structures can be likened to operating systems for IoT, they meet the criteria for platforms as they match different types of users against each other. (See also the section on vertical clouds on the previous page).

AI is being used as a management and administration tool within these platforms and can also be used to produce products in the form of smart apps. These can be tailored to different applications and are available for purchase with one click via corporate app stores. Specialist centres are also being created for various product and industry applications and being linked to production facilities to allow for online optimisation and troubleshooting by experts.

- Schneider Electrics' IoT platform is called EcoStruxure Platform and uses Microsoft Azure.

- In 2018, Emerson acquired GE's famous Predix Platform, which uses services from Microsoft Azure and Oracle, among others.

- Rockwell has Connected Enterprise and Factory Talk, which also use Microsoft Azure.

- Siemens' investment in this area is called MindSphere and rests on resources from Microsoft Azure, IBM Watson, SAP and even Amazon Web Services.

- Microsoft has taken a firm grip on the close-to-production IoT market through its partners. Microsoft Azure is its platform and includes services, tools, and infrastructures that can, among other things, simplify AI development.

- Service offerings include Microsoft Cognitive Services, a set of pre-built AI features, including vision, speech, language, and search functions. Everything is in the cloud and can be integrated into applications. Some features are customisable and can be optimised to transform and enhance organisational or industry-specific processes.

- ABB's venture is called Ability and is built on solutions from Microsoft Azure and IBM Watson.

The world market for IndTech – products and systems for industrial digitalisation and automation – is worth USD 340 billion and has a growth rate of 6–7%. The area can be divided into IT and OT (operational technology). The IT share is USD 100–110 billion, while operational technology for production and logistics accounts for USD 230–240 billion. Within OT, the distribution is 45 per cent for automation for the manufacturing industry and 55 per cent for process automation.

OT includes different types of industrial control systems and field equipment such as instrumentation, drive systems and robots. A particular growth area is industry's Internet of Things which complements traditional system environments. Several platform suppliers are now also launching dedicated and distributed systems for machine learning at a local level. Edge capacity on the factory floor can thus effectively be integrated into the cloud. For example, in 2018, Google released the third generation of the Tensor Processing Unit (TPU) chip.

This parallel development has led to several automation companies developing stand-alone AI modules with neural networks that can be plugged into the racks of the control systems. With this comes pre-custom type solutions for different processes or process objects.

2.3 Is AI expensive?

The computerisation of industry in the 1980s and 90s cost a large amount of money. Machines were replaced, and investment mainly focused on large process control systems and specialised computers, while thousands of kilometres of cable were laid. Air-conditioned computer halls were built, along with control rooms and cross-connection spaces. In short: it was expensive, but a good investment, nonetheless, as productivity and key quality figures skyrocketed. This type of primary investment will always be needed when rebuilding or when new investment is required, but the nature of digitalisation alters the equation. Consider the so-called "logic of small streams", where a large number of spread-out, smaller contributions can come together to produce great results.

Today, you do not need to build air-conditioned computer halls and to buy servers to bring good ideas to fruition. It's easy to order as much computing power and functionality as you require – including AI tools – from the cloud at comparatively low prices. The majority of the data monitoring and collection infrastructures one might require are already in place. If sensors and hardware are needed, there are – or will soon be – cost-effective IoT modules that will meet even the strictest precision and environmental requirements. Communication solutions are also expected to be wireless, reliable and inexpensive in the future.

These developments are leading to an evolving approach to change. It's becoming possible to be creative, to test the boundaries, and to think outside the box. Massive investments will not be required to unlock parts of the hidden value within plant and production processes and to outperform both competitors and customers' expectations.

The challenge with succeeding with AI is less about expensive investments in computer technology and more about obtaining the right skills in the right constellations. Experience shows that good AI projects are characterised by successful teams where domain and process knowledge, knowledge of analysis, and the right tools, all play a crucial role.

2.4 Concluding thoughts about the market as a framework for technology development

Our first aim was to examine the expected value-creating effects of AI and, based on the evidence we have studied, we have found that AI's impact on productivity development is likely to outpace by a factor of two previous, generic technology shifts, such as the introduction of steam, robotisation and IT. According to PwC, the global GDP in 2030 could be as much as 14 per cent higher due to the AI effect. Global impacts on industrial sectors leading up to 2030 could amount to as much as USD 2.3 trillion.

Our second ambition was to examine whether the development of AI and the associated supplier system could meet the demand for the technology that is being generated by the potential value creation it brings. Our view is that the centre of the development is now leaving the initial innovation phase and moving into the best-practice phase. This assessment is based partly on the fact that the large R&D investments being made need to yield profits, and partly on the fact that standardisation work is well on the way to show results. Besides, industry leaders around the world have come to realise the vast sums at stake in the coming transformation of the industry. Furthermore, there is also a dynamic that will arise when the three developmental hubs begin to work together, once development results reach the market. This is likely to produce an increase in torque for the entire system.

Thirdly, we wanted to understand whether this development was sustainable or whether we are currently seeing a hype effect that will wear off, with the actual market breakthrough coming much further down the track. Looking at Gartner's (often challenged) model Gartner's Hype Cycle for Emerging Technologies, there are different technical aspects of AI spread out across the different phases of the model. For example, deep learning using neural networks – which can be viewed as representative of the industrial use of AI – is currently at the "peak of inflated expectations" stage (Aug 2018) but could move to the "plateau of productivity" within 2–5 years.

Our assessment and our S-curve model suggest that digitalisation, in a broad sense, has now reached the beginning of the "plateau of productivity": best practice. It is challenging to make timing predictions around the introduction of AI to industry. On the one hand, there is a spread over different verticals with different conditions, and there are also various activities underway, ranging from management and administration to forecasting and foresight, to operational functions in production and logistics. In some areas, AI has already established technology, while in others, it is still in the developmental phase.

But we also found via our empirical evidence that machine learning is being quietly tested in many more places than you might imagine. Safety reasons dictate that the incubation period for new technology within heavier industries is much longer than in other commercial areas and longer still than the consumer area. With this in mind, 2–5 years seems not a long time, but rather a reasonable action period for translating ideas into operational benefits. A two-to five-year timeframe for the more permanent establishment of AI technology is also in line with our analyses, with the focus point shifting from a "best practice" situation to a commercial production breakthrough on the S-curve.

We conclude that the massive underlying forces driving both the demand for and supply of technology guarantee a stable development outlook for AI for industrial applications.

Within the period mentioned above, the timing seems right for a match between technology reaching the industry and the spreading of insight into the possibilities of this technology. This has the potential to create a significant industrial movement and thus deliver increased commercial demand at the company level and the achievement of previously unachievable results in production systems.

We also expect structural changes in the supplier system as entirely new concepts reach the market and previous industry and supplier limitations cease to be valid. There will be incentives for both extensive consolidation and repositioning in many areas.

Industry case study: Kone and ThyssenKrupp - Preventive Predictive Maintenance

Two hundred and sixteen centuries is a long time to wait for a lift. That figure is an estimate of the cumulative annual stopping time of the twelve million lifts in the world, moving about one billion people every day. To improve the maintenance of lifts, escalators and conveyor belts, the two suppliers Kone and ThyssenKrupp, have begun to use machine learning. According to the companies, it is just a taste of what is to come. The idea of the work is to anticipate errors before they happen, and the experience gained from these large-scale, global applications is expected to provide a valuable knowledge base for the broader development of AI in preventive maintenance – something expected to be required in most industries.

The challenge

Predictive maintenance is not a new concept. Industries that require high availability such as pulp and paper, chemistry, oil, gas and steelworks, to name a few, have long used statistical analysis tools to forecast interruptions and improve maintenance work. But machine learning provides a new level of accuracy and efficiency and makes predictable maintenance possible on a large scale on a large installed base. It is possible to identify common error patterns over hundreds of thousands of lifts, and at the same time, using algorithms, detect anomalies and specific behaviours for each lift plant. While two lift plants might be of the same model, their practical use will differ from day to day, as will the infrastructure around them.

It is simply not possible to apply simple sets of rules across such large and heterogeneous environments – which is why machine learning represents a real breakthrough in this context. Until now, predictive maintenance has involved identifying fault thresholds using a range of sensor data, which in turn can statistically indicate faults in lift plants. Machine learning involves using historical data in which fault events have been identified to allow the system to learn to find new faults – all without operators having to tell the AI what a fault pattern looks like.

The experience

In 2015, ThyssenKrupp launched a service called "Max" based on data from IoT sensors, control system data and data from the company's ERP environment and CRM systems from SAP and Oracle. In collaboration with Microsoft, a cloud-based data storage facility was created based on the Azure Cloud Platform. ThyssenKrupp currently provides the service to approximately 120,000 lifts and other systems, or 10 per cent of the installed base.

Open source code is used to build classification and regression models. A combination of models across different data streams and types of objects is compiled to achieve highly relevant and reliable results. The various predictive models are also gradually becoming outdated, due to lifts and escalators wearing out, being rebuilt and maintained, and so continuous re-learning is carried out.

The goal is to send field technicians to a facility before it fails. Although the maintenance system currently is not reaching that goal, the technician is often on the road when the call from the customer arrives. Once in place, the system has already done much of the troubleshooting work that would otherwise be started only once service staff are in place.

The introduction of AI technology at ThyssenKrupp has led to a review of some organisational boundaries between its service departments, IT and other functions. According to ThyssenKrupp's data, the system (used by 20,000 service technicians worldwide) has so far reduced the stoppage times of over 40,000 customers. It's not just the development of machine learning that has made this possible. Lower mobile data costs and the development of cloud technologies have also been enabling factors.

From an organisational development perspective, the project is not primarily technology-driven. From the outset, a broad group of different professions has been involved. Field technicians have been at the centre of the action, and they have been complemented with an IT team with skills in cloud and machine learning, as well as the skills to bring together and prepare the data. HR, legal, construction, production and other divisions have also become involved.

Kone – one of ThyssenKrupp's competitors – has developed its service offering along with IBM and Watson IoT systems. The partnership was launched in February 2017, and Kone has since equipped the facilities that use the service with IoT sensors to measure around 200 different parameters, such as movement, temperature, air pressure and forces within the machine.

Data is transferred to the IoT cloud platform as well as data and error status from the control systems. IBM Watson's natural language learning and machine learning processes have also been able to analyse information in maintenance logs and manuals accumulated over several decades. Watson can also be applied to images, sounds, and vibration patterns. Some generic components such as rotary machines have been modelled with general data from the installed base, and models have since been refined with more data specific to each device.

The result is that customers see significantly fewer stops and errors, and they experience a higher level of service. Plans include adding more data and expanded infrastructures that will further develop the customer offering. There is a plan to let people interact with the lifts so that the lifts, for example, sense when someone leaves a hotel room and then ensure that a lift is available on the right floor. Kone sees great quantifiable benefits from applications like these, which will drive AI to integrate into other applications as the technology improves.

Kone and ThyssenKrupp show that technical industries with large dispersed installation bases can use AI and machine learning technologies in the cloud to build predictive models that plan and make maintenance more efficient. But just as we have noted before, there are challenges when it comes to developing and applying the computer models, especially concerning data management. This means that these techniques are likely to be mainly used, where the return is most significant.

Sources: ThyssenKrupp, Kone, Computer Weekly.

2.5 The birth of modern AI

In the summer of 1956, a select group of researchers met at a seminar at Dartmouth College in the United States. The topic was Artificial Intelligence, and the optimistic goal of the summer meeting was to "achieve significant advances in the AI field".

Convening the Dartmouth Summer Research Project on Artificial Intelligence was a young assistant mathematics professor who would later become legendary. His name was John McCarthy, in his invitation, he wrote: "We propose that a two-month, ten-man study of artificial intelligence be carried out during the summer of 1956 at Dartmouth College in Hanover, New Hampshire. The study is to proceed based on the conjecture that every aspect of learning or any other feature of intelligence can, in principle, be so precisely described that a machine can be made to simulate it. .... We think that a significant advance can be made in one or more of these problems if a carefully selected group of scientists work on it together for a summer."

The seminar marked the beginning of a favourable period for American AI research. Support, in the form of considerable funding from the defence authorities during the Cold War, led to great academic freedom and a creative research climate. But the 1970s brought challenges. AI researchers had underestimated the challenges, and a series of setbacks followed. Both American and British-oriented AI-foundational research was deprived of its funding. The research was criticised for lack of realism and lack of results.

The 1970s came to be called the first AI winter. Several setbacks would come later, as the hype around the field repeatedly rose and fell. The so-called LISP machines and expert systems of the 1980s were market failures that once again reduced development grants and led to new freeze periods. However, in the late 1990s and early twenty-first century, AI development quietly began to make progress. Expert systems using the technology could be used commercially for logistics and medical diagnosis. These successes came from better methods and more, and cheaper, computational power. The twenty-first century, and especially the period after 2010, has shown that AI is now an established commercial field that is growing rapidly.

Consumer applications from Amazon, Google, Microsoft and Apple are being rolled out on a broad scale, while AI support is now built into finance, media, trade and industrial applications. AI for language management is expected to grab the largest share of the market in the coming years, while health care applications are forecast to have the highest growth rate. Meanwhile, industrial applications are also expected to grow rapidly, a trend that is supported by our analysis in the first part of this report.

The current commercial breakthrough of AI technology is the result of the simultaneous coming to maturity of several underlying fields. The rapid expansion of the Internet from the 1990s onwards means that large amounts of data are readily available today. Data is the raw material of AI technology and is transformed into money and growth using algorithms.

In parallel, other developments mean that the cost of computer capacity is now rarely a limiting factor. Algorithm technology has undergone a similar process. All this comes on top of unprecedented growth within tech companies and their large appetites for investing thousands of billions of dollars in AI development, as well as general conditions that have allowed for the spread of AI. The overall picture is clear: AI in the form of machine learning is an established field of commercial technology that is achieving significant breakthroughs within all verticals in all markets.

The purpose of this concluding section of the report is to provide a somewhat deeper technical perspective to complement the report's market focus in the first part.

We will start with an overall picture of IndTech and the scope of industrial IT, automation and digitalisation in general. This will be followed with a description of machine learning technology, and finally, a discussion of the data challenge, the concept of collaborative intelligence and the future of AI.

2.6 IndTech: an overview of the industry as an application area

The concept of IndTech brings together IT with both operational technology on the factory floor and digital development. It has a special significance in that it is where technologies from a range of different fields and periods come together (**Figure 2**). In addition to helping to transform the industry, the IndTech movement is creating a world market for industrial technology worth SEK 3.5 trillion per year [1]. IndTech is a hidden and yet giant industry and a field of excellence for Sweden, with numerous renowned companies operating in the area across the world.

The installed base of automation and industrial IT in the world is estimated to be SEK 50 trillion. This is where technology with roots in the 1980s meets with digital innovations generally not even developed for industry; something that's hardly surprising given that a range of other sectors encountered digitalisation far earlier. The picture of the field that is emerging is thus one of the great opportunities but also significant challenges.

The traditional view of system support for the industry has been a pyramid-shaped hierarchy, with operational technology closest to production, and IT

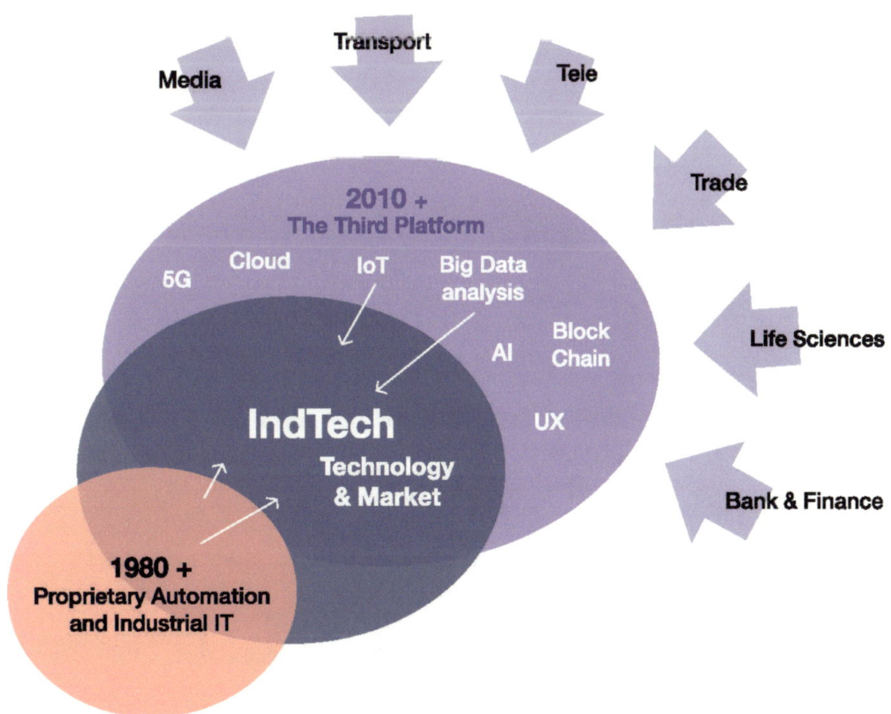

Figure 2.
The model for IndTech: traditional and new technologies come together and make "smart industry" possible. Classic automation and industrial IT meet digitalisation and create new digital platforms and business ecosystems. Source: Blue Institute, 2019.

for administrative processes located above it. The idea that this hierarchy, the Automation Pyramid, might be dissolved in favour of more flexible structures has long been the subject of discussion. How this might happen has been less clear.

Incremental change scenarios seem the most likely given industry's installed base of 1990s technology, much of which has a significant remaining life span, and the need for extensive standardisation work. In the short term, the focus may be on removing silos through better, more practical integration between computers and organisations, both within companies and in supply chains. In the longer term, the focus is likely to be on interoperability in the shape of the full interchangeability of information, without manual intervention, based on accepted industry standards.

To understand the general impact of digitalisation on the industry, it's essential to consider which existing structures could simply be replaced by new technologies (a less common scenario), and which are likely to go through incremental changes over a long period (the more common scenario). The challenge going forward will be to use digital platforms and information transparency to address market fluctuations with new organisational approaches and ways of doing business (**Figure 3**).

Industry's experience with previous technology shifts has demonstrated the importance of creating an overall conceptual picture, as well as having clear objectives from the outset and working towards them one step at a time. These objectives should include at a bare minimum: having digital infrastructure delivered through one, or several, specialised cloud services from different providers; using AI analysis for automation, augmentation and a collaborative approach between people and machines; using of the Internet of Things as a general application platform to lowers prices and simplify hardware and software.

Together, these three verticals form a digital platform with the potential to resolve information hierarchies over time. One of these verticals pertains to advanced analytics, an area in which machine learning if applied correctly, can be a potent tool. We will now examine this field in more detail.

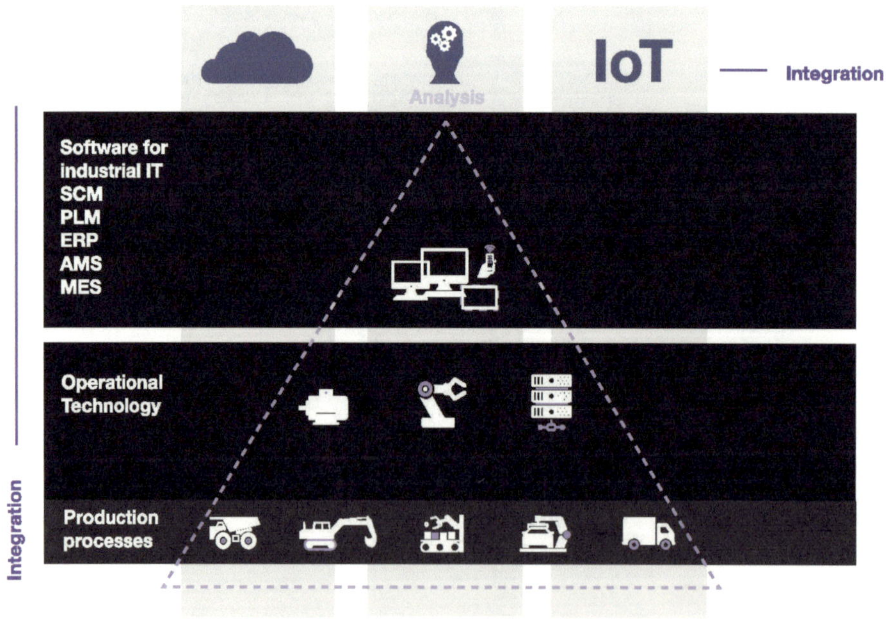

Figure 3.
Development can be summarised as integration in vertical and horizontal directions, and through new technology fields that both complement, improve and challenge the traditional environments and hierarchies. Source: Blue Institute, 2019.

DOI: http://dx.doi.org/10.5772/intechopen.93579

2.7 AI analytics with machine learning

Artificial intelligence is often seen as something almost supernatural, and the media is often prone to highlighting its more sensational aspects. But as we will see, machine learning might just as well be called data analysis or applied mathematical statistics. The principles are very logical, even if the calculation processes are wide-ranging and complex.

Advances in AI development typically based around machine learning being applied to larger and larger sets of data and the development and efficiency of learning algorithms. Machine learning is, therefore, the technology behind most types of AI we see today. While traditional computer programs adhere to predetermined explicit program instructions, machine learning algorithms scan data to detect patterns and then learn to make predictions. The algorithms adapt gradually, and the experience they gain is utilised and improves efficiency over time.

The mechanism behind machine learning centres on how tasks are presented as an input to a matrix-like structure; a neural computer network inspired by the functioning of the brain.

A machine learning algorithm expresses a function between the data it is fed, and the data produced by the model: $y = f(x)$. This function is always unknown, as it cannot be precisely determined mathematically, and this is where the finesse lies in machine learning: estimating the target function as accurately as possible. Correspondingly, if it is possible to determine the function in some other way, machine learning is not needed.

The output of the network, the prediction, depends on how the junction points in the network where the data meets during the process are given different values, called weights. These weights are the secret to the system's learning. (The junction points can be likened to the neurons of the brain (**Figure 4**)).

The problem lies in how to calculate the weights. The most common way is to start by giving them random values and seeing how significant the errors emerging from the model are. Each error is measured and then used to gradually change the weights and eventually approach a solution where the error is as little as possible—in other words, minimising the function's cost. A central part of the learning process is a mechanism called "backpropagation" that tells the network which mistakes it makes.

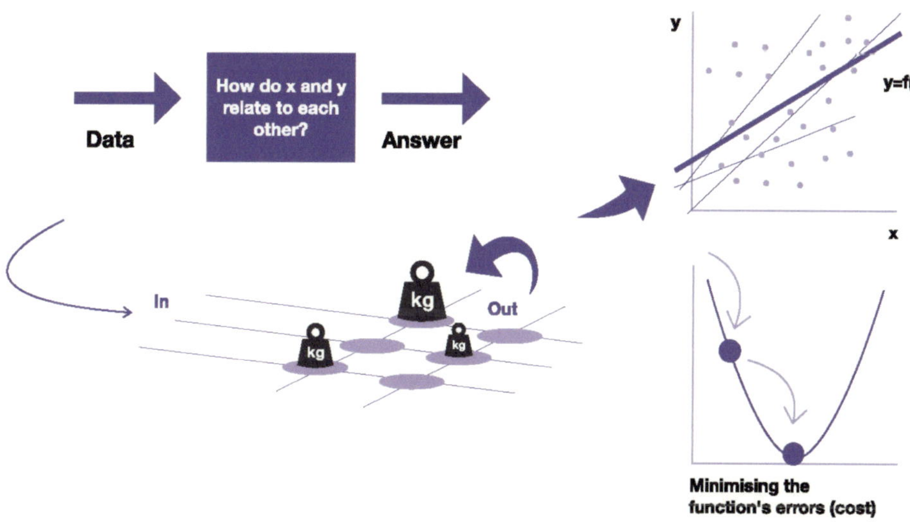

Figure 4.
A machine learning algorithm expresses a function between the data it is fed, and the data produced by the model: $y = f(x)$.

A tremendous amount of data is required to train and validate a model. Some models can automatically separate data into different clusters and see the context and patterns themselves, but many forms of neural networks require data with guidance. This includes examples of what should be entered and what the expected results should be. For this purpose, collections of open training and test data are created of various kinds, such as those for traffic images, with a label that classifies them as representing a traffic light, a pedestrian, etc.

As we have often returned to in this study, the amount, structure and quality of data are the most challenging parts of machine learning, which are both time consuming and costly.

For industry, the technology is useful in optimising the sourcing and the supply of materials; optimising internal and external logistics; planning production and forecasting demand and capacity utilisation; for process management and energy optimisation; for creating maintenance plans and working with preventive maintenance; for understanding customer behaviours; and for simulating cash flows. In summary, for progress in operational development.

The key to success using analysis as a method for operational development lies in good domain knowledge, that is knowledge of the company's operations and processes, and in the ability to create an analysis culture with a solid understanding of both mathematics and statistics. The tools needed are rapidly being commercialised and are becoming both cheaper and easier to use.

One of the simplest methods of classifying items through supervised machine learning, and also one of the most accurate, is called the "nearest neighbour" method. The technique is to measure the difference between two objects, or the distance between the objects. A large number of items are collected with each object labelled with a class affiliation. This is called the reference quantity. When a new unknown object is found, it is compared to the reference quantity until the object that differs the least from the new one is found. The unknown object is then considered to be of the same class as its nearest neighbour from the reference quantity.

Regression analysis, or regression, is a branch of statistics where the goal is to create a function that best fits the observed data. Linear regression is a method commonly used in machine learning contexts that has its limitations but compensates for these with simplicity, interpretability and efficiency. Simple linear regression assumes that a straight line can be adapted to the data and the regression equation can be described as $y = a + b\,x$. The intercept with the y-axis A and the slope B is calculated so that the error compared to the observed data is minimal. The error can be calculated using, for example, the least square method or maximum likelihood.

Logistic regression is an appropriate method of analysis when the dependent variable is binary. Like all regression analyses, logistic regression is a predictive analysis. Logistic regression is used to explain the relationship between a dependent binary variable and one or more independent variables.

With clustering, the aim is to divide the inputs into several groups. A difference between clustering and classification is that with clustering, it is not clear what the groups are in advance. This is typical of unsupervised learning.

2.8 How data becomes money: the machine learning process

A fundamental difference between neural networks and conventional computer programs is that the former develops in two stages. In the first stage, which can partly consist of regular programming, the width and depth of the network are determined, along with how it is to be provided with data and how it will be connected with the rest of the application and the process to be automated or optimised. The next stage is that the network begins to be trained.

DOI: http://dx.doi.org/10.5772/intechopen.93579

The machine learning process (**Figure 5**) – the pipeline – begins with data collection in a procedure called ingesting and includes the cleaning and normalisation of the data so that, for example, numerical scales of values are aligned with each other. This is a time-consuming part of the process and can take as much as 80% of the project time.

The data sets need to be representative, and it is essential to analyse how bias can affect the model. The critical issue is how data is selected and how it is normalised. Distortions and prejudices built-in by algorithms is one of the most significant risks of machine learning because these undermine the entire purpose of the technology. The old truth about "you put garbage in, you get garbage out" applies in the highest, amplified, degree to machine learning.

In many cases, the process involves working with streaming data. In that way, it is possible to choose to first save the data in a database or to collect the data continuously to fine-tune existing models. The alternative is to build new models and train them with new data occasionally. The decision affects the choice of algorithms, as some algorithms are suitable for fine-tuning and others are not.

The next phase is comprised of training the model, or to put another way: determining the weights in the function relationship so that the model delivers the best possible results. The procedure for setting the weights is called hyperparameterisation. A hyperparameter is a setting that controls how a model is to be created based on an algorithm.

In reality, the process of teaching a model by seeking the "correct" weights can include millions, perhaps billions of iterations. To increase performance during modelling, there need to be multiple, parallel work processes running. That is copies of a program that run simultaneously at different locations. The parallelisation calculations utilise special hardware. CPUs originally used for graphic drivers (GPUs) have proven to be excellent in these cases.

In the summer of 2019, there was an emerging discussion over the impact of machine learning technology on the environment and the climate, given the energy-intensive GPUs that run the learning processes. A recent article from the University of Massachusetts [2] has found it is the marginal fine-tuning of models, in particular, that consume energy, thus leaving an imprint on the climate if the computers are driven by, for instance, coal power. This is also one of the reasons why Sweden is a country of interest to the localisation of data centres.

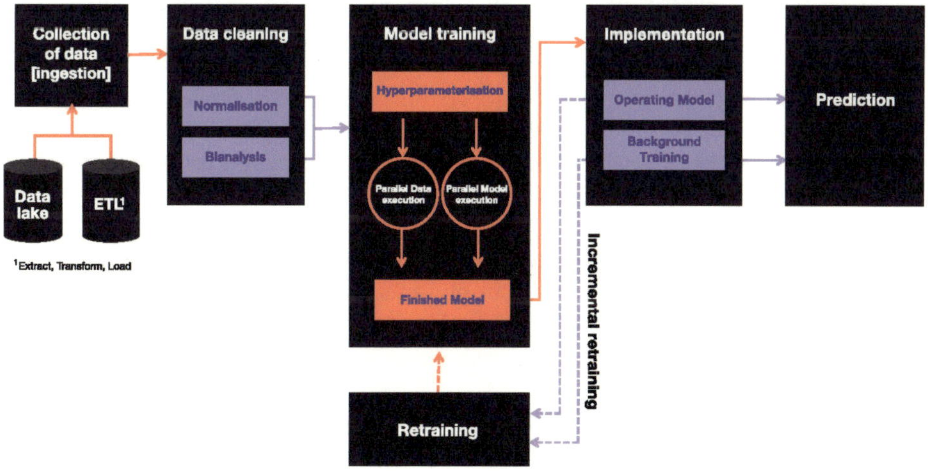

Figure 5.
The machine learning process or pipeline that begins with the collection of data and continues with data preparation and training of the model. The model is then implemented and run to provide accurate predictions. Source: InfoWorld.

The final phase of the process is to use the pre-trained model. The model is now run with new, live data to make predictions that can then be translated into intermediate values such as quality, time and efficiency, which in turn can be assigned a price. Data has thus been transformed into money.

2.9 Typical problem types and methods of analysis

1. Classification, which means that based on a set of training data, new input data is categorised into one of several different categories. An example of classification is identifying whether an image contains a specific type of object or product of acceptable quality from a manufacturing line.

2. Continuous estimates calculate the next numeric value in a sequence based on a set of training data. These types of problems are sometimes described as "predictions", mainly when applied to time-series data. An example of continuous estimates might be to forecast the sales demand for a product based on inputs such as previous sales, consumer preferences and the weather situation.

3. Cluster comparisons require systems that create sets of categories where the data instances have common or similar characteristics. An example of cluster formation is different consumer segments based on data from individual consumers, including demographics, general preferences and consumer behaviour.

4. Anomaly detection, which, with a set of training data, determines whether specific input data falls outside of a norm. For example, a system that has been trained with historical vibration data from a machine can determine whether a new data batch suggests there is a fault in the machine. Anomaly detection can be considered a subcategory of the classification problem.

5. Ranking involves algorithms being used for information retrieval problems where the results of a request need to be set against a criterion. Recommendation systems that, for example, suggest prioritised purchases of products use these types of algorithms to sort the suggestions by relevance before they are presented.

6. Recommendations are systems that provide recommendations based on a set of training data. A typical example is a system that suggests a "next purchase" for a specific customer based on the buying patterns of similar people and the observed behaviour of the particular person.

7. Data generation requires a system that can generate appropriate new data based on the training data. For example, a music composition system can be used to create music pieces in a particular style after being trained on pieces of music in that style.

    - Bad - means that the quality of available data is substandard, even though it has a clear physical significance. This makes it difficult to compensate for flaws in quality by adding more data of more or less the same type. The latter is a method that can work for applications using deep learning, such as image recognition.

- Broken - means that data that has been collected to train a machine learning model lacks the essential qualities of validity/relevance and contain error conditions. This then leads to false positives or negatives in the online implementation of the model. This is a serious problem because even a few or occasional erroneous statements can endanger the reliability of the system, and industrial AI applications typically have significant potential to impacts on assets and personal safety.

- Background – means the data patterns in industrial contexts can be transient. The process involved is volatile, fluctuating and fast. Interpreting such data often requires in-depth domain knowledge, and it's not enough to simply dig for more numerical data. In addition to precision around predictions and quality of performance, an ability to find the roots of possible anomalies is also required.

2.10 The data challenge

One of the biggest challenges with AI concerns the quality of the data needed to make predictions, create forecasts, and recognise patterns. It is a widespread issue, and a great deal of monotonous, routine work takes place behind the sometimes simplified depictions of AI that we see.

In autumn of 2018, BBC News [3] brought attention to a new concept: labelling farms. This is a rapidly growing global sector involving data centres that have been located in low-cost countries for economic reasons. Labelling farms today employ thousands of people whose only task is to help AI algorithms interpret data.

Pixel by pixel, the content of millions of images is classified; a car is identified as a car, a dog as a dog, a road sign as a road sign, and so on so that self-driving cars can recognise real-world objects. Similar data challenges are being encountered everywhere that AI is to be applied. The high cost of data preparation means that there are financial incentives to solve the data problem, and many projects are being carried out with the help of even more AI in a bid to find new solutions and better methods.

Industrial AI involves transforming raw data into "intelligent" predictions to make decisions. In industrial processes – in a steel mill or a paper mill – quick decisions are made in real-time at the millisecond level in models representing physical reality. Several challenges arise in such processes. Real-time requirements mean that the cost-effective and almost endless resources of the cloud need to be supplemented with locally distributed computational and storage capacity, also known as "edge". But the most fundamental challenges also concern the availability and the quality of the data.

Since the 1980s, industrial control systems have been producing enormous amounts of information. Industrial Big Data is available in every factory, and yet while industrial data is generally well structured, it often lacks quality. You sometimes hear talk of the "three B's" of Industrial Big Data: Bad, Broken & Background.

Teams of people who possess both process knowledge and computer science are required for the development of good adaptive models. There is also a need for method development, with experience teaching us that data preparation demands a disproportionate amount of work. This is a serious issue that needs to be continuously addressed and prioritised, lest it becomes an obstacle in releasing industrial value.

2.11 Solutions to data deficiency and manual intervention

Much of the success of modern AI applications are based upon bottom-up strategies within which models are trained using large, well-structured data sets typically collected via the Internet. For example, the GPT-2 text bot was trained using a data set of eight million web pages. Intelligent assistants like Apple's Siri or Amazon's Alexa use thousands of terabytes of data to perform their tasks, and self-driving cars consume about forty terabytes per eight hours of driving, according to INTEL.

For operational industrial applications, large amounts of information are being collected. However, critical processes, in particular, lack the volumes needed to train good models. There is a lack of data in marginal or edge cases, and it is not always easy to deliberately address such deficiencies (by inducing errors in physical processes). The errors they represent correspond to high costs due to significant production disruptions. This is a problem that also applies to other, normally data-rich applications. One of the considerable challenges in the development of autonomous vehicles is managing the most unusual of operating cases. Another characteristic of today's AI technology is that it tends to easily become "confused" if circumstances deviate significantly from what is expected.

Methods are in development to overcome these weaknesses. Similar to human intelligence, they involve working in a more flexible, top-down manner, which allows for reduced data requirements and enhanced speed. There are a number of trends related to the development of more natural systems worth keeping an eye out for soon.

2.12 Eight trends

1. The first trend involves giving robots conceptual properties (both physical and artificial) that in turn give them a greater ability to perceive themselves – and their environment. See the text box on page 45, describing how researchers at the University of Columbia have succeeded in giving a robot such properties.

2. Another developmental avenue involves something of a renaissance of the concept of "expert systems" within which computers become better at doing what human process operators do by making adjustments in real-time to optimise processes.

   Siemens has developed data-efficient methods such as these based on "reinforcement learning" to control the company's gas turbines. In this area, traditional neural networks would take up to a hundred years to learn the complex combustion processes. The method has subsequently been developed to increase the efficiency of the company's wind turbines. Google is also using technology to reduce the energy consumption of its data centres successfully.

3. A third way to address the weaknesses of today's AI algorithms is to give computers more common sense. According to an article in Harvard Business Review [4], the Allen Institute for Artificial Intelligence is working on developing test data that can be used to verify what common sense means to a machine. Meanwhile, DARPA is investing USD 2 billion in AI research through, among other things, creating models that mimic human cognition. And Microsoft and McGill University have jointly developed a system for distinguishing ambiguities in natural language; a challenge needs to be solved if, among other things, computers are to be able to communicate with human beings in a human way.

DOI: http://dx.doi.org/10.5772/intechopen.93579

4. A fourth track is the possibility of letting computers make similar balances of probability assumptions to those that humans intuitively make. This is being made possible through stochastic Gaussian processes that can function and recognise patterns within limited data sets and learn from experiences. Another feature of this method is that processes are traceable if something goes wrong, unlike with the black boxes of neural networks.

5. Yet another method of advancement is Probabilistic Programming for the applications described above. This method brings together the best practices for mimicking human intelligence such as probability theory for modelling, statistical methods for drawing conclusions, and neural pattern recognition networks, along with symbolic program languages that hold the system together.

6. "Explainable AI" is an adjacent developmental track. The black box phenomenon of machine learning can be problematic. Therefore, systems must be able to justify how they have arrived at their conclusions. It's also essential to ensure that human beings can have trust in the way that such systems arrive at their results and decisions when, for example, traffic situations, legal support or medical diagnosis become automated.

7. Federated machine learning is another method showing promise. The idea was launched in 2017 by Google as a concept within which the ability to train a model is decoupled from the up-until-now necessary central storage of data in the cloud. The method can train a single machine learning algorithm over several decentralised servers that store data, without actually exchanging data with other servers. It allows multiple actors to build a common, robust machine learning model without sharing data, thereby addressing critical issues such as data privacy, data security, data access rights, and access to heterogeneous data. It also enables capacity in distributed applications.

   This way of working is based on the idea that a distributed device, such as a phone, downloads an existing shared model, improves it by learning the data that is locally available on the phone/device and then summarises the changes as a small concentrated update. Only the update is sent to the cloud, via encrypted communications, where it is immediately computed and integrated with other user updates to improve the shared model. All training data remains on the local unit, and no individual updates are stored in the cloud.

   The technology can contribute to breakthroughs for industrial operational applications based on conventional automation or IoT, where distributed capacity is both a prerequisite and natural in the current concept of control and monitoring.

8. Finally, AutoML or Automatic Machine Learning looks like the holy grail for solving the many and long routine steps found in today's state of the art technology. Automated machine learning involves automating the process from end-to-end. As we have noted, typical machine learning projects involve extensive pre-processing before the dataset can be made available for actual machine learning.

Pre-processing is followed by a selection of an algorithm, hyperparameterisation and fine-tuning to maximise predictive performance in the final model. In addition, many of these steps require both experience and specialist knowledge. What could be more logical, then, than to suggest AutoML as an artificial intelligence-based solution to these growing challenges? Automating the process would be an effective productivity-enhancing method, which besides would be likely to provide solutions and models that exceed manually designed ones.

AutoML solutions with drag-and-drop-based user interfaces, and that do not require any coding in the ordinary sense are now on the market and are offered by all major platform providers such as Google, MS Azure and IBM, along with many specialised smaller companies. The technology is evolving rapidly and will further lower the **threshold for users.**

Industry case study: BillerudKorsnäs in Gävle, Sweden - Deep Process Learning.

DEEP is a project that will show how deep process learning (deep learning) can be used for the next step in process automation. The project takes advantage of data that already exists in process control systems and uses it to suggest the measures required to improve selected key performance figures. The project, which is a collaboration with PiiA, consists of a consortium between BillerudKorsnäs, Peltarion, PulpEye and FindIT.

"The forestry sector has an advanced supply chain with multiple levels of complexity and difficult, resource-intensive processes...".

The process industry accounts for almost half of all industrial production in Sweden. So, the achievement of efficiency and productivity improvements within it is certain to have a significant impact on the Swedish economy. The forestry sector has an advanced supply chain with multiple levels of complexity and difficult, resource-intensive processes: from felling to barking and chipping the wood, to boiling, washing and bleaching the pulp before it reaches the paper machine to be refined to produce paper and cardboard of various grades.

Process industries produce vast amounts of data and have a high degree of automation, but they also face a variety of challenges. These challenges cannot always be addressed through traditional analysis methods. As such, the data produced can be a valuable asset, capable of being refined through AI to generate insights, predictions and automation algorithms – thus creating the next stage of productivity, quality and automation.

BillerudKorsnäs is a forestry company that supplies packaging materials and packaging solutions. The company has three divisions: Division Board, which manufactures and sells liquid and non-liquid packaging board, as well as fluting and liner; Division Paper, which produces and sells high-quality kraft and sack paper; and the Solutions Division, which meets the needs of brand owners for efficient packaging solutions and systems.

During a feasibility study for DEEP, BillerudKorsnäs and Peltarion jointly led a machine learning project to predict the kappa number of pulp after boiling. The kappa number is a measure of residual lignin in the pulp and determines the boiling process required for different pulp qualities. The project was successful and resulted in a useful technique for predicting the kappa number. This success encouraged further development of the approach in other process steps.

"The successful use of machine learning as a tool is based upon a deep understanding of the processes that are to be optimised".

The challenge

Paper machine 4 in Gävle is a cardboard machine that manufactures liquid packaging board for juice and milk packaging, among other things. The purpose of the DEEP project has been to realise the efficiency potential identified in the manufacturing process by proposing optimal machine operational parameters. An essential feature of the finished liquid packaging board is the carton's bending stiffness. This property is determined by complex relationships between the different stages of the manufacturing process, not least by the pulp's fibre properties. The goal is to produce strong packaging using less raw material.

To meet the quality objectives at optimum production speed, process settings must be continuously evaluated and adjusted. In the DEEP project, data is being collected to support the online optimisation of such decisions. The data used in the project consists of high-resolution microscopy images from PulpEye's analyser which provides information about the pulp's fibre properties and camera images from the drying cylinder which provides information about the dewatering of the pulp, in combination with measurement values from different sensors in the system. In the next step, data will be used to develop a suitable model to predict quality properties.

During the DEEP project, many different attempts were made using various methods, including deep learning with Peltarion's self-developed platform.

The experience

BillerudKorsnäs has formed a digitisation team with different competences from different parts of the organisation, and that initiates and runs transformation projects. The company's various AI initiatives are part of that transformation process.

BillerudKorsnäs' experience shows that deep learning technology is ripe for use in various types of classification problems and for further increasing the degree of process automation. The process industry is characterised by a combination of large amounts of data and a high degree of automation, which partly produces conditions that differ from other fields that apply deep learning and machine learning. Over time, the technology will find its place in process analysis and control and will solve many more problems that affect efficiency, quality and logistics.

One of the essential takeaways from BillerudKorsnäs' AI projects is the need for domain knowledge and the ability to formulate the right problems. The successful use of machine learning as a tool is based on a deep understanding of the processes that are to be optimised.

BillerudKorsnäs is continuing its work on developing processes with the help of AI, and another project will be launched in collaboration with PiiA in the spring of 2019. This time, it will be led together with Finnish Quva OY as a data analytics provider.

Sources: PiiA, BillerudKorsnäs, Peltarion.

2.13 Man and machine: collaborative intelligence

It was once said that we should, "Let the machine take care of the details and let the man think and dream". And, as Anders Ynnerman, a professor at Linköping University, states, "for every AI system that we have where we add on the human aspect, we get a much better system."

At the same time, there is a fear that AI will eventually push people out of the labour market. The latter is hardly inevitable or even the most plausible outcome. Never before have digital tools been better suited for collaboration with people. And while AI will surely change the way work tasks are performed, and who performs them, the role of machines in future will be to reinforce and supplement human abilities rather than to replace them.

The concepts of collective and collaborative intelligence are also worth bearing in mind. Models where people's intellectual capacity can be increased through smart, collaborative methods, either working with other people or with machines, will have a significant impact on industrial development. Man's abilities in leadership, teamwork, creativity and social interactions will complement AI's speed, scalability and quantitative ability to keep track of large complex data sets. Industrial activities require both.

But the above line of reasoning also demonstrates the need for changed processes and in many cases, radical transformations of both business activities and the way people and machines interact on a practical level. An article in Harvard Business Review [5] notes that the business effects of artificial intelligence depend on the ability to "rethink" activities so that they both incorporate AI and cultivate related abilities in human employees, in addition to allowing creative experimentation and having clear AI strategies. Last but not least, it is crucial to managing data in both a relevant and responsible manner.

> *"For every AI system that we have where we add on the human aspect, we get a much better system."*
>
> *Anders Ynnerman, Professor at Linköping University*

AI will lead to more automation and more advanced automation. One of the significant advantages of automation is avoiding errors caused by people not being able to repeat tasks efficiently. A robot that is asked to do the same motion a thousand times makes the same motion a thousand times – as long as the sensors and the mechanics work. A person might be able to perform it three times but is at the same time a master at interpreting their senses and dealing with new, unexpected situations.

The process of how this might happen is not yet clear and making machines that act in a human-like manner is a complex matter. The recent accidents involving one of a highly advanced Boeing aircraft model have, in a frightening way, also shown that for every human mistake that a machine eliminates, there is a risk that a new one will be introduced. There are endless possibilities for misunderstandings to occur between human intelligence and machines. In the industrial context, the challenge boils down to establishing collaborative intelligence, and how well the interface between human and machine works. This developmental field is known as UX – user experience – or in the AI context, it's perhaps more appropriately called MMC, man-machine communication.

Issues with misunderstandings and mistakes have the potential to intensify as the degree of automation increases further. Humans will no longer have full control over machines. Overall, this will lead to a decrease in those parts of industry domain knowledge that include artisanal process knowledge. At an operational level, the challenge for the machine operator will be to monitor a process over a significant amount of time and to be prepared to take over the moment something goes wrong. Problems in this area have the potential to be costly in the process industry and utterly catastrophic within aviation.

One conclusion that can be drawn is that machines that do not allow people to keep up with the processes they are managing aren't optimal in events where people are forced to take over. Another conclusion is that the best kind of automation is not necessarily where the computer automatically does most of the work, but rather where there are an optimal distribution and a realisation that people and machines will probably never understand each other perfectly. We have two pilots in the cockpit and two operators in the control rooms, and unfortunately, both can sometimes be expected to do unexpected things.

2.14 Looking ahead

In this section, we have skimmed over some of the concepts and constructs that may come in handy from an applied industrial perspective. Of course, there are countless other aspects of AI that could potentially be taken into account when assessing a technology which proponents claim to be "intelligent." Many of these issues relate to morals and ethics. As society and industry move ahead, we will likely encounter machines with questionable intentions and distorted development, whose intent is to benefit individual stakeholders. AI will influence people's attitudes; false correlations and self-reinforcing feedback will eventuate – and algorithms may influence reality to gain even more influence, even though their base assumptions are false. The origin and quality of data will continue to be an issue and, last but not least, we will face uncertainty around what is real and true: will we, in future, be able to trust what we see and hear? Will we be able to trust pictures, movies and sounds?

From an industry development point of view, our hopes for AI and machine learning might be for them to provide greater flexibility than that currently found in our simple neural networks which are only capable of performing one task at a time and are expensive and arduous to retrain. We might also hope to see significant productivity gains in system development, while there is also room for improvement in the deployment of models.

But we can rest assured that these are areas that are currently being addressed by research. Likewise, the substantial data requirements, the need for manual intervention, and the problems with edge data all need to be addressed. The actual learning process, with its hyperparameterisation, needs to be further automated. Another potentially growing concern is the lack of transparency in neural networks, which, for the most part, resemble black boxes.

It's impossible to know how and when these issues will be addressed. It could take years, or there may be sudden breakthroughs, such as when the AlphaGo defeated one of the world's best Go players with the help of reinforcement learning. But it does not change the fact that AI and machine learning are already powerful enough tools to change the industry, and that those who acquire knowledge, experience and an upper hand when it comes to applying the technology have everything to gain.

> *"Will we in the future be able to trust what we see and hear? Will we be able to trust pictures, movies and sounds?"*

2.15 Glossary

The article contains some terms that may need clarification. Key terms include:

Artificial Intelligence (AI)

The term "Artificial Intelligence" (AI) does not have any clear definitions or delineations. AI research itself is both specialised and spread across many subfields. For this analysis, we have chosen the definition also used by Vinnova in the study of Artificial intelligence in Swedish business and society, 2018".

This is: "The ability of a machine to mimic intelligent human behaviour. Artificial intelligence is also the designation of the science and technology field that aims to study, understand and develop computers and software with intelligent behaviour."

When we talk about AI in an industrial context, we are primarily referring to machine learning technology with neural networks.

Algorithmisation

Algorithmisation is a mega-trend within which more and more value-adding activities are managed and controlled by algorithms instead of human beings.

IndTech

IndTech is used to describe the development, companies and markets that arise when traditional automation and industrial IT meet digitisation. IndTech companies include:

- Suppliers of industrial automation, such as ABB or Siemens.
- Suppliers of industrial IT software, such as SAP or IBM.
- Providers of digital platforms, such as Microsoft or Amazon Web Services.
- IoT providers, such as Ericsson or Nokia, and operators, such as Telia or Telenor.
- System integrators and machine suppliers who base their process or mechanical engineering offerings on digital technology. These include companies such as Sandvik, Epiroc, Valmet and many more.

Platformisation

Platformisation can be used to describe the general movement of various companies' automation and IT support to the cloud, and also to describe the movement of platforms created by open standards to platforms owned and controlled by a particular actor. Because the value of a platform tends to increase for all involved as more people use it, there is a tendency for already-large platforms to grow even bigger.

Operational Development - OD

We have chosen to use the term operational development to encompass the operational changes in processes or in organisations that lead to increased efficiency or increased customer values. Within this area, AI can be a potent tool.

Author details

Örjan Larsson
Blue Institute and Strategic Innovation Program PiiA, Västerås, Sweden

*Address all correspondence to: orjan.larsson@blueinst.com

References

[1] Blue Institute. PiiA Insight. Västerås, Sweden: PiiA; 2018

[2] Strubell E et al. University of Massachusetts. In: Energy and Policy Considerations for Deep Learning in NLP. Ithaca, NY, USA: Cornell University; 2019

[3] BBC News. Why Big Tech Pays Poor Kenyans to Teach Self-Driving Cars. United Kingdom: BBC; 2018

[4] James Wilson H et al. The future of AI will be about less data. In: Not More. Brighton, Massachusetts: Harvard Business Review; 2019

[5] Collaborative Intelligence: Humans and AI Are Joining Forces. Brighton, Massachusetts: Harvard Business Review; 2018

Chapter 5

Artificial Intelligence and ISO 26000 (Guidance on Social Responsibility)

Weiwei Zhao

Abstract

With the rapid development of artificial intelligence, it has a more and more far-reaching impact on social, economic, cultural, and other fields. At the same time, artificial intelligence faces ethical, moral, privacy, and security issues. In order to realize the healthy development of artificial intelligence, it is urgent to apply the social responsibility management system to artificial intelligence. Based on the seven core subjects of social responsibility proposed by ISO 26000: organizational governance, human rights, labor practices, the environment, fair operating practices, consumer issues, and community involvement and development. In this chapter, the possible risks of artificial intelligence in these seven aspects are analyzed, and the corresponding countermeasures are discussed according to the causes of these problems. The final conclusion is the aspects that artificial intelligence should pay attention to when fulfilling its social responsibility.

Keywords: artificial intelligence, ISO 26000, social responsibility, seven core subjects, issues

1. Introduction

Artificial intelligence has become the focus of social concern. At present, various countries are expanding and strengthening new industrial clusters, implementing big data's development actions, strengthening the research and development of the new generation of artificial intelligence, and promoting the "Internet" in many fields such as medical care, old-age care, education, culture, sports, and so on. At present, the innovation rhythm of artificial intelligence technology is accelerating constantly, the application scene also changes with each passing day. The realm of artificial intelligence is not limited to playing go and acting as a smartphone assistant; manufacturing, warehousing, transportation, car, education, health, finance, home, escort, entertainment, services, and many other industries can see the presence of artificial intelligence. And what's more, it's. We are now from industrial water testing to large-scale application stage. The global market for artificial intelligence will reach $643.7 million in 2016 and $36.8 billion in 2025, according to a study by the US Tractica [1].

Artificial intelligence is the subversive technology that leads the future industrial change. It may mean as much to human society as the Internet, electricity, and steam engines. The more powerful technology is, the more obvious the tool

IntechOpen

attribute is in practical application: it can greatly improve social productivity, bring well-being to human society; it may also be used improperly; and it brings new challenges to social management. Therefore, it is necessary to standardize the application of artificial intelligence by strengthening social responsibility. Recently, the dispute over Facebook and data privacy, Uber auto-driving accident of, which is relate to the artificial intelligence security. It is wise to consider the social responsibility of artificial intelligence in a timely manner.

2. Artificial intelligence and industry

Artificial intelligence (AI), as a leading and strategic technology in the future, has become an important driving force for the new round of scientific and technological revolution and industrial transformation. Artificial intelligence can run through the design, process, production, management, service, and other links of industry, so that the industrial system has the mode and result of intelligent functions such as description, diagnosis, prediction, decision-making, and control. The impact of the COVID-19 outbreak in 2019 on industrial production has become apparent. Some enterprises have accelerated the application of intelligent industrial robots and other new ways to carry out intelligent production. Industrial enterprises can make use of AI to digitize and intellectualize manufacturing, supply, sales, and other information in production, and finally achieve the purpose of providing consumers with fast and effective personalized product supply.

Through artificial intelligence learning, industrial enterprises can assist maintenance personnel and engineers with more accurate identification and diagnosis from massive historical maintenance records, technical data, drawings, experience, and other data, so as to shorten the maintenance time and improve the accuracy of predictive maintenance. Artificial intelligence analysis can inform business decision makers of the data performance reflected in what is the most impact on profitability, what is the reason? By using AI technology, manufacturing will achieve higher engineering efficiency, shorter time to market, and production flexibility in the future.

Industrial enterprises form a sustainable industrial ecological chain through the Internet, virtual economy, personalized marketing, and hardware manufacturing. The relationship between the business and the consumer will change the original custom relationship. Take the smart refrigerator as an example. As the pace of life quickens, people have much less free time to do housework. The advent of smart refrigerators will be necessary to improve people's quality of life and save resources. When people buy food and put it in the refrigerator, it often causes waste. On the one hand, they buy too much food at one time; on the other hand, they forget it after putting it in the refrigerator. Smart refrigerators can not only process expired food by themselves and purchase fresh food but also make overall arrangements to reduce food waste and make personalized recipes. It adjusts the location of the ingredients in the fridge to remind people to eat them in time, depending on whether they are fresh or not. In addition, the smart refrigerator can also analyze the rationality of users' meals and make recipes. At the same time, it is suggested that once the user has determined the ingredients that need to be supplemented, the smart refrigerator will automatically select e-commerce for home delivery, directly realize the automation and intelligence of food distribution and delivery, and realize the remote viewing and control of the mobile phone. The use of artificial intelligence in industrial products can greatly enhance user experience and product value.

3. Artificial intelligence and social responsibility

Since 2016, the issue of social responsibility for artificial intelligence has received increasing attention. In September 2016, The British house of commons science and technology committee issued a report, robotics and artificial intelligence, calls for greater ethical research on artificial intelligence to maximize the benefits and try to minimize its potential threat. In the report, they recommended a commission on artificial intelligence should be established to identify principles to govern the development and application of artificial intelligence, provide advice to the government, and foster public dialog [2]. In September 2017, the World Commission on Ethics of Scientific knowledge and Science and Technology (COMEST), which is an advisory body and forum of reflection that was set up by UNESCO in 1998, issued the report on Robotics Ethics. Considering that robots not only need to respect the ethical norms of human society but also need to incorporate specific ethical norms into robots. There are seven relevant ethical principles and values, which are human dignity, value of autonomy and privacy, do not harm principle, principle of responsibility, and value of beneficence and justice [3]. In January 2017, Future of Life Institute, an artificial intelligence research institute, convened Asilomar Conference. A large number of experts and scholars in the fields of law, ethics, philosophy, and so on convened the 2017 Asiloma Conference and formed 23 AI principles as a guide to the research, development, and utilization of artificial intelligence. They believe designer and builder of stakeholders in the moral implications of their use, misuse, and actions, with a responsibility and opportunity to shape those implications [4]. In addition, governments and industry have invested heavily in setting up research funds to advance the social responsibility of artificial intelligence. In December 2016, the Institute of Electrical and Electronic Engineers global initiative for ethical considerations in artificial intelligence and autonomous systems, a standard-setting organization, released ethical design: a vision for prioritizing human well-being with artificial intelligence and autonomous systems (version 1: for public discussion). The aim is to encourage scientific and technological personnel to prioritize ethical issues in the research and development process of artificial intelligence [5].

Figure 1.
Social responsibility seven core subjects bases on ISO 26000.

Based on ISO 26000, social responsibility means"responsibility of an organization for the impacts of its decisions and activities on society and the environment, through transparent and ethical behaviour that contributes to sustainable development, including health and the welfare of society; takes into account the expectations of stakeholders; is in compliance with applicable law and consistent with international norms of behaviour; and is integrated throughout the organization and practised in its relationships" [6]. To identify relevant issues and set its priorities, artificial intelligence should address the following core subjects (see **Figure 1**): organizational governance, human rights, labor practices, the environment, fair operating practices, consumer issues, and community involvement and development. This chapter comprehensively analyzes the problems of artificial intelligence on social responsibility in theory and practice by using the research framework of ISO 26000.

4. Challenges of artificial intelligence faces on social responsibility

4.1 Organizational governance

Organizational governance is the most crucial factor in enabling artificial intelligence to take responsibility for the impacts of its decisions and activities and to integrate social responsibility throughout the organization and its relationships [7]. The White House's October 2016 report on artificial Intelligence, "preparing for the Future of artificial Intelligence," said organizational governance should aim to ensure public safety and a fair market. Some characteristics in the process of artificial intelligence research and development and operation challenge the organizational governance of artificial intelligence social responsibility. Firstly, the covert nature of artificial intelligence research and development, which little visible infrastructure investment, is required. Secondly, the decentralization nature of artificial intelligence research and development, which developers, participants, etc., may be distributed in different countries or regions. Thirdly, the discontinuity nature of artificial intelligence research and development, which involving many components and elements. Fourthly, the opacity nature of artificial intelligence research and development, which stakeholders do not know the artificial intelligence system. It may be difficult to define the artificial intelligence system that needs to be managed by the organization.

Artificial intelligence-related products, such as robots and self-driving cars into the human society, when it causes damage, how to allocate responsibility? Existing governance frameworks, such as product liability and fault liability, have limitations in managing the damage caused by artificial intelligence systems. Designers may also be unable to foresee their follow-up actions, making it unfair for designers to take responsibility for unforeseen damage. Because of the autonomous learning ability of artificial intelligence, when the user loses control of the artificial intelligence system, the user or anyone is no longer responsible for the artificial intelligence system.

On the one hand, what kind of organizational governance can contribute to the development of safe, reliable artificial intelligence systems? On the other hand, when artificial intelligence systems cause personal and property damage, how should responsibilities be allocated? These two aspects are not only realistic problems that need to be studied in depth, but also challenges faced by existing management systems.

4.2 Human rights

Human beings influence everything on earth. They have invented many different tools to improve human productivity. Artificial intelligence is also a tool invented by humans; it has its uniqueness; it may be smarter than us in many ways. A huge paradigm shift is taking place. We will develop faster than before. Artificial intelligence will solve problems that we cannot solve in the future, such as cancer and traffic accidents. But artificial intelligence has also raised a range of questions about human rights. Since 2016, celebrities such as Stephen William Hawking, Elon Musk, and Eric Emerson Schmidt have expressed concern about the development of artificial intelligence. It is even thought that the development of artificial intelligence will open the door to human destruction.

Robots and artificial intelligence systems are becoming more and more human-like, both in their external form and in their internal mechanisms. What on earth should man do with artificial intelligence? Can artificial intelligence enjoy a certain moral or legal status? The rights of machines have been paid more and more attention and become an unavoidable problem in human society. With the increasing popularity of robot applications, this problem will become more and more important. In addition, what is the law of artificial intelligence systems such as robots? Natural person? Legal person? Animal? Or is it a new subject of law? Answering this question may involve agency, tax payment, liability, and so on [8].

Although artificial intelligence provides a wide range of information, gives people a lot of choices, and offers many opportunities, its essence is to liberate human knowledge and loosen control, and in practice, if it is controlled by certain power organizations, humans will be restricted in the reception of certain information and narrow their horizons opportunities. This will result in a more serious centralization, which is easily ignored and difficult to detect.

Artificial intelligence has an impact on human privacy, freedom, and dignity. It challenges human privacy and data protection by collecting, utilizing, automating, and intelligentizing data analysis. At present, artificial intelligence based on machine learning needs a lot of data, many of which are personal data. How to protect human privacy? If human beings are surrounded by artificial intelligence systems everywhere in the future, how can individual freedom be realized? Gender discrimination. Artificial intelligence always associates women with certain elements, a "sexist" artificial intelligence, which thinks the person standing in the kitchen should be a woman. Gender bias is not only common in databases but also magnified by artificial intelligence.

4.3 Labor practices

Artificial intelligence has many beneficial effects on labor practice. Artificial intelligence can help companies select suitable candidates, avoid the risk of false resumes, and reduce recruitment and training costs. At the same time, it can also create a more honest and fair competitive environment for job seekers. Like any science and technology in human society, the application of artificial intelligence in many fields such as transportation, medical health, manufacturing, service industry, and so on will inevitably bring safety problems. In July 2015, for example, a robot at the Volkswagen factory in Germany suddenly "shot" a worker in the chest, killing him instantly.

Artificial intelligence replaces human beings on a large scale in low-skill, low-creative work or tasks, which is the trend of the times. With the advent of more versatile artificial intelligence systems, even highly skilled, highly creative occupations, such as doctors and teachers, may be partially replaced by artificial intelligence.

4.4 Environment

Artificial intelligence can promote the environmental protection, such as the related technology can better control the air pollution. But at the same time, the development of artificial intelligence also brings more and more electronic garbage. Electronic waste exacerbates environmental degradation and brings environmental radiation; with the rapid development of artificial intelligence technology, the garbage generated by the replacement of related products will be even more severe [9]. Abandoned mobile phones, computers, etc. can be regarded as electronic waste. These electronic wastes contain precious metals such as gold, silver, copper, and so on, which can benefit from the recovery, but also contain a lot of toxic substances such as lead, mercury, and so on. If it is not properly treated, it will cause great pollution harm to the environment.

The emergence of artificial intelligence has given us too many reasons for upgrading electronic devices. Not only phones, sockets, cooking machines but also surveillance cameras, traffic lights, and so on, which are widely used as urban infrastructure, are also being updated. Artificial intelligence reduces the cost of related products. Low-cost products mean shorter service life and higher wastage rates. The resulting mass of discarded hardware could become a far-reaching burden on the environment.

4.5 Fair operation practices

Artificial intelligence automated decision-making systems are increasingly widely used in many fields such as education, employment, advertising, medical care, criminal justice procedures, and so on. From speech assistant sex discrimination to crime assessment software discrimination against blacks, the unfairness of artificial intelligence system decision-making has spread to many fields. In the concept of artificial intelligence, the wedding dress is a white western wedding dress and does not "know" the culture of the third world. The search engine responds not to reality but also to its user's understanding of reality. Search for "black," but come up with "black criminals." When identifying image content in the United States and the third world, artificial intelligence is always "selective blind." Today's machines, of course, do not have the ability to experience emotions or deliberately impose prejudices, but honestly reflect the real prejudices in the database and even in society, which are sometimes not what we want.

Artificial intelligence will replace human beings in taking on more and more decision-making, and the question is how can fairness and justice be guaranteed? How can we ensure that there is no discrimination and injustice in the algorithm? When an individual is implicated in such a decision, how can he or she be provided with a complaint mechanism in order to achieve fairness to the individual? If you have artificial intelligence technology, you will have wealth.

Tech giants will enjoy rich resources of big data, which may pose a threat to mankind, although it cannot be inferred. Big date's monopoly will become a barrier and a tool for industry giants to pursue their personal interests. It may lead to the accumulation of wealth in the hands of a small number of people, resulting in new inequalities.

4.6 Consumer issues

Artificial intelligence provides consumers with a lot of convenience. The merchant recommends the next item according to the consumer's shopping history and interest. Through the network taxi, greatly reduce the travel cost of consumers and provide consumers with a more reliable time security. Self-driving can reduce the death rate in traffic accidents because machines do not suffer from fatigue, road rage, drunk driving, speeding, and congestion problems, which can reduce many traffic accidents. At the same time, due to the remote internet operation of unmanned cars, the average speed of the vehicle can be greatly improved, and it can also be of great help to the urban traffic. Artificial intelligence can help people solve the problem of parking better. Also with its help, you can better find parking spaces, through positioning, to avoid the trouble of forgetting the location of parking.

Artificial intelligence also brings a series of problems to the rights and interests of consumers. Many businesses promote artificial intelligence face recognition payment. Face recognition payment does provide a quick and convenient payment experience for consumers, but how to use the consumer portrait data obtained by businesses? Artificial intelligence system for personal data automation, intelligent analysis, and decision-making affect consumers' personal rights and interests, and consumers may be completely unaware of this, and how to achieve the balance between data commercial utilization and consumer rights and interests? In recent years, there have been many accidents in driverless cars. Who should bear the responsibility for the accidents?

4.7 Community involvement and development

In addition to offline communities, there are online communities. Cyberspace is a real virtual existence, an independent world without physical space. The leakage of Facebook user data reflects the fact that personal privacy has been embezzled in the era of big data. People do not keep their information secret anymore, they volunteer to share their information online, but it just means people are adopting new definitions and new privacy rules. Keeping information private does not mean keeping it secret; people want to control the information share to whom and how to use the shared information.

5. Causes of problems

5.1 Technical limitations and culture factor

Intelligent robots, however, lack the unique conscience of mankind. The reason for this concern is essentially due to the limitations of artificial intelligence technology. Today's artificial intelligence does not have the ability to be emotional or deliberately biased, but honestly reflects the biases that exist in the database and even in society. Artificial intelligence recombines and pushes the contents through the algorithm; however, the original data are not completely fair; some of the original data have been biased, and the algorithm will further expand the original data. In addition, the machine is prone to make another mistake, is to treat most of the characteristics of the data as general features; this can be very unfair for a small number of data. Artificial intelligence for data mining and understanding always has a variety of limitations.

5.2 Lag of policies and regulations

Artificial intelligence technology as the emerging development technology, and it is not a long time, which causes the related social responsibility problem that are gone beyond the scope of the existing law, such as the environment, labor practice, and community involvement and development. These problems are more constrained by the social responsibility consciousness of researchers and developers. Since they are not mandatory, the effect is not very obvious. At present, many countries have made efforts; however, these policies are still relatively simple, need human being to improve constantly.

5.3 Stakeholders' social responsibility consciousness is insufficient

How stakeholders manage conflicts of interest, including current and long-term interests, local interests and overall interests, anthropocentrism and sustainable development? [10]. If developers, designers, and users have poor motives to treat artificial intelligence, such as for nuclear testing, inventing intelligent weapons for war will eventually destroy the harmony and stability of human society. If people use artificial intelligence technology to carry out illegal and criminal activities, it will pose a threat to the public interest. In addition, the random dissemination of false information about artificial intelligence will make people panic when people do not know enough about artificial intelligence. The scientific and cultural qualities of the developers and users of artificial intelligence technology can greatly affect the sustainable and healthy development of artificial intelligence technology.

The mistakes, discrimination, and prejudices of artificial intelligence also come from technical staff problems. They lack consideration of the data and the social realities behind them, and groups and regions that do not have the right to speak are likely to be more severely marginalized in the future.

6. Suggestions

By regulating the development of artificial intelligence through social responsibility, it can be avoided from being improperly used, so as to deviate from the original intention of letting technology benefit human society and to build up consumers' trust in artificial intelligence, thereby reducing unnecessary suspicion, panic, and exclusion. To strengthen the social responsibility management of artificial intelligence, the characteristics of artificial intelligence should be deeply considered.

6.1 Organizational governance

A social responsibility committee for artificial intelligence could be set up to guard the design and development of artificial intelligence to ensure that robots conform to ethical, legal, and other norms of human society. The committee should be interdisciplinary, involving both male and female participants. Conduct cross-disciplinary monitoring of artificial intelligence applications, identify industry best practices, and propose regulatory measures in due course. On the one hand, it promotes the application and innovative development of artificial intelligence by clearing some existing rules and institutional barriers. On the other hand, it ensures the security and reliability of artificial intelligence systems in these fields through the perfection of rules and standards, maintain public safety, enhance public trust, and accelerate the popularization of new technologies.

To promote the standardization of social responsibility in artificial intelligence at the international level, we should continue to work toward the international unification of social responsibility standards of artificial intelligence. It is necessary to establish the corresponding artificial intelligence social responsibility standardization framework, so as to avoid the conflicts and disputes arising from the inconsistency of standards among countries.

6.2 Human rights

When artificial intelligence enters human society, it is necessary to abide by the legal, moral, other norms and values of human society, and to act legally and in accordance with human morality. Because artificial intelligence system is the subjective design of research and development personnel. On the one hand, it is necessary to embed all kinds of norms and values into artificial intelligence system in a technically feasible and effective way, so that the system can make ethical behavior when it is running. On the other hand, it is necessary for research and development personnel to avoid subjective bias, preference, discrimination, and so on in the process of artificial intelligence system research and development.

The policy makers, researchers, and consumers of products need to take responsibility toward artificial intelligence, because once the artificial intelligence technology is developed, it is unlikely will fall back. It is necessary to integrate the social responsibility concept into the artificial intelligence technology.

6.3 Labor practices

Automation driven by artificial intelligence is disrupting the labor market. In the future, it is necessary to strengthen the education and training of talents in artificial intelligence, so as to ensure that workers can adapt and transform to the new employment and work paradigm brought about by artificial intelligence.

In the environment of artificial intelligence, the working hours and rest of workers are being blurred. The organization should create the conditions as far as possible so that the employees can realize the balance between work and life. To prevent the stress and anxiety of employees caused by the speed-up of the work rhythm caused by the intelligence of the workplace, and to create the appropriate professional environment to adapt to the physical and mental health of the employees.

6.4 Environment

E-waste can be recycled in a more rational and environmentally friendly way. Establish professional electronic waste disposal institutions to provide workers with protective measures to separate precious metals and toxic substances from electronic waste in a safe environment and to dispose of toxic substances in a specified manner. Educating stakeholders, advertising, setting up e-waste collection sites on the streets, and so on is to make people aware of the dangers of private disposal of e-waste.

6.5 Fair operation practices

The degree of automation of artificial intelligence decision-making is increasing day by day. It is necessary to ensure transparency, participation, and accuracy in artificial intelligence decision-making process. When an algorithm-based artificial intelligence system makes important decisions, it ensures transparency, participation, and accuracy. Transparency and accountability should be the primary goal in the design of artificial intelligence system.

6.6 Consumer issues

Artificial intelligence is a series of related technologies, which is widely used and involves all aspects of society. It should be taken into account in the process of social responsibility management. Moreover, the development of artificial intelligence technology relies heavily on big data, which greatly improves data acquisition.
And the efficiency of processing also increases the risk of improper application and disclosure of data, so it is necessary to pay special attention to data application and privacy protection in social responsibility management.

6.7 Community involvement and development

Pay attention to the network community. The Internet is not an extrajudicial place. Behavior in network community, including business innovation, must be on the rule of law track. For a long time, there has been a view that the Internet should be completely self-disciplined, to survive the fittest by itself. Now it has been found that this is not the case. Regulate the developers, users, and owners of artificial intelligence; do not harm the legitimate interests of the state, society, and individuals.

7. Conclusion

The development of artificial intelligence has had a profound and long-term impact on human production and life. While making people enjoy the good life, it also allows people to feel its negative effects, such as infringing on human privacy and bringing new inequalities to human beings. There are a lot of problems in theory and practice of artificial intelligence on social responsibility. However, these problems will not be solved in a short time. This chapter discusses the social responsibility problem of artificial intelligence with some related suggestions. We will proceed from the seven core subjects to enhance the social responsibility of artificial intelligence and ultimately achieve the sustainable development.

Acknowledgements

This work was funded by the National Key R&D Program of China under grant No.2017YFF0207603 and No.2016YFF0202502.

Author note

Some ideas of this chapter are from my three published article: 1. "Research on social responsibility of artificial intelligence based on ISO 26000" Source: Advances in Intelligent Systems and Computing, v 856, p 130-137, 2019; 2. "How to improve corporate social responsibility in the era of artificial intelligence?" Source: IOP Conference Series: Earth and Environmental Science, v 186, n 6, October 11, 2018; 3. "Improving Social Responsibility of Artificial Intelligence by Using ISO 26000" Source: IOP Conference Series: Materials Science and Engineering, v 428, n 1, October 1, 2018.

Author details

Weiwei Zhao
China National Institute of Standardization, Beijing, China

*Address all correspondence to: zhaoww@cnis.ac.cn

References

[1] Tractica. Artificial Intelligence Revenue to Reach $36.8 Billion Worldwide by 2025 [Internet]. 2016. Available from: https://www.businesswire.com/news/home/20160825006052/en/Artificial-Intelligence-Revenue-Reach-36.8-Billion-Worldwide [Accessed: 05 April 2019]

[2] House of Commons Science and Technology Committee. Robotics and artificial intelligence: fifth report of session 2016-17. 2016. Available from: https://publications.parliament.uk/pa/cm201617/cmselect/cmsctech/145/145.pdf [Accessed: 05 April 2019]

[3] Corporate author: World Commission on the Ethics of Scientific Knowledge and Technology. Report of COMEST on robotics ethics. 2017. Available from: https://unesdoc.unesco.org/ark:/48223/pf0000253952 [Accessed: 05 April 2019]

[4] 2017 Asilomar Conference. Asilomar AI Principles. 2017. Available from: https://futureoflife.org/ai-principles/?cn-reloaded=1 [Accessed: 05 April 2019]

[5] Committees of The IEEE Global Initiative for Ethical Considerations in Artificial Intelligence and Autonomous Systems. Ethically Aligned Design: A Vision for Prioritizing Human Wellbeing with Artificial Intelligence and Autonomous Systems (AI/AS). 2016. Available from: https://standards.ieee.org/content/dam/ieee-standards/standards/web/documents/other/ead_v1.pdf [Accessed: 05 April 2019]

[6] International Organization for Standardization. ISO 26000:2010 - Guidance on social responsibility; 2010

[7] Zhao WW. Improving social responsibility of artificial intelligence by using ISO 26000. In: IOP Conference Series: Materials Science and Engineering, Vol. 428, no. 1, October 1, 2018, 3rd International Conference on Automation, Control and Robotics Engineering. CACRE; 2018

[8] Jianfeng C, Na S. Global ai safety and ethics research. In: Zhiyi P, Zhun Z, Ling QQ, editors. Annual Report on Development of Cyberspace Security in China. China: Social Sciences Academic Press; 2017. pp. 58-77. ISBN/ISSN: 978-7-5201-1723-4

[9] Zhao W-W. Research on social responsibility of artificial intelligence based on ISO 26000. In: Advances in Intelligent Systems and Computing, Vol. 856. pp. 130-137, 2019, Recent Developments in Mechatronics and Intelligent Robotics - Proceedings of International Conference on Mechatronics and Intelligent Robotics. ICMIR; 2018

[10] Zhao W-W. How to improve corporate social responsibility in the era of artificial intelligence? In: IOP Conference Series: Earth and Environmental Science, Vol. 186, no. 6, October 11, 2018 International Conference of Green Buildings and Environmental Management. GBEM - Sustainable City and Regional Development; 2018

Chapter 6

Operationalizing Heterogeneous Data-Driven Process Models for Various Industrial Sectors through Microservice-Oriented Cloud-Based Architecture

Valdemar Lipenko, Sebastian Nigl, Andreas Roither-Voigt and Zelenay David

Abstract

Industrial performance optimization increasingly makes the use of various analytical data-driven models. In this context, modern machine learning capabilities to predict future production quality outcomes, model predictive control to better account for complex multivariable environments of process industry, Bayesian Networks enabling improved decision support systems for diagnostics and fault detection are some of the main examples to be named. The key challenge is to integrate these highly heterogeneous models in a holistic system, which would also be suitable for applications from the most different industries. Core elements of the underlying solution architecture constitute highly decoupled model microservices, ensuring the creation of largely customizable model runtime environments. Deployment of isolated user-space instances, called containers, further extends the overall possibilities to integrate heterogeneous models. Strong requirements on high availability, scalability, and security are satisfied through the application of cloud-based services. Tieto successfully applied the outlined approach during the participation in FUture DIrections for Process industry Optimization (FUDIPO), a project funded by the European Commission under the H2020 program, SPIRE-02-2016.

Keywords: industrial optimization, model predictive control integration, machine learning model integration, Bayesian network integration, enterprise resource planning (ERP) forecast model integration, prediction model integration, model calculation graph, microservice-oriented architecture, cloud computing

1. Introduction

In the area of industrial manufacturing performance optimization, prediction models are often used to predict future plant outputs in order to increase product quality and energy efficiency or optimize planning of plant maintenance activities.

Forecasts and predictions are utilized to generate lead time to plan and operate the manufacturing processes in a highly optimized way. To achieve predictive

IntechOpen

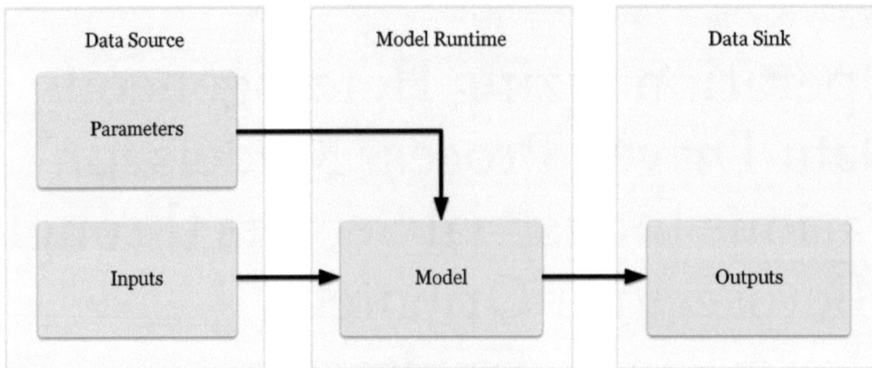

Figure 1.
Most basic representation and elements of a model.

operation on optimized performance, analytical calculation models are developed and operationalized to calculate the future behavior of manufacturing plants. This chapter describes the techniques and methods for operationalizing the prediction models in industrial manufacturing environments.

A model is defined as a simplified mathematical representation of a real natural process. Such process can be an industrial plant (e.g., a heat exchanger) or a complex plant like a continuous production machine (e.g., large-scale waste incineration steam boiler). Models are also called digital twins in applications where a model represents the real plant in a very high degree of details as in finite-element-method (FEM) model of a system (machine) or its components.

A model in the context of this chapter is characterized by three main features:

1. The model is a representation of a real plant.

2. The model is simplified "version" of the real plant that does not include all properties or behavior.

3. The models can be generalized, so they represent a type or a class of a real plant and not necessarily a real existing instance of a plant at a specific time.

Models are typically operationalized calculation (software) modules that allow a causal calculation of outputs from inputs and parameters. A model is defined as a calculation module that can predict the future industrial plant behavior. Such "plant behavior" can be the quality of produced product and the consumption of energy such as steam, electricity, raw material, and chemicals. Operating industrial plants based on predictive analytics allows optimized planning and real-time optimization. An industrial manufacturing or process industry plant has usually strictly separated information technology (IT) and operational technology (OT) systems (**Figure 1**).

2. Requirements and convergence of industrial system architecture

Industrial manufacturing industry continuously seeks for performance optimization strategies and ways of operationalizing new methods for improving product quality, production efficiency, energy efficiency, emission reduction, and, of course, cost reduction techniques. Digital transformation programs need to support these performance optimization goals. By driving convergence of OT and IT on all levels of enterprises (people, processes, interfaces, and system architecture), beneficial effects from integrated and consistent data and information utilization can be achieved.

Furthermore, a state-of-the-art operational architecture should be open and scalable for a whole ecosystem of internal or external partners to benefit from a learning culture on how to operate all production and business processes at maximum performance levels under changing market, raw material, and environment conditions (**Figure 2**).

Typical areas of industrial OT systems are as follows:

- machines and parts of machines (pumps, mixers, valves, tanks, etc.);
- automation systems [distributed control system (DCS), programmable logical controller (PLC), supervisory control and data acquisition (SCADA) system]; and
- operator user interfaces of SCADA system and DCS.

Typical areas of industrial IT systems are as follows:

- business applications like customer relationship management (CRM) system;
- business intelligence and data warehousing;
- enterprise resource planning (ERP) system;
- data analysis;
- computing systems and technology; and
- data warehouse and storage systems.

The main questions relevant in the design process of an industrial prediction model operationalizing framework development can be as follows:

- How to make exchange of prediction models as easy as possible for an industrial ecosystem with partners from process industry and scientific community and commercial partners?
- What framework elements have impact on the calculation performance of large-scale models and what are the performance requirements based on the dynamic behavior of the processes of interest?
- What architectural security elements are required to ensure a safe operation of the model calculation runtime system and user interfaces.

2.1 Requirements by heterogenous industries

Various industries have significantly different requirements for the integration of prediction and optimization models.

Examples of different industries that benefit from prediction/forecast model integration are shown in **Table 1**:

Industry category	Model value/benefits
Continuous process industry (e.g., pulp and paper mill, continuous waste water treatment plant, oil and gas refinery)	• Production rate of continuous production machines • Tank level utilization for stable production • Optimized quality stability
Original equipment manufacturer (OEM), e.g., micro-combined-heat-power unit manufacturer	• Improved overall equipment efficiency • Machine flexibility in conditions with changing raw material properties • Enable predictive maintenance use cases
Energy and utilities (e.g., waste incineration power plant, district heating network)	• Increased process stability (temperatures, pressure, steam flow) from predictive boiler operation using feedforward model predictive control • Increased energy efficiency by predictive control of excess combustion air (= O_2) control
Discrete manufacturing (e.g., automotive)	• Increased overall equipment efficiency (productivity time quality acceptance ratio) • Reduced costs due to lower reject rates

Table 1.
Industry categories and model benefits (by examples).

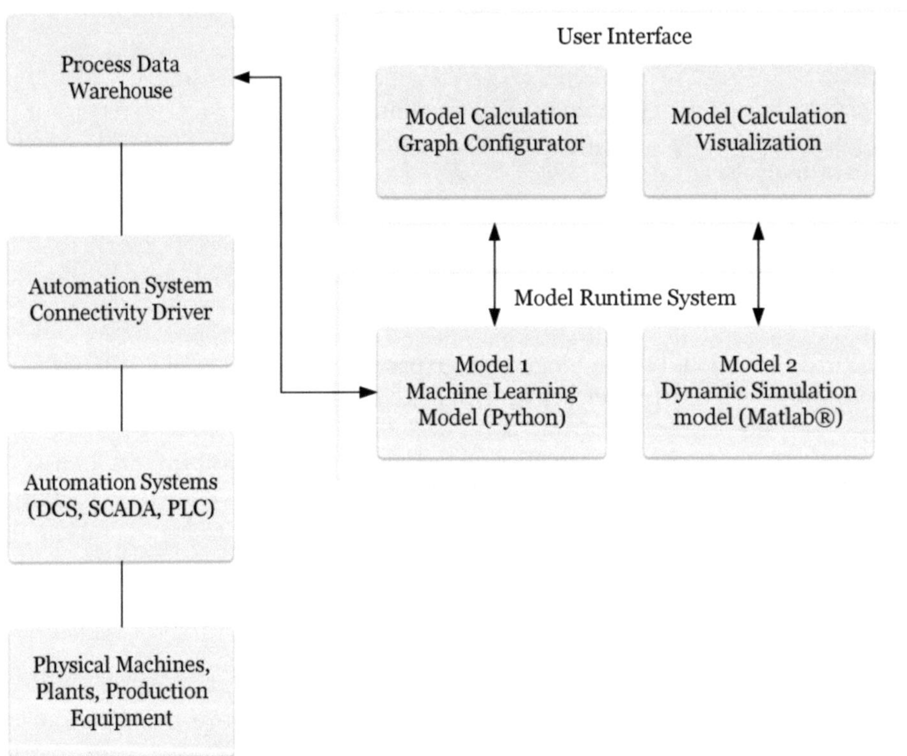

Figure 2.
Model operationalization framework overview (examples of).

2.2 Requirements by heterogenous models

Forecast and prediction models can be of various types and purposes of integrated usage by a smart manufacturing system. The huge spectrum of very specific requirements makes it hard for practitioners to find models from existing libraries that fulfill the requirements of the specific application. An ontology-based approach to classify and identify such model application usage scenarios can be found in Ref. [1]. The challenge of model exchange can be solved by using standards for model exchange such as functional mockup interface (FMI) or functional mockup units (FMU), as specified in Ref. [2].

Model examples illustrating the range of requirements are as follows:

- prediction of future process information based on production plan;
- prediction of future process output based on (just) recent process inputs, process states, and predicted future disturbance variables;
- prediction models for use in model predictive control (MPC). Dynamic MPC models can be linear or nonlinear; and
- first principle physical models (dynamic, static).

Models can be developed in various simulation software tools and have very different requirements when it comes to the runtime environments. As models are software components, many dependencies need to be fulfilled to run the models. Running models in this context means execution with actual inputs and user-specified parameters to calculate the predictions for a specific period (usually a time period in the future).

In order to avoid a complicated or even contractionary model runtime system architecture, a containerization technology can be used. In a software container (like Docker [3]), each model gets the required dependencies installed in the container rather than on the global runtime system. Utilizing container technologies

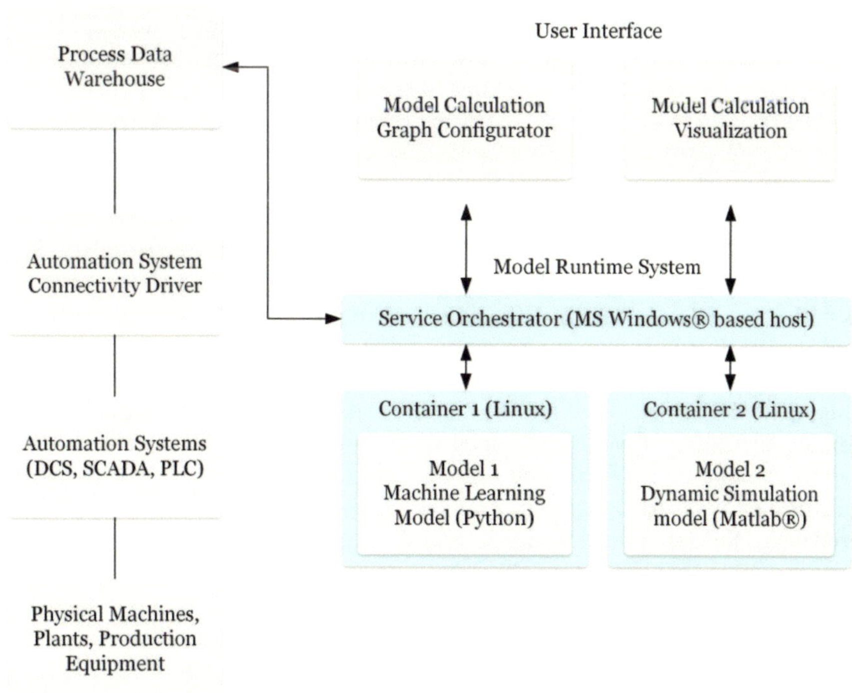

Figure 3.
Microservice-based architecture for operationalized model integration.

allows a clean deployment of operational models in their own containerized runtime system avoiding any additional requirements on the main application level as a container provides all required dependencies.

The resulting system architecture is called microservice architecture as it contains smaller independent software components packed into smaller application units (containers). An additional service orchestrator application is required to handle the messages between the OT systems and the microservices (containers) and the user interface.

Figure 3 shows the system architecture for an example with two containers where Linux is used as container operating system and the model container host server can be Microsoft Windows® based. This scenario allows the integration of models for all commonly used operating systems and can therefore provide all dependencies of other software components needed by the models to run/execute properly.

3. Calculation graph configurator

This section will cover why it is beneficial to use a calculation graph configurator when integrating one or more models into one software solution. Furthermore, it will be explained how node-RED can be used as a calculation graph configurator [4].

3.1 Calculation graph configurator overview

Since executing models often requires multiple calculation steps, for example, preprocessing the model input data or apply filters when selecting input data, a visual tool to connect and modify these calculation steps is beneficial. For example, in the FUDIPO project Node-RED, a flow-based programming tool is used as a calculation graph configurator. One big advantage of a tool like Node-RED as a calculation graph configurator, is that it saves a lot of cost due to the fact that Node-RED is highly customizable and open source. Therefore developing a custom calculation graph calculator, is not necessary. Furthermore, Node-RED is cross platform compatible as it runs on Node.js. There is also an official Docker image for Node-RED [4].

3.2 Node-RED as a calculation graph configurator

In Node-RED, calculation steps are called nodes. The nodes communicate via JavaScript Object Notation (JSON) messages. Node-RED provides a base set of nodes with a special functionality like nodes making a HTTP-Request or executing a JavaScript code. In addition to the base nodes, the community develops and contributes nodes that are available, thus providing very versatile functionalities. Node-RED also offers a dashboard where the model results could be displayed, although the default dashboard of Node-RED does not support multiple users, it should just be used for debugging or if only one user accesses the user interface (UI). Another great feature of Node-RED is that it provides a node to map data. Data mapping is required when the data source variables and the model input variables are not in the same structure. In some cases, the data must be mapped twice, once before the model execution, and a second time after the model has been executed to store the model results, as shown in **Figure 4**.

Node-RED's dashboard offers some basic UI elements, like charts, input forms, buttons, switches, and slides; however, it is also possible to write HTML, cascading style sheet (CSS), or JavaScript code directly, and this offers the ability

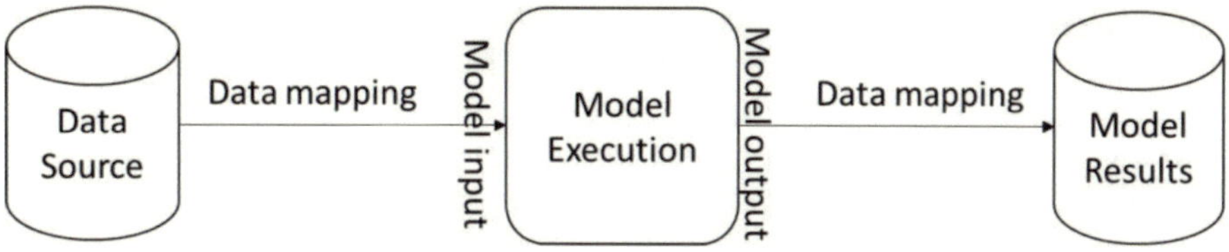

Figure 4.
Data mapping for normalized model execution.

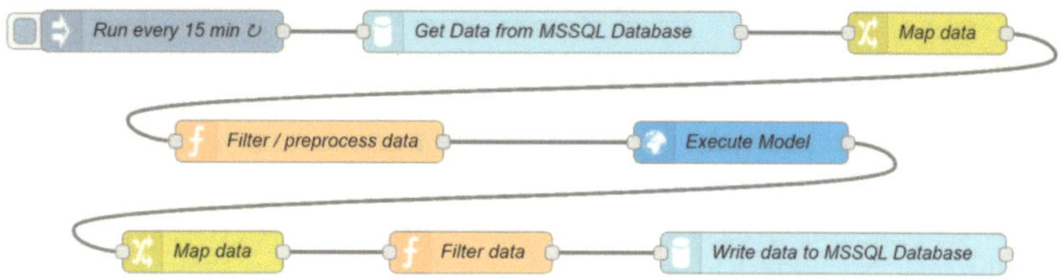

Figure 5.
Calculation graph example in Node-RED.

to create a highly customizable UI within Node-RED. Furthermore, Node-RED offers the possibility to customize its theme to match the company's color scheme. Additionally, Node-RED also offers the ability to secure the editor and dashboard user interface with a username and password, so if you have sensible data or do not want your Node-RED to be publicly available, it can be secured with a login mask [4].

3.3 Node-RED calculation graph example

An example calculation graph would be to retrieve data from an MicroSoft Structured Query Language (MSSQL) Server → map the data to match the model input data → filter or preprocess the input data → execute the model → map data back to match the schema of the MSSQL database → filter the model result data → save the filtered model results in a MSSQL server. In **Figure 5**, the example just explained is realized in Node-RED [4].

The first node runs the calculation graph every 15 minutes. At the time of writing, the interval can range from 1 second to 596 hours. The "Get Data from MSSQL Database" is a contribution node, which means that a user created this node and provided the source code, so everyone can use it. This node retrieves data from an MSSQL Server using Structured Query Language (SQL) statements. The yellow "Map data" node uses JSONata to change the structure of a JSON object. In the "Filter/preprocess data," node JavaScript is used to apply custom-made filter and preprocessing algorithms. The "Execute Model" in this example sends a HTTP POST request to an Application Programming Interface (API), where the model is being executed and returns the model results as an HTTP response. The next node maps the model result data to the schema of the MSSQL database. Afterward, these data are filtered again before writing it back to the MSSQL database in the last node. An advantage of this solution is that there is only little effort needed to integrate models into the solution. The only thing that needs to be developed is API that executes the model. Node-RED also offers the possibility to execute terminal commands, so if a model can be executed via the terminal, even less effort is needed to integrate the model. This calculation graph could also be used for multiple systems or machines if the model is designed to do so [4].

4. User interface

In this section, it will be discussed how a user interface for displaying the model results and for uploading or tuning the models could be developed. The key features, a user interface should have, are as follows:

- visualizing model results with charts;
- allowing the model developer to upload a new version of the model; and
- allowing the model developer to test and tune the model.

Node-RED allows the model developers to tune and test their models with the live data; however, Node-RED does not offer the feature to upload new models. A possible solution for this would be to use Node-RED's dashboard to visualize the model results, to use Node-RED's editor to make it possible for the model developer to verify and tune his model(s) with live data, and to build a separate website enabling the model developers to upload newer and improved versions of their models. As already mentioned, Node-RED's dashboard has the big advantage that it is very easy to use, even people with little to no knowledge about HTML, CSS, or JavaScript can make a simple user interface in Node-RED. However, the major flaw in the provided dashboard is that it does not support multiple users; for small use cases or debugging purposes, this might be enough, but in large-scale applications where multiple people are going to use the dashboard, another solution must be used. In such cases, it is either possible to develop a new website and access the model results either from an API or from a database, or use a fork of Node-RED's dashboard, developed by the community, that supports multiple users [4].

Another possible solution would be to display the model results in the company's current software solution. If there is already a software that acts as a user interface for the machine data, it might be possible to display the model results too.

If the goal is to continuously integrate multiple models, a possible solution would be to develop a website where the model developers are able to upload and maybe even verify, possibly with real live data from the machines, their models. Even though such an automated process might be convenient, it requires a lot of development effort and it should be considered if the initial effort is worth spending. Another possibility is to initially integrate the model manually in the solution and provide the model developers the possibility to update their models, by uploading a new version to a website. This is especially useful if there are models that require frequent updates. For example, training a machine learning model requires a lot of computing power, due to this it is suboptimal to train the model on the server where the model is executed. Thus, training the model on a separate computer or server that is suited for such high loads is better; therefore, a web interface could be used to upload the newly trained model, providing the model needs to be trained continuously. Another possible feature such a website could have would be allowing model developers to test the newly uploaded model with live data. So, model developers are able to ensure that their models are working correctly.

To be able to display the model results in a website in most cases, a charting library is needed. One of the many JavaScript charting libraries is Highcharts®. The advantage of Highcharts® is that it is very configurable, well documented, and feature rich. Highcharts®, for example, is able to export the data that are displayed in the chart as an Excel sheet or download the chart as an image. Highcharts® has many feature add-ons, called modules. Another big advantage of Highcharts® is that it has a boost module, which significantly boosts the performance of the charts making it possible to display well over 500,000 datapoints in one chart while keeping about the same performance as a chart with 1000 datapoints [5].

5. Pulp and paper use case

Processes in the pulp and paper industry are considerably complex with significant time delays (up to few hours) resulting in major difficulties for appropriate process optimization and control. An example of such process is continuous cooking in pulp digester, aiming at removing lignin from wood chips [6]. The most widely used index for measuring residual lignin present in the pulp is kappa number [6]. The digester primary control objective consists in minimizing the variability of kappa number, keeping it in a small range within few percent of target value (too low and too high kappa numbers negatively both impact quality and production stability).

The current situation is shown in **Figure 6**.

Through utilization of various process-specific analytical data-driven models, it is possible to substantially improve the process control, hence also the product quality and production stability.

One of such data-driven approaches is to conduct the forecast of future observations, such as based on certain characteristics of wood chips to predict the resulting kappa number (which is otherwise known only with the delay of ca. 4 hours from the time when respective wood chips came into cooking process). Measurement of wood chips with near-infrared (NIR) provides the capabilities to extract data on lignin content, moisture, and reactivity of wood chips. Subsequent utilization of specific physical and/or machine learning models enables to forecast the resulting kappa number.

MPC integration is another important approach in this context, in particular accounting for multivariable specifics of the underlying process control. The presence of measurement noise and various complex chemical process uncertainties, which are common in the pulp and paper industry, can also effectively be addressed through MPC [7]. Here, data from NIR-based soft sensors coupled with various dynamic process models are the main enablers of a feedforward MPC [7].

Process diagnostics (such as identification of digester hang-ups, screen clogging, and channeling of liquor inside the digester [8]) can effectively be done by input of certain information on current system status to a causality tree, such as Bayesian Networks (BN), with subsequent inference for computations of probability of different faults. Hence, the resulting decision support system is a valuable tool to improve diagnostics and fault detection capabilities in pulp and paper applications.

A comprehensive view on model-based control and diagnostics for pulp digester provides [9] the possibility of feedforwarding the lignin content of incoming wood chips based on NIR measurement under the incorporation of modeling and simulation studies. Additionally, Rahman et al. [9] proposed a simple Bayesian network-based diagnostic approach to detect pulp digester faults. Rahman et al. [7] provided an approach for feedforward MPC as applied to the pulp and paper industry.

Moreover, the analytical data-driven models need continuous updating to utilize process changes and to learn from experience. The feedback can be done automatically, by process operators, maintenance staff, and so on.

Thus, **Figure 7** provides an overview of the potential process improvement as compared to the current situation.

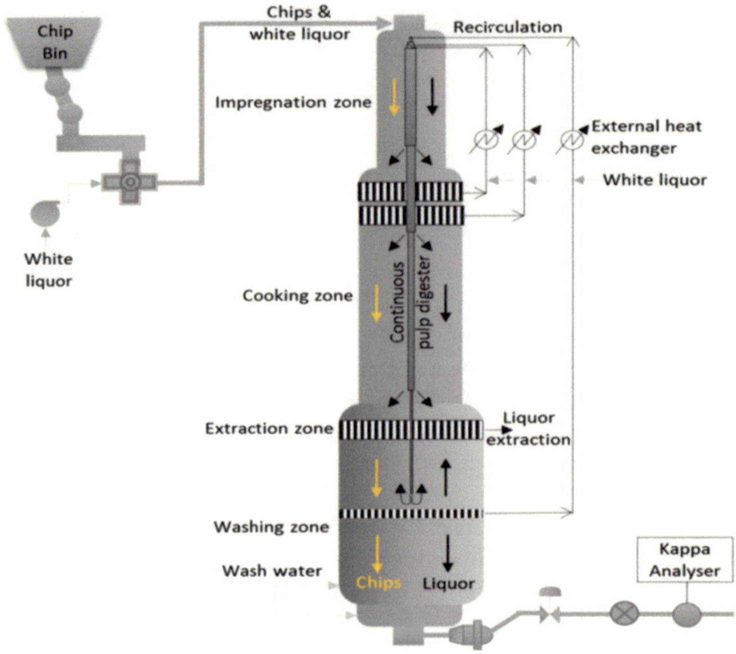

Figure 6.
Pulp digester [8].

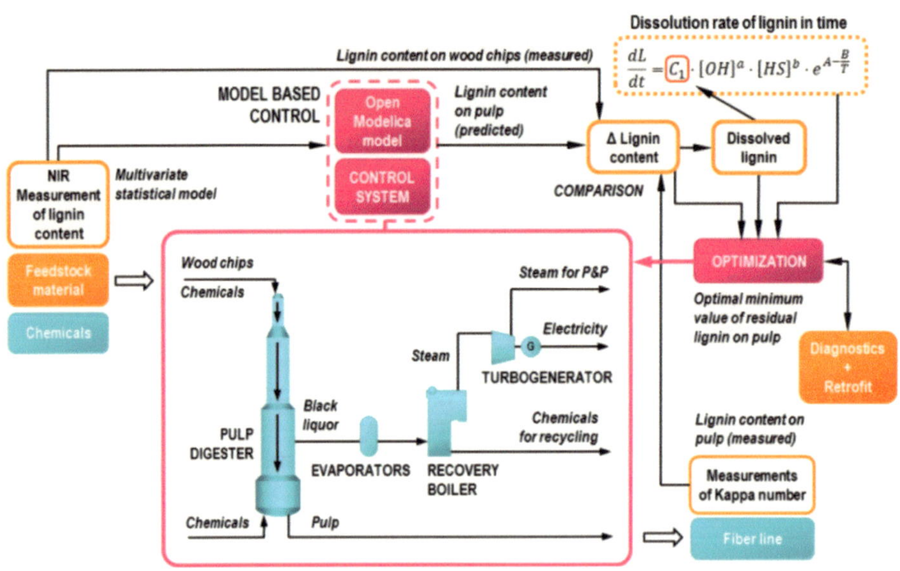

Figure 7.
Pulp digester improved process [8].

6. Micro turbine powered combined cycle

A key factor to reduce the operating cost of a micro gas turbine (MGT) powered combined heat and power fleets is to improve the maintenance strategy. The goal is to stop maintaining the micro gas turbines on a timely basis, instead maintaining MGT based on anomaly detection (condition-based maintenance) with specifically developed models that are able to detect MGT faults. In addition to lowering maintenance and operating costs using such models can also increases the safety of micro gas turbines. Two ways to develop such models would be either by using a physics-based approach or by using a data-driven approach, and the best results are achieved when combining a physics-based model and a data-driven model. The challenge when developing the models for a fleet of micro gas turbines is that often models are tuned for a specific machine and cannot be applied to a similar machine; however, when having a fleet of thousands of micro gas turbines, it would require way too much effort to develop a model for each system; therefore, the model must detect anomalies for multiple MGTs. To sum it up models should predict the maintenance actions long before the MGT fails, this will likely increase the safety, reduce the maintenance cost, and possibly increase the availability of MGTs [14, 15].

Physics-based models are based on mathematical formulas and constraints between sensor data. A physical model for MGTs basically simulates a micro gas turbine and compares the simulated values with the real sensor data. With the results, a degradation of a MGT can be calculated. The main disadvantage of such models is that they require an expert knowledge of the machines they are developed for. In contrast, data-driven models do not require such comprehensive knowledge of the field. Data-driven models use big quantities of sensor data and known failures. These data-driven models are able to classify and predict the future failures based on learning (machine learning) from the historical data. The drawback of data-driven models is that it is often not comprehensible how the model results are exactly calculated.

6.1 Example system architecture for micro gas turbines

This subsection will illustrate an example system architecture for micro gas turbine powered heat and power fleets.

Figure 8 represents an example system architecture for MGT fleets. Every MGT in the fleet writes its sensor data to a database, which is in the Azure [16] cloud for maximum scalability and availability. The SQL server has two databases, one for the sensor data and a separate database for the model results. The databases are separated because if one database crashes, the other will keep running. If even more reliability is needed, the two databases can also be in separate SQL servers. The virtual

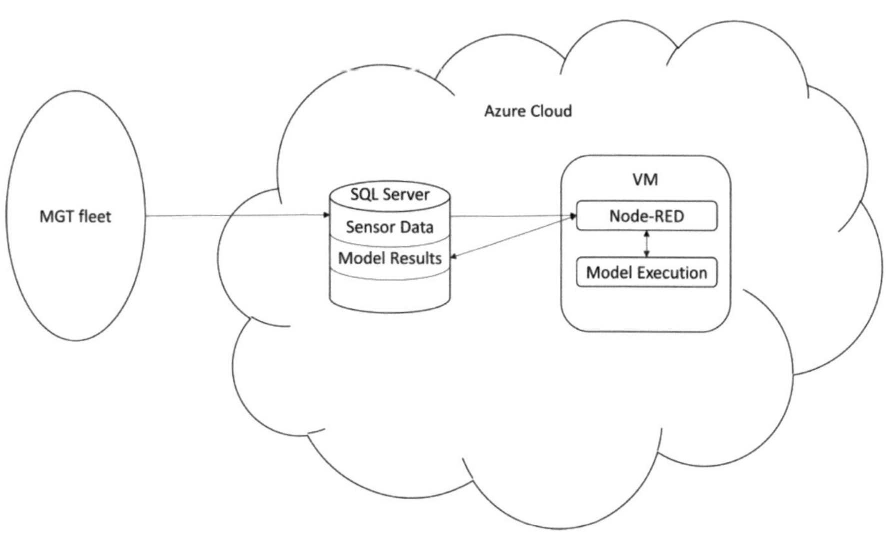

Figure 8.
Example system architecture for MGT fleets.

machine (VM) runs Node-RED as a calculation graph configurator and allows the model developers to verify and tune their models. Node-RED then executes the calculation graph as explained in Section 3.3. In this example, Node-RED's dashboard is being used to validate, tune, and debug the models. Additionally, the model results (e.g., future maintenance actions) are visualized in the company's own UI. Furthermore, the model results and sensor data could be processed and displayed in the UI of the customer, so the owners of the MGT(s) could see how good their MGT(s) perform, or even allowing them to plan for the future maintenance costs [4, 5].

7. System environment

In the present setting, the view on the overall system architecture may not be limited to IT aspects but is necessarily extended to OT. Hahn [10] provides an in-depth analysis of the differences between IT and OT in the domain of industrial control systems (ICSs). On the one hand, OT is traditionally seen as the heart of ICS. On the other hand, modern ICS increasingly utilizes IT capabilities, creating a convergence of OT and IT domains [10]. For an in-depth understanding of the impact and posed challenges of OT and IT consolidation, it is therefore important to align on different principles to handle data in these both domains, together with differences in how the overall systems are operated and managed, which technologies are used to support them, and so on. Hahn [10] provides the respective explanations as follows:

- Operational domain. In order to control and monitor physical processes (such as power grids, pulp production, and oil refinery), various sensors, actuators, controllers, and human operators are altogether utilized in ICS. The resulting unique operational requirements are much different from the traditional IT environments, whose focus is more strongly on controlling and managing the data, retrieved here from the underlying OT.

- Technical domain. The unique technical requirements for the software used to support the operations of ICS mainly result from or are closely related to specific communication protocols and architectures; real-time performance demand; domain-specific device manufacturers and integrators; complex integration of digital, analog, and mechanical controls; and so on.

- Managerial domain. From the management of OT system's point of view, their underlying complex physical infrastructure usually requires much larger capital investments than it is the case for the IT systems. Hence, the operation of ICS is subsequently usually planned for decades, in order to recuperate the cost.

Thus, in order to successfully integrate specific solutions from the IT domain (in particular, analytical data-driven solutions) with OT in a holistic ICS, the following requirements should be, respectively, satisfied [11]:

1. Data source requirements. Support for data produced by various industrial machines and sensors, usually connected in one network, without the possibility to access over a common application programming interface (API) is crucially important. The unique characteristics of single heterogeneous elements should be harmonizable for further processing.

2. Processing requirements. Data from sensor instrumented industrial devices commonly possess typical Big Data characteristics, so-called four Vs:
   - *velocity*: data arrive rapidly from many different sources;
   - *volume*: huge amounts of data produced in seconds or milliseconds;
   - *variety*: structured, semi-structured, and unstructured data appear; and
   - *veracity*: occurrence of noisy or otherwise of a poor-quality data.

   This makes such data very challenging to manage, integrate, and analyze.

3. Human interface requirements. The insightful results from assessed data are necessary to be made available to operations professionals in convenient manner for them. In other words, the respective process database should eventually contain all the data required to accordingly inform process responsible (in form of visualizations, notifications, alerts, etc.).

4. Security requirements. Integration, development, deployment, extension, and customization are required to take place in secure environments with ensured security standards on at least the same level as already in-place.

An important extension to the considerations regarding IT and OT is the runtime environment itself, or more precisely the decision between installing and running single solution components on the premises of demonstrators (on-prem) as opposed to remote facilities (cloud-based). The outlines of advantages and disadvantages of on-prem and cloud-based solutions are, respectively, depicted in **Tables 1–3**. At the same time, since in this case the single advantages and disadvantages are often overlapping, the summarized information should be seen as a very general overview only.

Since on-prem and cloud-based solutions have own strong advantages and disadvantages, there is naturally no single best choice for every different demonstrator. In other words, it is crucially important to enable an efficient system architecture setup for on-prem the same as cloud based. Then, the decision for a particular setup is left to be met solely in accordance with respective internal policies, with posing

Advantages of on-prem solutions	Disadvantages of on-prem solutions
Great possibilities for customizations	Solution implementation takes longer
Possibility for specific security policies	High system maintenance effort
High solution design flexibility	High system management cost
In-house system and data knowledge	High upfront investment

Table 2.
Advantages and disadvantages of on-prem solutions.

Advantages of cloud-based solutions	Disadvantages of cloud-based solutions
Short implementation time	Limited customization possibilities
No need for system maintenance	Security depends on cloud provider
Low upfront investment	Low solution design flexibility

Table 3.
Advantages and disadvantages of cloud-based solutions.

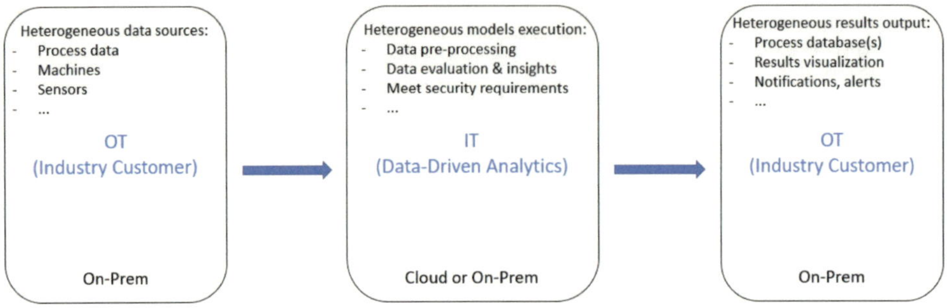

Figure 9.
Simplified generic system overview.

no technological limitations whichsoever. Hence, disregarding the differences between the outlined system architectures, they can be consolidated as depicted in **Figure 9**.

Hence, the consolidated system environment may easily be generalized as of a cloud-based nature, differencing solely in certain administrative responsibilities of the underlying technical infrastructure.

8. Microservice architecture

The decision in favor of microservice architecture (MSA), which is a variation of service-oriented architecture (SOA), naturally results from the cloud-based (either a public cloud or an on-prem configuration) specifics of the underlying solution. In a nutshell, a solution adopting microservice-based architecture consists of a large number of small services, each responsible for a single-specific aspect (e.g., data access, execution of certain model, and specific data preprocessing step) [12]. Hence, the main benefits of solutions implemented based on MSA include their major scalability capabilities (both scaling up and scaling down depending on the present circumstances), reusability, loose coupling, and their advanced technology agnostic nature, which are probably the most important in the discussed context. The latter advantage of MSA enables easily utilize different runtime environments and programming languages in single microservices of one mutual solution, hence also adapting to technological changes to avoid technology lock-in, and so on [12].

Coming back to the previously elaborated requirements in the context of OT consolidated with IT to successfully implement analytical data-driven solutions, MSA solves these as follows:

1. Data requirements. The heterogeneous nature of the underlying data produced by industrial machines and sensors is best supported through the creation of separate microservices responsible for particular types of data, retrieving data from specific sources, and so on. Moreover, every new type of data or data source can further be easily supported by simply adding a new and independent respective microservice to the overall solution. Through further microservice, heterogeneous data can be harmonized and brought to common format and structure for the subsequent processing stages.

2. Processing requirements. The architecturally unlimited number of microservices in-place best support the requirement to process even data produced in near real time through simply extending the number of instances of respective

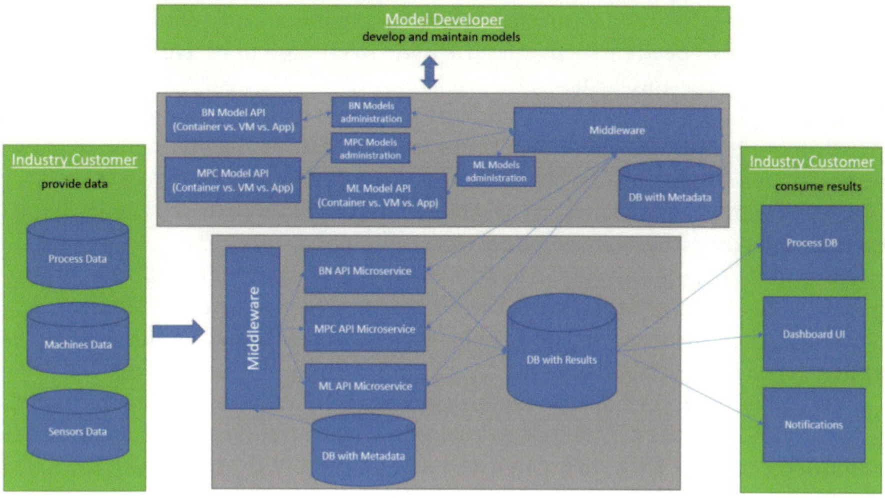

Figure 10.
Microservice-based system outline.

microservices. The only bottleneck could be imposed through requests dispatching entity, such as a certain API management microservice, which is nevertheless efficiently replicable nowadays as well. Thus, all of the V's of Big Data can efficiently be approached by utilizing MSA.

3. Human interface requirements. The way of how results of data processing and analysis are made available to process professionals (process operators, process engineers, etc.) is again not limited by the architecture and only requires the implementation of appropriate single-independent microservices for writing the data to respective type of process database, or even simply visualizing in-place in the underlying solution. Moreover, the initially chosen implementation is easily adjustable during the whole solution lifecycle based on, for example, specific usability requirements.

4. Security requirements. Single microservices made responsible for security act as respective middleware, thus ensuring conformity to require security standards. Again, customization and adaptability are major benefits of choosing MSA as opposed to monolithic solutions.

An exemplarily overview of how MSA can successfully be utilized for operationalizing the heterogeneous data-driven process models is depicted in **Figure 10**.

9. ERP integration of prediction models

In today's times, the ERP system has a central role in almost every company. Its capabilities help the business to act, plan, and decide on the data, which will be collected in such systems. In all firms, such decision-making processes are happening every day. The question is how a company could become more competitive than its competitors. What if, company leaders could predict future events or would be able to connect the dots beforehand. This and maybe other reasons are the motivation to integrate the prediction models in ERP systems.

The challenge is, to bring the best parts of business know-how, software engineering, and data science together, to be able to predict the future.

The first step is to understand the company's business. Only with that knowledge, problems can be identified, which are worth enough to be avoided to save leaders in firm's time and costs.

One example of such problem could be the question: When and how much will my customer order which products? There are too many parameters to answer this question with a deterministic approach, which means the only way to solve this problem is to get an empiric way of thinking. So, companies can work with that what they already gathered in so many ways, with data.

This leads to the second step to get the understanding of that data. The advantage of having an ERP system is that the information is already collected and described. To get one step further, the challenge is to find out or explore where these data are stored and verify the quality.

Once this is done, the content of the third step is to prepare the data for modeling. Simply spoken, the task here is to bring the data in a correct format that prediction models can work with. Every model needs a different format, which means in this step, it is also needed to know which model should be used. If the formatted data and prediction model fit together, the next step can be done.

Step 4 is about modeling. In this phase, data and the selected prediction model will be evaluated, tested, and improved in a couple of iterations or even exchanged. The quality of a prediction model can be measured in predictive power and predictive robustness. The power measures the capacity of the input variables to explain the target. The robustness is the ability to display the same level of performance on new data sets as training ones.

The final step is to deploy and integrate the prediction model in the ERP system. To show how this can be accomplished, the following paragraphs will rely on the ERP solution Systems, Applications, and Products (SAP).

A typical SAP on-premise installation always comes with an application server called SAP S/4HANA and a database management system called SAP HANA [13–16].

When and how much will my customer order which products?

In the paper business, the products can be distinguished by different characteristics, but one major characteristic is the grade. So, for every grade, different raw materials are needed to be ordered and purchased. Also, a major factor is that for every grade, the paper machine needs a different setup, which leads also to effort and costs. So, for the management of a paper producing company, it would be beneficial to predict the customer's behavior for the next month to save time and costs.

The information which customer orders which product with amount and point of time is maintained the so-called customer sales order. Information to the product is available in the material master. The following database tables are meant: VBAK (Sales Document Header), VBAP (Sales Document Position), MARA (Material Master), and AUSP (characteristic values).

To be able to work with that data, it is needed to bring all tables in view to aggregate the information on a monthly basis. This can be accomplished with a so-called core-data-service (CDS), which is nothing else than a view with special functionalities. The CDS-View plays a major role when it comes to data preparation and analysis.

As there is the need to predict the future time events, the auto regressive integrated moving average (ARIMA) model will be chosen for this scenario. The ARIMA model is delivered within the HANA database by SAP Predictive Analytics Library (SAP PAL) and can be accessed via ABAP Managed Data Procedures (AMDP) from the application layer. It must be mentioned that there are also different integration scenarios possible. For example, a R- and/or Python-Integration can be implemented with SAP as well. This approach will be used for high sophisticated models where the functionalities of SAP PAL cannot match for this requirement.

To integrate this predictive function in existing applications or reporting tools, an Advanced Business Application Programming (ABAP) will be written for this purpose.

The ABAP executes the training of the model and the forecast for the next period at one point of time.

To get an overview how this solution with its different integration scenarios could like the following illustrations should help to get a better understanding what has been described before (**Figure 11**).

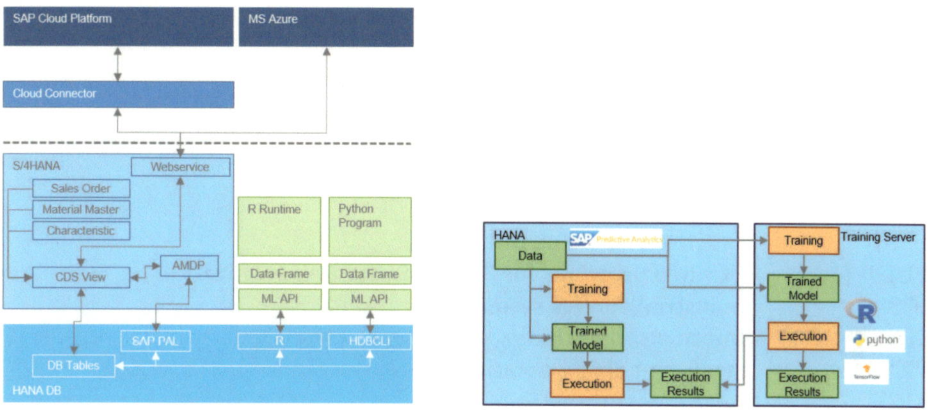

Figure 11.
Architecture for SAP ERP integration of predictive models.

10. Conclusions

Operationalizing and deployment of industry process prediction models can be achieved efficiently by setting up a microservice-based architecture that gives not only the industry companies but also the scientific community full flexibility in using software components to run the prediction models. Successful integration of heterogenous models for different types of industries has shown high added value for operating the prediction models for business performance optimization. Process industry and OEMs benefit by the proposed architecture from a very fast and cost-efficient implementation of models into their OT and IT environment.

Acknowledgements

The work presented in this paper is carried out in connection to the FUture DIrections for Process Industry Optimization (FUDIPO) project. This project has received funding from the European Union's Horizon 2020 SPIRE-2 (Sustainable Process Industry Research) program under Grant Agreement No. 723523.

Conflict of interest

The authors declare no conflict of interest.

Abbreviations

ABAP	advanced business application programming
AMDP	ABAP managed data procedures
API	application programming interface
ARIMA	auto regressive integrated moving average
BN	Bayesian networks
CDS	core-data-service
CRM	customer relationship management
DCS	distributed control system
ERP	enterprise resource planning
FMI	functional mockup interface
FMU	functional mockup unit
FUDIPO	FUture DIrections for Process industry Optimization
ICS	industrial control systems
IT	information technology
JSON	JavaScript object notation
MGT	micro gas turbine
MPC	model predictive control
MSA	micro service architecture
MSSQL	MicroSoft structured query language
NIR	near-infrared
OEM	original equipment manufacturer
OT	operational technology
PLC	programmable logical controller
SAP	systems, applications, and products
SCADA	supervisory control and data acquisition
SOA	service-oriented architecture
SPIRE	sustainable process industry through resource and energy efficiency
SQL	structured query language
VM	virtual machine

Author details

Valdemar Lipenko, Sebastian Nigl, Andreas Roither-Voigt* and Zelenay David
Tieto Austria GmbH, Vienna, Austria

*Address all correspondence to: andreas.roither-voigt@tieto.com

References

[1] Novák P, Šindelář R. Applications of ontologies for assembling simulation models of industrial systems. In: OTM Confederated International Conferences on the Move to Meaningful Internet Systems; 17 October 2011. Berlin, Heidelberg: Springer; 2011. pp. 148-157

[2] FMI-Standard [Internet]. Available from text [Accessed: 13 August 2019]

[3] Docker Container Technology [Internet]. Available from: https://www.docker.com/ [Accessed: 13 August 2019]

[4] Node-RED [Internet], 2019. Available from: https://nodered.org/ [Accessed: 13 August 2019]

[5] Highcharts [Internet], 2019. Available from: https://highcharts.com/ [Accessed: 13 August 2019]

[6] Correia FM, d'Angelo JV, Almeida GM, Mingoti SA. Predicting kappa number in a Kraft pulp continuous digester: A comparison of forecasting methods. Brazilian Journal of Chemical Engineering. 2018;**35**(3):1081-1094

[7] Rahman M, Avelin A, Kyprianidis K, Jansson J, Dahlquist E. Model Based Control and Diagnostics Strategies for a Continuous Pulp Digester. InPaperCon, 2018

[8] Pulp and paper (Billerudkorsnäs)—FUDIPO [Internet], 2019. Available from: https://fudipo.eu/pulp-and-paper/ [Accessed: 12 August 2019]

[9] Rahman M, Avelin A, Kyprianidis K, Dahlquist E. An approach for feedforward model predictive control for pulp and paper applications: Challenges and the way forward. In: InPaperCon 2017; 23-26 April 2017. Vol. 10. Minneapolis, Minnesota, USA: TAPPI Press; 2017. pp. 1441-1450

[10] Hahn A. Operational technology and information technology in industrial control systems. In: Cyber-Security of SCADA and Other Industrial Control Systems. Springer Champions; 2016. pp. 51-68

[11] Morris HD, Ellis S, Feblowitz J, Knickle K, Torchia M. A Software Platform for Operational Technology Innovation. International Data Corporation.; 2014. pp. 1-7

[12] Balalaie A, Heydarnoori A, Jamshidi P. Microservices architecture enables devops: Migration to a cloud-native architecture. IEEE Software. 2016;**33**(3):42-52

[13] Färber F, May N, Lehner W, Große P, Müller I, Rauhe H, et al. The SAP HANA database—An architecture overview. IEEE Data Engineering Bulletin. 2012;**35**(1):28-33

[14] Aslanidou I, Zaccaria V, Rahman M, Oostveen M, Olsson T, Kyprianidis K. Towards an integrated approach for micro gas turbine fleet monitoring, control and diagnostics. In: Global Power and Propulsion Forum 2018; Zurich, Switzerland. 2018

[15] Rahman M, Zaccaria V, Zhao X, Kyprianidis K. Diagnostics-oriented modelling of micro gas turbines for fleet monitoring and maintenance optimization. PRO. 2018;**6**(11):216

[16] Azure—Microsoft [Internet], 2019. Available from: https://azure.microsoft.com/en-us/ [Accessed: 13 August 2019]

Section 2

Industrial Applications of AI

Chapter 7

Machine Learning Models for Industrial Applications

Enislay Ramentol, Tomas Olsson and Shaibal Barua

Abstract

More and more industries are aspiring to achieve a successful production using the known artificial intelligence. Machine learning (ML) stands as a powerful tool for making very accurate predictions, concept classification, intelligent control, maintenance predictions, and even fault and anomaly detection in real time. The use of machine learning models in industry means an increase in efficiency: energy savings, human resources efficiency, increase in product quality, decrease in environmental pollution, and many other advantages. In this chapter, we will present two industrial applications of machine learning. In all cases we achieve interesting results that in practice can be translated as an increase in production efficiency. The solutions described cover areas such as prediction of production quality in an oil and gas refinery and predictive maintenance for micro gas turbines. The results of the experiments carried out show the viability of the solutions.

Keywords: machine learning, prediction, regression methods, maintenance, degradation prediction

1. Introduction

The amount of data accumulated by man's activity is uncountable. Millions of tuples are registered daily in the databases, each of them constitutes an observation, an experience to learn from, and a situation that could reoccur in the future in a similar way. Learning from experience is something that humans do naturally and constantly, but what happens if the number of experiences exceeds our ability to process it? What happens if a fact is repeated millions of times and never happens again in exactly the same way?

Machine learning (ML) is the area of artificial intelligence, which deals with learning from the experience, that is, to extract automatically implicit knowledge in the information (stored in the form of data) [1].

In this paper we will describe two real-world industry problems that have been solved using ML. The first of these consists in predicting the quality of the final products of an oil and gas refinery, described in Section 2. The second consists of a model for estimating the degradation of a fleet of micro gas turbines, described in Section 3.

In the next section we offer the theoretical elements necessary for the development of our solutions. Any interested reader can find in Section 1.1 the description of the ML methods we have used. We also describe the general working scheme of the ML applications.

1.1 Machine learning

There are countless examples of complex real problems solved through ML, such as [2–6]. In ML, one important subarea is inductive learning; this type of method assumes that a set of examples or instances is known [7]. Formally, **learning** is defined as:

Theorem 1.1. Let a set of examples $(x_i, y_i), x_i \in D$ be a domain space state D and $y_i \in S$ be a solution domain space state S, or, let $(x_i), i = 1, 2, \ldots, n,$ be a domain space state D where the solutions are not defined, that is, S is undefined. The task of creating a system that can learn the input-output pairs $\{(x, y)\}$ or learn the characteristics inherent to $\{x\}$ is defined as learning.

The first case refers to **supervised learning**, where there is a solution y_i (the class label) for each input vector x_i, these examples are known as "classified" or "labeled" [8]. The second case refers to **unsupervised learning**, in which a system learns characteristics, traits, groups, and concepts from unlabeled data.

The supervised learning is a technique to deduce a function from training data. One component of the pair is the input data and the other, the desired results. The output of the function can be a numerical value (as in the regression problems) or a class label (as in the classification ones).

Formally **supervised learning** is defined as:

Theorem 1.2. Let T a training set, which is formed by pairs $(x_i, y_i), i = 1 \ldots n,$ where n represents the number of features, x_i is defined as input vector, y_i is the output value (the target). If, y_i is numeric then it is a regression task, and if, y_i is discrete then it is a classification task.

The need for supervised learning arises from the requirements of having an automated procedure that is much faster than a human supervisor and that, at the same time, can avoid biases and prejudices adopted by an expert [9].

There is also another area in ML known as semi-supervised learning (SSL) [10, 11]. SSL uses both labeled and unlabeled data for training. Reinforcement learning (RL) is an example of SSL. In RL, the model learns how to act in changing environments. It is about taking suitable action to maximize reward in a particular situation. It has been widely used in games, autonomous driving, and many industrial applications. **Figure 1** summarizes the previous definitions.

However, to get to the learning process, it is necessary to go through some preparing phases first. **Figure 2** shows these phases.

The first phase is the data collection; the data can come from multiple sources, be in different formats, etc. The second phase is the data preprocessing; in this

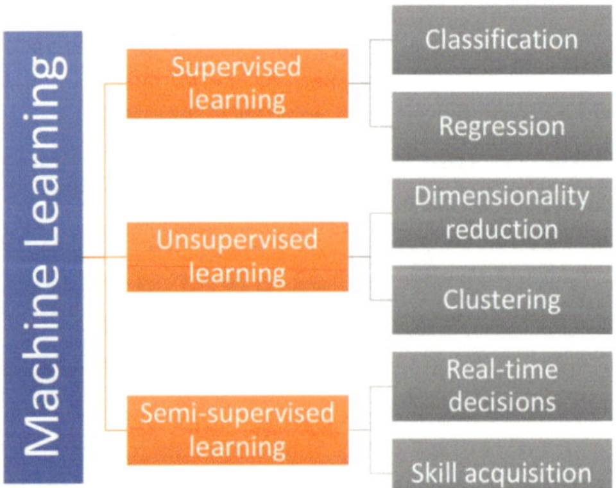

Figure 1.
Machine learning tasks.

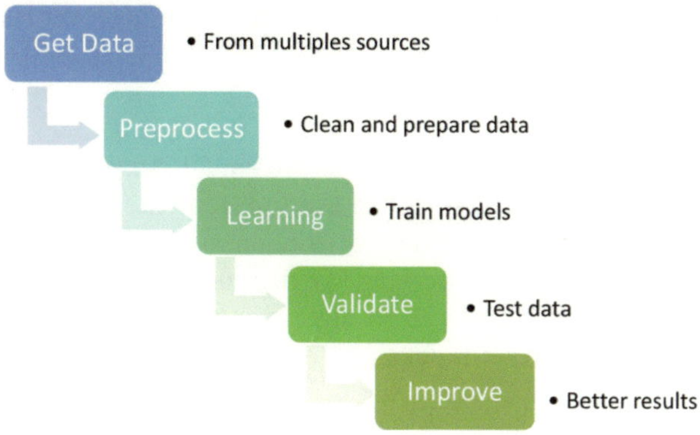

Figure 2.
Learning phases.

phase numerous tasks are performed in order to prepare the data for the learning stage. These tasks can remove noise; normalize, discretize (is needed for the learning phase), and remove/replace missing values; select features; and select instances. When the data is ready, the learning phase can start. The data is partitioned in:

Training set: is the set of examples used to build the model to train the ML model.

Testing set: is the set of examples used to test the models. The ML model will assign an output to each example in the test set. In classification, if the output value assigned by the ML model matches the label that has the example in the test set, then it is true classified. In regression, an error is computed using the difference between the real value and the predicted. **Figure 2** shows the phases described above.

Given the relevance of preprocessing to our study in the following subsection, we will describe in detail some of the preprocessing techniques.

1.2 Preprocessing steps

In real-world applications, especially the industrial ones, data is rarely clean and homogeneous. Most often we find data that tends to be incomplete, redundant, noisy, or inconsistent. The area of ML that deals with the above problems is known as preprocessing. The preprocessing task consists of the set of techniques that are carried out before the learning process. Its objective is to obtain a higher-quality dataset. These techniques can be divided in the following groups:

1. Handling missing values: missing values occur for various reasons: human errors, errors in sensor measurements, data is merged from various sources, etc. Some learning methods can deal with missing values internally, while others do not. The most common techniques to deal with missing values are:

    a. Remove the variables or remove the examples with missing values. This technique can reduce the data sample and cause loss of information.

    b. Replace with an "estimated value." There are several methods to estimate a missing value, such as the mean value of the variable, the median, the most frequent value, and so on.

2. Handling noise: a noisy value is a value that is not the correct one. It is also known as corrupt data. The noisy value may be very close to the true signal.

3. Handling outliers: an outlier is a value that is much different than the other values. Most of the time, the outliers are noise, but sometimes a data point that is true signal can be an outlier.

4. Instance selection (IS): not all instances are equally important. IS consists of the selection of the most appropriate examples for the learning stage. It is also known as dataset reduction or dataset condensation. During the IS process, you can select those most representative instances, free of noise, outliers, or missing values. Some of the used FS algorithms are those based on rough set theory and fuzzy rough set theory.

5. Feature selection (FS): not all features are equally important. FS consists of the selection of the most representative variables or features for the learning stage. Selecting the right subset of features to be used for the learning phase has proven an improvement in the performance of supervised and unsupervised learning.

1.3 Learning algorithms background

In this section, we will describe the most significant learning algorithms from the state of the art, emphasizing those that were used in the present research. First, we will describe some classifiers and then some regressors.

1.3.1 Classification task

As we defined in previous sections, classification is the learning task where each input vector corresponds to a discrete output value, known as a class. Next we will describe the most representative classifiers in the state of the art.

- Decision tree C4.5 [12]: In 2008, it was ranked as #1 in the Top 10 Algorithms in Data Mining pre-eminent paper published by Springer LNCS. C4.5 builds decision trees (DT) from a training set, using the concept of information entropy. At each node of the tree, C4.5 chooses the attribute of the data that most effectively splits its set of samples into subsets enriched in one class or the other. The splitting criterion is the normalized information gain (difference in entropy). The attribute with the highest normalized information gain is chosen to make the decision. The C4.5 algorithm then recurses on the partitioned sublists.[1] Decision trees are predictive models that use a set of binary rules to calculate a target value.

- k nearest neighbors' classifier (kNN) [13]: It is a non-parametric algorithm. Its purpose is to use a dataset in which the instances are separated into several classes to predict the classification of a new instance. This method, for a new example to be classified, finds its k nearest neighbor using Euclidean distance, and then the example is classified by a majority vote of its neighbors. In a similar way, this method is used in a regression task. The numeric output is mean of the nearest neighbors.

- Random forest (RF) [14]: It is an ensemble method formed by decision trees. During the training phase, the method builds *n* decision trees from randomly

[1] https://en.wikipedia.org/wiki/C4.5_algorithm.

sampled datasets with replacement and randomly selected subset of features, where *n* is an input parameter. In the testing phase, each individual tree spits out a class prediction and the class with the most votes is then predicted. RF avoids the overfitting of the traditional decision trees.

- Multilayer perceptron (MLP) [15]: It is one of the most used artificial neural networks. It consists in a set of layers (minimum 3): one input layer, one or more hidden layer, and one output layer. The input layer has as many neurons as features in the training set. The number of hidden layers and the number of neurons in these layers are parameters defined by the user. The number of neurons in the output layer corresponds to the number of classes in training set. MLP used backpropagation for the learning process. MLP can be used for classification and regression task.

- Support vector machine (SVM) [16]: It is a discriminative classifier defined by a separating hyperplane. This algorithm performs as follows: given a labeled training set, it outputs an optimal hyperplane which categorizes new examples.

1.3.2 Regression task

Regression is a widely used task in the world of industrial applications. It learns from the data and then when facing a new entry, is able to predict an output value. The most used regression algorithms are:

- Linear regression (LR) [17]: is a linear method that models the relationship between a group of dependent variables and one or more independent variables. In LR the relationships are modeled using linear predictor functions.

- Partial least square (PLS) [18]: is also similar to linear regression but that at the same time projects the data into a lower dimensional space, so that less variables are used in reality in the prediction model.

- Decision tree regressor: is regression method that works in the same way as the DT as a classifier; it was introduced in [19]. A decision tree arrives to an estimation by asking a series of questions to the data. Every node of the tree represents a binary question to be answered. Each question is further restricting our possible values until the model has enough confidence to make a single prediction. In this way, it is possible to build very accurate rules about the data.

- Ridge [20]: is a method of regularization also known as Tikhonov regularization that puts weighted l_2 norm penalty on the regression coefficients. This method has shown very good results in regression problems, specifically in those of linear regression with the problem of multicollinearity. Multicollinearity, correlated independent variables, is very common in problems with a large number of features.

- LASSO [21]: is another regularization method that puts weighted l_1 norm penalty on the regression coefficients. The least absolute shrinkage and selection operator, known as LASSO, is a method that performs variable selection and regularization in order to enhance the prediction accuracy and interpretability of the statistical model it generates.

- Gaussian process regression (GPR) [22]: is a non-parametric, Bayesian method for regression that infers all possible values over a probability distribution. In recent years, GPR has gained popularity among the machine learning researchers because of its robustness and performance in terms of classification and prediction accuracy. The advantage of using GPR is that it can be utilized in the exploration and exploitation scenario [22].

2. Use case 1: predicting output quality in Tüpras

Obtaining high-quality products is a fundamental objective of the Turkish Refinery Tüpras. Its four main products (diesel 95%, diesel sulfur, HSRN 95%, and LSRN 95%) must meet certain quality parameters dictated by the customer. In practice, achieving the quality required by the client is not a simple task, since during the distillation process the oil is subjected to many physical and chemical processes. However, taking into account that (a) in each of the phases of distillation of the crude oil, many variables are measured (in the laboratory or using sensors), (b) the initial chemical properties of the crude oil are known, and (c) the company have historical data on the final quality of the products, in this investigation, we will use machine learning techniques to predict the final quality values of Tüpras products.

2.1 Problem description

The main task of the Tüpras refinery is to convert crude oil into usable final products, satisfying the specifications established by consumers. To achieve the quality specifications, it is necessary to take many decisions, which means in our context change the manipulated parameters in the distillation process. **Figure 3** shows how the crude oil goes through a distillation process.

As can be seen in **Figure 3**, crude oil goes through several processes before becoming a final product. When we analyzed the historical data we have, we observed that in only 7 of the 254 days of which we have information, the three products were in the desired range. This gives us the measure of how hard it can be to achieve the desirable quality. Predictions based on historical values, using ML, can help achieve the desired quality. Knowing in advance the quality value, it will be possible to take decisions and make changes in the distillation process that allows to reach the desired value.

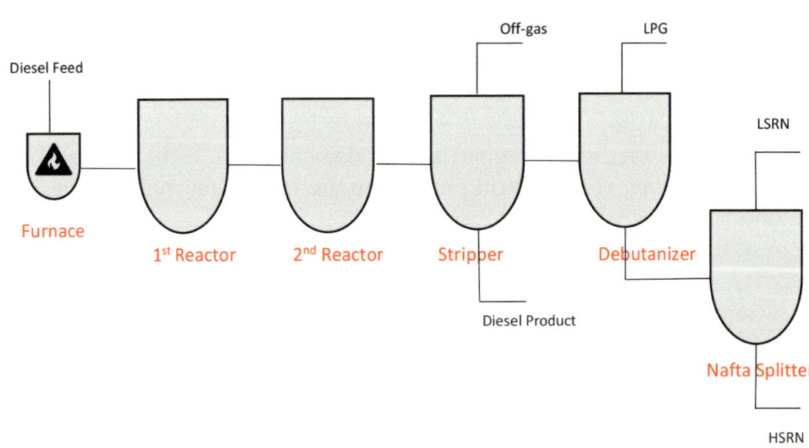

Figure 3.
Tüpras refinery process scheme.

2.1.1 Variables and the frequency in measuring

The complete cycle, starting with crude oil until transforming into a high-quality diesel, lasts approximately 240 min (4 hours). Having a large amount of data that comes from different sources, measured with different frequency, our first task is to create a dataset that logically and consistently unifies the complete distillation cycle.

Data was collected from 260 days, but after removing missing data, we have 254 days left. The data consists of:

- 17 raw input feed characteristics measured once a day where the timepoint was not specified.
- 272 process-related parameters measured every minute each day, in total 1440 measurements each day.
- 44 output feed characteristic variables where we only predict four of them (diesel 95%, diesel sulfur, HSRN 95%, and LSRN 95). The output variables were measured at 8 am every day and are valid for process measurements from 4 am to 8 am.

For the creation of the dataset, we consider:

- The dataset was created in the form $x_1, x_2, \dots, x_n \rightarrow y$, where n is the number of independent variables and each x_i represent a variable measured during the distillation process. These variables can be sensor measurements, manipulated variables, control variables, and the 17 raw input feed characteristics. The output variable or the dependent variable (y) is the final quality. In this way we have a decision system ready for learning task.
- We take into account the time delay of the process.

Thus, in total we have 279 (17 feed +272 process) independent variables that are used to predict four dependent variables. However, since the output variables are only valid for 4 hours, that is, 240 minutes in total, there are 240 × 272 measurements plus 17 input variables for each output variable sample. Thus, there are many more independent variables than dependent variables, and therefore, we use the mean and standard deviation of each process parameter over each 4-hour period as

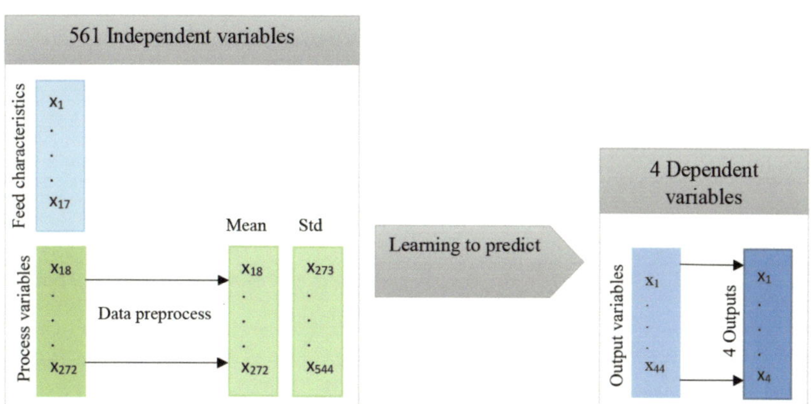

Figure 4.
Dependent and independent variables in the learning process.

features. Consequently, we have 17 + 272 × 2 = 561 independent variables for each 4-hour period. A graphic description can be found in **Figure 4**.

Notice also that only the above 4 hours of each day have labels—that is, have valid output variables (data is labeled)—while the remaining 20 hours lack valid labels (data is unlabeled).

Now we have the data ready for the learning process; in the next subsection, we will describe learning process.

2.2 A first approach: predicting final quality

In this subsection, we will describe the different experiments we used to evaluate the prediction performance for the output variables. First, we will report results from learning only from the labeled data. Next, we will present an analysis that uses learning curves to understand the learning problem, whether more data or more features would help improve the performance. Finally, we will describe results from applying semi-supervised learning where also the unlabeled data was used.

2.2.1 Experiment 1: prediction with only labeled data

In the first experiment, we use four regression methods described in previous section. We use LOO[2] cross-validation to investigate which method works best for predicting the four output variables when only trained on the labeled data. In LOO, we use $N - 1$ (where N = 254 days) data points for training the machine learning method that is then tested on the remaining data points. This is repeated N times resulting in N different predictions. For evaluating the prediction performance, we use root mean squared error (RMSE) that takes the square root of the mean of the square of the difference between the predictions and the true values.

In **Table 1**, the result is shown, and as can be seen, ridge regression has the overall best result with the smallest error (RMSE) for three of the output variables, while random forest has the smallest error for two of the output variables. We also notice that the errors of the two first output variables are not improved much by any of the methods compared to PMEAN, while the two last are improved quite a lot. Thus, in the following section, we will try to improve the performance for ridge regression.

2.2.2 Experiment 2: learning curves

In order to investigate whether we can learn some more from collecting more data or whether more features are needed, we can plot learning curves. Learning

Methods	**Output variables**			
	Diesel 95%	**Diesel sulfur**	**HSRN 95%**	**LSRN 95%**
PMEAN[a]	2.50	1.00	8.22	5.31
Ridge	2.36	0.79	3.69	3.68
PLS	2.44	0.79	4.53	4.34
RF	2.36	0.73	4.05	3.96

[a]As baseline, PMEAN is a simple algorithm used as a base of comparison. PMEAN uses the mean of the training data as prediction.

Table 1.
RMSE of the LOO cross-validation result for labeled data.

[2] Leave-One-Out

curves show the number training examples on the x-axis and the accuracy on the y-axis for both training data and test data that was not used for training. As accuracy we use the negative mean squared error (negative MSE), that is, the negative square of the RMSE. The learning curves for the output variables using ridge regression are shown in **Figure 5**.

The upper blue curve shows accuracy for training data, and the lower orange curve shows the accuracy for test data. Higher value means better performance, and as can be seen, the accuracy is better for the training curve than for the test curve, which is natural since the test curve should indicate the generalization performance of the algorithm. By extrapolating the curves, we can draw some conclusions from them.

The number of training examples is quite limited so what can be learned from the curves is also quite limited. However, we notice that the learning curves for the two upper output variables—**Figure 5a** and **b**—are quite similar, while the same can be said for the two lower learning curves, **Figure 5c** and **d**. We also observe that for the two upper learning curves, the test curves reach a plateau around −6 and −0.65, respectively, after which no more improvement is seen. Neither do we see much of an improvement for the training curves. This indicates that more training examples will not likely improve the performance, but instead we need more features or a more complex algorithm. For the lower left learning curve (c), we do not see the plateau that clearly, while the lower right curve (d) shows increasing performance with more data. So, for the lower curves, more training examples might improve the performance. In the next experiment, we will investigate this by using a semi-supervised approach that also uses the unlabeled data for training.

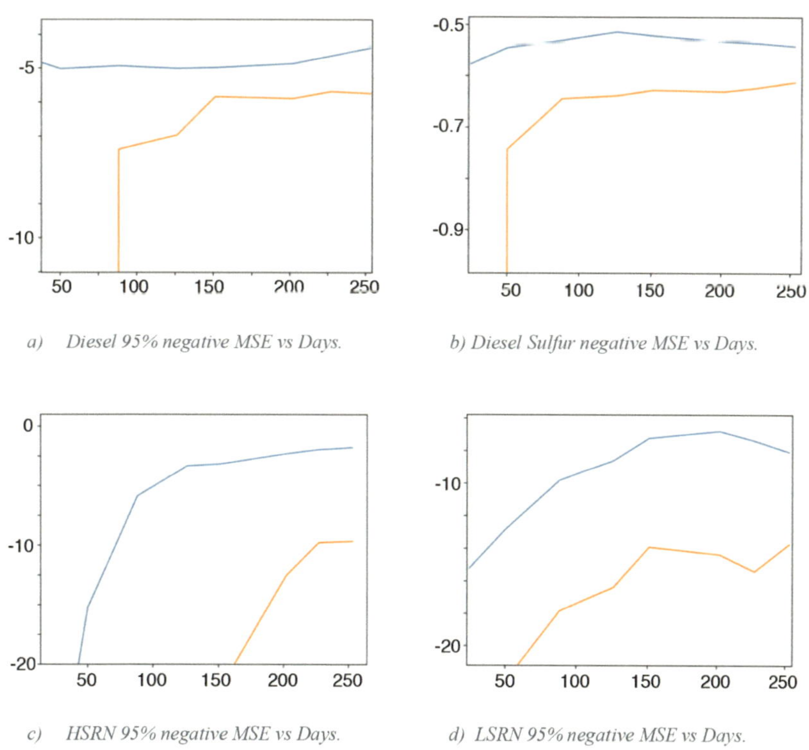

a) Diesel 95% negative MSE vs Days.

b) Diesel Sulfur negative MSE vs Days.

c) HSRN 95% negative MSE vs Days.

d) LSRN 95% negative MSE vs Days.

Figure 5.
The learning curves for the four different output variables.

2.2.3 Experiment 3: prediction with a semi-supervised approach

A semi-supervised approach uses both labeled and unlabeled data in the training phase [11]. In essence, we achieve this by creating more training examples and by using the ML algorithm to label them. We create unlabeled data by moving a sliding window of length 4 hours over each day with time step of 1 hour from 4:00 am to 0:00 pm. This created 20 periods of 4-hour data with 1 labeled and 19 unlabeled time periods. This increased the number of training examples from 254 to almost 5000 ($\approx 20 \times 254$). The algorithm we use to train on both labeled and unlabeled data has the following steps:

1. Train the learning method using only labeled data.

2. Predict the labels (output variables) of the unlabeled data.

3. Train the learning method using both labeled data and the data with predicted labels.

4. Repeat step 2 and step 3 until the difference between the old and new predicted labels becomes small.

The algorithm uses the maximum likelihood principle in that it converges toward the values with maximum likelihood, similar to how the expectation-maximization (EM) algorithm works [23].

For evaluation, we use also LOO cross-validation. That is, we used only labeled data for evaluation but used all unlabeled data in the training phase as described above and tested the trained method on the left-out labeled data. The result is shown in **Table 2**. The overall best approach is clearly ridge regression with semi-supervised learning. This confirms the observation from the learning curve analysis that the first and second output variable would not improve with more training examples, while the two last output variables we can indeed see improved performance with more data.

2.3 A second approach

After concluding a first stage in which we carried out the study shown in the previous section, we obtained new data from Tüpras. With the new data with a total 269 samples, we designed three new experiments. The objective of the following three experiments is to find with which dependent variables the best predictions of the variables are achieved.

Methods	Output variables			
	Diesel 95%	Diesel sulfur	HSRN 95%	LSRN 95%
PMEAN	2.50	1.00	8.22	5.31
Ridge	2.36	0.79	3.69	3.68
Ridge (SEMI)	2.34	0.82	2.54	3.31

Table 2.
RMSE of the LOO cross-validation with semi-supervised learning.

2.3.1 Experiment 1: not using the controlled variables

In our first experiment, we will predict the quality of the output variables without using the controlled variables. As in previous section, we will use LOO validation. **Table 3** shows the results. As we can observe, best results are obtained in all cases for LASSO, while ridge performs much worse for diesel 95% than in the first approach.

2.3.2 Experiment 2: using controlled variables

In our second experiment, we will predict the quality of the output variables including the controlled variables as independent variables. **Table 4** shows the results. Again, LASSO gets the best results in all cases, while ridge is the second best.

2.3.3 Experiment 3: using only controlled variables and diesel feed

In our third experiment, we will predict the quality of the output variables using only controlled variables and diesel feed characteristics. **Table 5** shows the results. LASSO gets the best results in all cases.

Methods	Output variables			
	Diesel 95%	Diesel sulfur	HSRN 95%	LSRN 95%
PMEAN	2.47	0.99	8.21	5.31
Ridge	2.58	0.85	2.64	3.49
PLS	2.50	0.82	3.32	3.91
LASSO	2.26	0.78	2.21	2.68

Table 3.
RMSE of the LOO cross-validation for experiment 1.

Methods	Output variables			
	Diesel 95%	Diesel sulfur	HSRN 95%	LSRN 95%
PMEAN	2.45	1.00	8.24	5.34
Ridge	2.41	0.88	2.79	3.56
PLS	2.72	0.88	3.30	3.92
LASSO	2.29	0.79	2.26	2.81

Table 4.
RMSE of the LOO cross-validation for experiment 2.

Methods	Output variables			
	Diesel 95%	Diesel sulfur	HSRN 95%	LSRN 95%
PMEAN	2.46	1.00	8.23	5.32
Ridge	2.18	0.85	5.68	4.42
LASSO	2.07	0.82	5.64	4.43

Table 5.
RMSE of the LOO cross-validation for experiment 3.

2.4 Partial conclusions

In this section we have described the use of regression methods to predict the four output variables of the Tüpras refinery.

In our first approach, we have described the evaluation of using ridge regression, partial least squares, and random forest to the problem of predicting the four output variables, where the ridge regression had the best performance. We have also shown that using a semi-supervised approach, we could improve the performance for two of the variables, which also indicates that more data collected from the process would most likely further improve the performance. However, for two of the variables, the learning methods did not improve the performance much compared to the baseline using the mean value of the training data, and neither did semi-supervised learning. For those two variables, we need to consider other relevant features than the mean or standard deviation.

When using more data (second approach), we constantly get the best results using LASSO regressor for the prediction of the four output variables. From our results we conclude that:

- For the prediction of diesel 95 it is better to use only the controlled variables and the diesel feed characteristics.

In contrast, ridge regression shows varying performance in the experiments, while being many times the second best. Thus, ridge seems to be less stable than LASSO for this problem.

3. Use case 2: predictive maintenance model from micro gas turbine

The need to predict maintenance intervals is a problem that currently has great relevance in the field of ML applied to the industry. Predicting in advance if a device needs maintenance can result in significant savings in time and money. With predictive maintenance, important failures and breakdowns in production time can be avoided [5, 23]. It is a fact that the maintenance intervals recommended by the manufacturers almost never correspond in practice with reality. This is largely due to the fact that local conditions vary a lot from one environment to another and manufacturers operate with generic measurements that do not take into account specific conditions.

In this section we will describe a proposal to estimate the performance degradation of a fleet of micro gas turbines. An important issue to consider is that there is no explicit degradation measure, which therefore must be estimated.

3.1 Problem description

The existing method for estimating degradation uses a linear model fitted to data from a reference system which then is used to correct the generated power from an installed system. Thus, the values are corrected and normalized so that they can be compared. In **Figure 6**, we can observe an example of the current approach. The yellow curve is the corrected power which shows the current approach to measuring degradation.

In addition to that, there are some conditions that make the problem unusual. These conditions are as follows:

- The systems are small-scale and low-cost installations, so there is only a small number of sensors available.

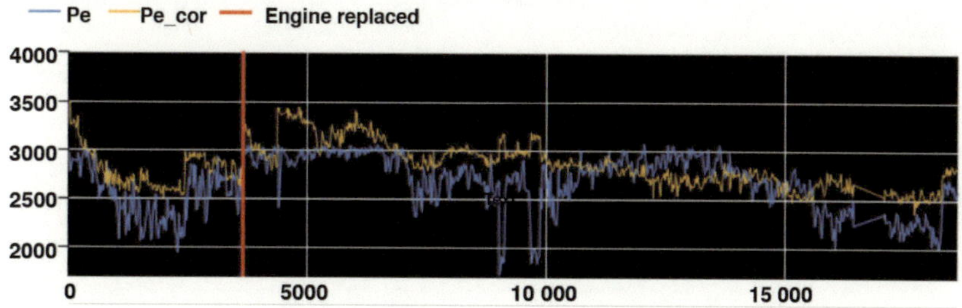

Figure 6.
Pe *is measured and* Pe_cor *is corrected power; engine replacement indicates start of currently installed engine life.*

- At the time of development of our method, there were not many systems installed and not many failures recorded, and thus, a traditional supervised ML approach could not be used.

- Each system is always running at full capacity, where the ideal power is the maximum power generated when there is no loss due to degradation and no effect of ambient conditions.

- Finally, there are recordings of maintenance actions, but the effect of an action on remaining degradation is not known.

Given the above list of circumstances, the design goal of the proposed method is to measure degradation:

1. Using only data from real systems and removing the need for a reference system

2. More smoothly than the existing method

3. Relative to the ideal power generation

3.1.1 Data collection and preprocessing

Data was collected from five different micro gas turbines with system ID 24, 27, 28, 29, and 30. The data was sampled every minute, but we used only samples from every second hour, since it was deemed to be sufficient for long-term degradation modeling. We use data from the parameters shown in **Table 6**.

3.2 Approach: power degradation model

The proposed method uses a regression model where physical properties are taken into account. As we said before, we are not facing a classic problem of supervised learning, since degradation cannot be measured. Thus, instead we let both the degradation and ideal power be properties of the model, and the model is trained to predict the measured electric power.

Now we introduce our model: let y be the measured power, $\vec{x}$ be a column vector with the ambient parameters like weather, pressure, etc. Then, let $\vec{t}$ be a column vector with time-dependent variables and n and m be the number of systems and number of maintenance periods, respectively. We use $1 \le i \le n$ to denote a specific

Variables	Parameters	Unit
Predicted variable (y)	• Net electric output power	Watts
Ambient (contextual) variables (x)	• Measured return water temperature	Kelvin
	• Inlet air temperature	Kelvin
	• Ambient pressure	Bar
	• Ambient pressure at stand still	Bar
	• Measured turbine speed	Rpm
	• Set point requests based on heat demand	—
	• The internal set point for desired speed and turbine exit temperature	—
Ambient pressure variable*	• Ambient pressure is missing	Dummy
Time-dependent variables	• Total number of running hours	Hours
Affecting the degradation trend (t)	• Total number of starts and stops	Frequency
Maintenance actions (M)	• Total number of running hours when action was taken	—
The ideal power per system (k)**	• Net electric output power during installation	Watts

**To handle missing values of the ambient pressure variable, we add a dummy variable that is 1 when the variable is missing and 0 when it is present.*
***The ideal power was measured when a system was installed and corresponds to the power that would be generated without disturbances from ambient variables and degradation due to wear.*

Table 6.
Parameters used in the prediction model.

system and $1 \leq j \leq m$ for a specific maintenance period. Now, we define the generic model of degradation as follows:

$$y = k_i + g\left(\vec{x}\right) + e\left(\vec{t};i,j\right) \tag{1}$$

where k_i is the known ideal power generation for system i, function g is the effect of ambient variables, and function e is the degradation over time due to wear.

In the above equation, the signal is divided into to two parts: a variation caused by ambient conditions and a degradation trend given by the time-dependent variables. It is also assumed that the variation due to ambient variables is the same for different systems, while the degradation is dependent on both the individual system and the maintenance period.

Let us assume a linear model for both functions g and e as follows:

$$\begin{aligned} g\left(\vec{x}\right) &= \vec{c}^{\,T}\vec{x} \\ e\left(\vec{t};i,j\right) &= a_i + b_j + \vec{e}_o^{\,T}\vec{t} + \vec{e}_i^{\,T}\vec{t} \end{aligned} \tag{2}$$

where $\vec{c}$ is a column vector to model ambient conditions and a_i and b_i model remaining degradation at start or after a maintenance action for system i and maintenance period j, respectively, and $\vec{e}_0$ and $\vec{e}_i$ are column vectors modeling the degradation common for all systems and for each individual system, respectively.

System	r	RMSE	MAPE
24	0.91	81.80	2.35
27	0.95	72.13	2.20
28	0.82	71.31	1.95
29	0.92	62.85	1.83
30	0.95	52.49	1.50
All	0.92	70.59	2.01

Table 7.
Estimation model results over five systems.

In order to get a feasible solution, we put a l_1 regularization on the remaining degradation coefficients in $\vec{a}$ and $\vec{b}$ so that the coefficients are kept close to zero. We also assume that the degradation is monotonic so that $\vec{e} \leq 0$. The solution was implemented using a machine learning framework called Keras[3] together with TensorFlow[4] as backend.

3.3 Experimental results

In this experiment, we use the existing method that uses corrected power derived from a reference system to validate the new method. **Table 7** shows, for each of the five systems we have tested, the Pearson correlation coefficient (r) between the estimated negative degradation and the corrected power, the root mean squared error, and mean absolute percentage error (MAPE in %) for predicting the measured power.

As can be seen, the correlation coefficients are above 0.9 for all but one system (28), which indicates that the proposed method is indeed a good replacement for the corrected power. Also, the RMSE and MAPE are of reasonable sizes.

3.4 Partial conclusions

In this use case, we presented a machine learning approach that incorporates physical properties into the model in order to estimate the degradation of a fleet of gas systems. In addition, we show that it was a good replacement of the existing approach to measuring degradation that was based on data from a reference system.

4. Conclusion

In this chapter, we presented an overview of machine learning and presented example use cases where we applied machine learning. In the first use case, we predicted the diesel product quality using common regularized linear regression, while in the second use case, we used a more customized regularized regression to implement predictive maintenance.

---

3 https://keras.io.

4 https://tensorflow.org.

As general conclusions, we can summarize:

- LASSO and ridge regressor were very efficient methods in predicting diesel quality at UC 1.

- For the prediction of diesel 95, it is better to use only the controlled variables and the diesel feed characteristics.

- The incorporation of physical properties into the degradation model in use case 2 is very useful for the final maintenance prediction.

A summary of general approach to solving a problem with machine learning is to:

1. Start by defining the learning problem: what variable should be predicted? If there is no explicit variable, it might be an unsupervised problem, but as in use case 2, it can also be a variable that is not measured. Thus, the sought variable needs to be extracted from or part of the estimated model.

2. Next, chose a performance metric that measures the desired outcome. In use case 1, it was quite simple since the diesel quality was measured directly, while in use case 2, the desired outcome—the time when the degradation of a system is too bad—was not measured directly.

3. Then, start out with a simple model, like a linear regression model, which also can be used as a baseline for comparison of more complicated models used in the next step.

4. Plot and analyze the learning curves (see Section 2.2.2). If the curves indicate potential of using a more complex model, then try with a more complex model like a random forest or a neural network. However, the selection of model is also dependent on the size of the dataset. If there is only small dataset as in use case 1, it is not possible to use a too complex model since more model parameters need more data for training.

5. Finally, test the models on a dataset not used for training above. This is to ensure that the performance measures the generalization power of the model and to avoid overfitting to the training data.

As an overall conclusion, we can see that we ended up with quite simple variants of linear models in both use cases, which is not uncommon given the authors experience from industrial problems. Another general comment is that in most cases each industrial problem is quite unique and there is no single solution that fits every problem. So, it is important to understand the problem domain and chose methods that fit that particular problem. Hence, machine learning is not a silver bullet that will solve all problems. If there is a good physical model, a machine learning model will probably not be a better choice. However, it might be a benefit to create a hybrid model combining the physical model with a data-driven machine learning model.

Acknowledgements

The research of Dr. Enislay Ramentol has been funded by the European Research Consortium for Informatics and Mathematics (ERCIM) Alain Bensoussan

Fellowship Programme. This project has received funding from the European Union's Horizon 2020 research and innovation programme under grant agreement No 723523.

Author details

Enislay Ramentol[1†], Tomas Olsson[2*†] and Shaibal Barua[2,3†]

1 Fraunhofer Institute for Industrial Mathematics, Kaiserslautern, Germany

2 Research Institute of Sweden, Västerås, Sweden

3 School of Innovation, Design and Engineering, Mälardalen University, Västerås, Sweden

*Address all correspondence to: tomas.olsson@ri.se

† The first two authors contributed equally; the third author contributed description of the state-of-the-art algorithms.

References

[1] Mitchel T. Machine Learning. New York: McGraw Hill; 1997

[2] Mazurowskia MA, Habasa PA, Zuradaa JM, Lob JY, Bakerb JA, Tourassib GD. Training neural network classifiers for medical decision making: The effects of imbalanced datasets on classification performance. Neural Networks. 2008;**21**:427-436

[3] Merschmann L, Plastino A. A lazy data mining approach for protein classification. IEEE Transactions on Nanobioscience. 2007;**6**:36-42

[4] Milletari F, Navab N, Ahmadi S-A. V-net: Fully convolutional neural networks for volumetric medical image segmentation. arXiv:1606.04797v1 [cs.CV]. 2016

[5] Ramentol E, Gondres I, Lajes S, Bello R, Caballero Y, Cornelis C, et al. Fuzzy-rough imbalanced learning for the diagnosis of high voltage circuit breaker maintenance: The SMOTE-FRST-2T algorithm. Engineering Applications of Artificial Intelligence. 2016;**48**:134-139

[6] Ronneberger O, Fischer P, Brox T. U-net: Convolutional networks for biomedical image segmentation. arXiv: 1505.04597v1 [cs.CV]. 2015

[7] Kasabov NK. Foundations of Neural Networks, Fuzzy Systems, and Knowledge Engineering. Cambridge, Massachusetts, London, England: Massachusetts Institute of Technology; 1998

[8] Cherkassky V, Mulier F. Learning from data. In: Concepts, Theory, and Methods. 2nd ed. Hoboken, New Jersey: John Wiley & Sons, Inc.; 2007

[9] Orriols-Puig A, Bernadó-Mansilla E, Sastry K, Goldberg DE. Substructural surrogates for learning decomposable classification problems: Implementation and first results. In: Conference Companion on Genetic and Evolutionary Computation: GECCO 07. 2007

[10] Triguero FHI, García S. Self-labeled techniques for semi-supervised learning: Taxonomy, software and empirical study. Knowledge and Information Systems. 2015;**42**:245-284

[11] Chapelle O, Schlkopf B, Zien A. Semi-Supervised Learning. 1st ed. Cambridge, Massachusetts, London, England: The MIT Press; 2010

[12] Quinlan JR. C4.5 Programs for Machine Learning. CA: Morgan Kaufmann; 1988

[13] Peterson LE. K-nearest neighbor. Scholarpedia. 2009;**4**(2):1883. Revision 137311

[14] Ho TK. Random decision forests. In: Proceedings of the Third International Conference on Document Analysis and Recognition (Volume 1). Washington, DC, USA: IEEE Computer Society. 1995

[15] Gardner MW, Dorling SR. Artificial neural networks (the multilayer perceptron)—A review of applications in the atmospheric sciences. Atmospheric Environment. 1998; **32**(14):2627-2636

[16] Hearst MA. Support vector machines. IEEE Intelligent Systems. 1998;**13**(4):18-28

[17] Seber GAF, Lee AJ. Linear Regression Analysis. 2nd ed. Hoboken, New Jersey: John Wiley & Sons, Inc.; 2003

[18] Vinzi VE, Chin WW, Henseler J, Wang H. Handbook of Partial Least Squares: Concepts, Methods and

Applications. 1st ed. Berlin, Heidelberg: Springer Publishing Company, Incorporated; 2010

[19] Breiman L, Friedman J, Olshen R, Stone C. Classification and Regression Trees. Monterey, CA: Wadsworth and Brooks; 1984. New edition

[20] Hoerl AE, Kennard RW. Ridge regression: Biased estimation for nonorthogonal problems. Technometrics. 1970;**12**(1):55-67

[21] Tibshirani R. Regression shrinkage and selection via the Lasso. Journal of the Royal Statistical Society, Series B. 1994;**58**:267-288

[22] Schulz E, Speekenbrink M, Krause A. A tutorial on Gaussian process regression: Modelling, exploring, and exploiting functions. Journal of Mathematical Psychology. 2018;**85**:1-16, ISSN: 0022-2496. DOI: 10.1016/j.jmp.2018.03.001

[23] Moon TK. The expectation-maximization algorithm. IEEE Signal Processing Magazine. 1996;**13**(6):47-60

Chapter 8

Consensus Control of Distributed Battery Energy Storage Devices in Smart Grids

Javad Khazaei and Dinh Hoa Nguyen

Abstract

One of the major challenges of existing highly distributed smart grid system is the centralized supervisory control and data acquisition (SCADA) system, which suffers from single point of failure. This chapter introduces a novel distributed control algorithm for distributed energy storage devices in smart grids that can communicate with the neighboring storage units and share information in order to achieve a global objective. These global objectives include voltage regulation, frequency restoration, and active/reactive power sharing (demand response). Consensus theory is used to develop controllers for multiple energy storage devices in a cyber-physical environment, where the cyber layer includes the communication system between the storage devices and the physical layer includes the actual control and closed-loop system. Detailed proof of designs is introduced to ensure the stability and convergence of the proposed designs. Finally, the designed algorithms are validated using time-domain simulations in IEEE 14-bus system using MATLAB software.

Keywords: consensus control, battery energy storage, smart grids, distributed control, droop control

1. Introduction

Integration of highly distributed renewable energy sources has introduced significant challenges to the resiliency and efficiency of the smart grid systems. This is mainly due to uncertain behaviors of these renewable energy sources and their dependency to weather conditions [1, 2]. Battery energy storage has been introduced as a solution to solve the intermittency and uncertain behaviors of renewable energy sources in smart grids. The energy storage is normally connected to the electric power system through a voltage source converter and by controlling the charge/discharge rate of the storage units, output power regulation of renewable energy sources can be achieved [3, 4].

Currently, energy storage units are distributed throughout the grid. Given the centralized structure of supervisory control and data acquisition (SCADA) system, it cannot meet the requirements of highly distributed renewable energy and storage devices of future smart grid systems. In addition, the centralized controllers suffer from single point of failure and are not a suitable choice for energy storage control when the grid resiliency is significantly important [5–7]. As an example, an energy storage device which is permanently out of energy is not able to use its power capability to support the load until a charge event is scheduled. Moreover, any charge discrepancies will result in battery degradation due to the increased depth of discharge.

Distributed control of distributed energy storage units has recently been introduced [8–11]. For example, a coordinated control is proposed for low voltage distribution networks in [8], which mitigates voltage fluctuations in a distribution feeder using distributed storage units. As another example, the voltage regulation issue of distribution feeders is resolved using a droop-based distributed controller to cooperatively charge/discharge the storage units to regulate the feeder voltage considering the state of charge of the batteries [9]. In Hammad et al. [10], a virtual inertia-based distributed controller is designed for transient stability of a power system using distributed storage units. The authors then used a feedback linearization control algorithm to evaluate the proposed virtual inertia and its effectiveness in stability of the system.

Consensus control of energy storage units has also recently been proposed as an emerging technique for synchronization of distributed storage devices [12–20]. In Khazaei and Miao [13], the authors introduced a state of charge balancing algorithm for distributed storage devices in AC microgrids using consensus theory and validated the results using a real-time simulator. In Guan et al. [15], a dynamic consensus approach was introduced to balance the discharge rate of energy storage devices in AC microgrids. The proposed model achieved power regulation by adjusting the virtual resistance of voltage-controlled inverters. A novel distributed controller was designed for load management in distribution networks using distributed battery storage systems. The proposed methodology used limited communications to coordinate multiple storage units with solar power energy penetration. The authors also have significantly studied the consensus design for storage devices for power sharing and energy synchronization [13], power sharing of heterogeneous storage units with droop control [14], voltage and frequency regulation of storage devices in smart grids [16], output power regulation of double-fed induction generator-based wind farms [19], and power sharing of storage devices with different droop schemes [20].

This chapter summarizes the findings of the authors in the distributed control design of energy storage devices in smart grids to provide ancillary services including: (1) voltage regulation, (2) primary frequency support, (3) equal active power sharing between storage units based on their capacities, (4) equal reactive power sharing based on storage capacity, and finally, and (5) controlling the load in both islanded and grid-connected modes. Time-domain simulations on a modified IEEE 14-bus system are performed to validate the effectiveness of the proposed designs.

The rest of the paper is as follows: Section 2 covers the battery energy storage model. Consensus design for heterogeneous storage units is considered in Section 3. Section 4 includes multiple case studies to validate the designs, and Section 5 concludes the chapter.

2. Battery energy storage model

A basic schematic of an energy storage device, which is connected to the grid through a DC/AC converter and an LCL filter is illustrated in **Figure 1**. The energy storage control uses the well-known synchronous reference frame control, where three cascaded control loops are adopted. The main objective is to control the active and reactive demand using a droop control method in dq reference frame. The droop control receives active and reactive power measurements from a sensor installed at the point of common coupling (PCC) as

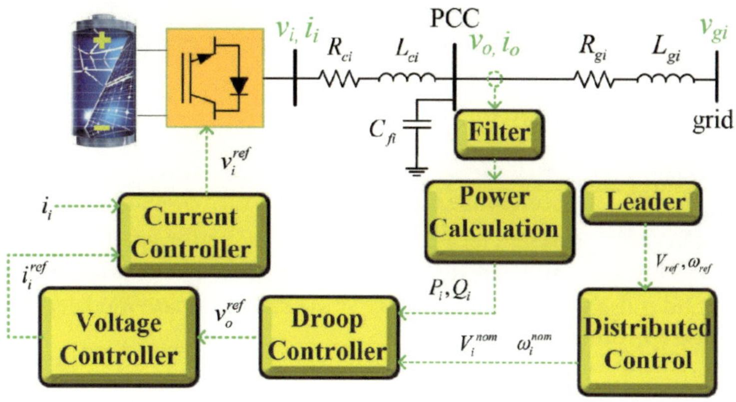

Figure 1.
Control structure of a battery energy storage system.

$$P_{mi} = \frac{3}{2}\left(v_{od}i_{od} + v_{oq}i_{oq}\right) \tag{1}$$

$$Q_{mi} = \frac{3}{2}\left(-v_{od}i_{oq} + v_{oq}i_{od}\right) \tag{2}$$

where P_{mi} and Q_{mi} are measured active and reactive powers, respectively. In addition, v_{od}, v_{oq}, i_{od}, and i_{oq} are measured converter voltages and currents at the point of common coupling, as illustrated in **Figure 1**. The measurements will then pass through a low-pass filter,

$$P_i = \frac{\alpha_p}{s + \alpha_p} P_{mi} \tag{3}$$

$$Q_i = \frac{\alpha_q}{s + \alpha_q} Q_{mi} \tag{4}$$

where α_p and α_q are the bandwidths of the low-pass filters. The AC-side dynamics of the ith energy storage system in dq frame is expressed as:

$$L\frac{di_{id}}{dt} - \omega L_{ci}i_{iq} + R_{ci}i_{id} = v_{od} - v_{id} \tag{5}$$

$$L\frac{di_{iq}}{dt} + \omega L_{ci}i_{id} + R_{ci}i_{iq} = v_{oq} - v_{iq} \tag{6}$$

where v_{od} and v_{oq} are dq voltages at the point of common coupling (PCC), v_{id} and v_{iq} are the dq frame converter output voltages, and i_{id} and i_{iq} are the dq reference frame currents flowing from the PCC to the converter.

2.1 Inner current controller

The most inner control loop in the energy storage system is the current controller, which is in charge of regulating the converter current in a decoupled manner. The inputs to this controller come from the voltage controller and are the reference dq frame current setpoints. Two proportional integral controllers are utilized which regulate the d and q axis currents with feedforwarded loops. Dynamics of the current controller for ith battery storage system are presented as [14]:

$$
\begin{aligned}
v_{id}^{ref} &= -\left(k_p + \frac{k_i}{s}\right)\left(i_{id}^{ref} - i_{id}\right) + \omega L i_{iq} + \frac{1}{\tau s + 1} v_{od} \\
v_{iq}^{ref} &= -\left(k_p + \frac{k_i}{s}\right)\left(i_{iq}^{ref} - i_{iq}\right) - \omega L i_{id} + \frac{1}{\tau s + 1} v_{oq}
\end{aligned} \tag{7}
$$

where v_{id}^{ref} and v_{iq}^{ref} are reference converter voltages to be sent to the pulse width modulation (PWM) controller, k_p and k_i are the PI regulator gains, and τ is the time constant of the low-pass filter for voltage measurement. The controller computational delay and pulse width modulation (PWM) switching are modeled by τ_s which can be ignored for simplicity [14].

2.2 Voltage control

Voltage controller receives inputs from the droop controller and provides reference currents for the inner current control loops. Similar to the current controller, two PI regulator are used for the voltage controller. Dynamics of the voltage control loop can be modeled by [14]:

$$
i_{id}^{ref} = \left(v_{od}^{ref} - v_{od}\right)\left(k_{pv} + \frac{k_{iv}}{s}\right) - \omega_0 C_{fi} v_{oq} + \frac{1}{\tau_v s + 1} i_{od} \tag{8}
$$

$$
i_{iq}^{ref} = \left(v_{oq}^{ref} - v_{oq}\right)\left(k_{pv} + \frac{k_{iv}}{s}\right) + \omega_0 C_{fi} v_{oq} + \frac{1}{\tau_v s + 1} i_{oq} \tag{9}
$$

where $v_{od}^{ref}, v_{oq}^{ref}$ are dq frame reference converter voltages, v_{od}, v_{oq} are the dq frame measured voltages at the PCC passed through a low-pass filter, and i_{od}, i_{oq} are the dq frame converter output currents. In addition, τ_v is the time constant of the filter that is used in voltage controller.

2.3 Droop control

Droop control is used when multiple converters are installed in parallel to support the load based on their capacities. The principle of the droop control is based on the fact that a converter with higher capacity should share more load compared to a converter with lower capacity. This will be taken care of by designing droop gains properly. The droop controller receives measurements from active and reactive powers at the point of common coupling and provides reference voltage and frequency to that shape the reference voltages in dq frame for the voltage controller loop such that

$$
v_{od}^{ref} = |V_i| \cos \omega_i t \tag{10}
$$

$$
v_{oq}^{ref} = |V_i| \sin \omega_i t \tag{11}
$$

The droop controller sends the reference voltage magnitude and frequency setpoints of the converter to the voltage controller. Dynamics of the droop controller can be expressed by:

$$
\omega_i = \omega_i^{\text{nom}} - D_i^P P_i \tag{12}
$$

$$
|V_i| = |V_i^{nom}| - D_i^Q Q_i \tag{13}
$$

where ω_i is frequency setpoint of ith battery storage, ω_i^{nom} is the nominal frequency to be developed using the consensus theory, D_i^P is the active power droop gain, $|V_i^{\text{nom}}|$ is the nominal voltage magnitude of the ith storage to be designed by consensus control, and D_i^Q is the reactive power droop gain. The droop gains can be found by:

$$D_i^P = \frac{P_i^{max} - P_i^{min}}{\omega_i^{max} - \omega_i^{min}} \tag{14}$$

$$D_i^Q = \frac{Q_i^{max} - Q_i^{min}}{|V_i|^{max} - |V_i|^{min}} \tag{15}$$

The droop control design is similar to the primary frequency and voltage control of synchronous generators, where the voltage and frequency will not be regulated to their nominal values. To restore the voltage and frequency to their nominal values, a distributed controller is designed in this work using a consensus theory. The proposed controller receives signals from neighboring storage devices and modifies the nominal frequency/voltage in the droop equations to regulate the voltage and frequency to their setpoints.

3. Heterogeneous consensus design

The main objective of this section is to supplement a secondary controller to the droop controller of the storage devices. The controller receives information from neighboring storage units and shares the power between storage devices to regulate the voltage and frequency of the system at the point of common coupling. Furthermore, a virtual leader is considered, which can be assigned to one energy storage in the system, or a few storage devices. The leader will have the setpoints of the voltage and frequency in the system and will share the information with its neighboring storage units. To develop such a control design, the battery energy storage model needs to be developed. In our recent work, a simplified battery energy storage model was developed. The model accounts for the dynamics of the droop controller and active power/energy relationship of the battery. Such model can accurately incorporate the dynamics of energy storage devices in smart grids. Dynamics of the energy storage devices can be represented by [14]:

$$\begin{aligned} \omega_i &= \omega_i^{\text{nom}} - D_i^P P_i, \\ |V_i| &= |V_i^{\text{nom}}| - D_i^Q Q_i, \\ \dot{E}_i &= \frac{-D_i^P}{3600} P_i, \\ \dot{P}_i &= u_i^P. \end{aligned} \tag{16}$$

To develop such simplified model, it is assumed that dynamics of voltage controller and current controller are much faster than the droop controller, therefore, their dynamics can be ignored. In the above model, u_i^P is the input for distributed active power sharing, and D_i^P reflects the heterogeneity of batteries. To achieve equal power sharing, $D_i^P P_i$ should be regulated among batteries so that a battery with higher capacity (lower droop gain, D_i^P) contributes more to the power sharing. To minimize the number of communication links between the storage devices, this paper regulates the nominal voltage and frequency of neighboring storage units. In this method, there will be no need to receive measured voltage and frequency signals from neighboring storage units and the control design only requires the nominal frequencies ω_j^{nom} and nominal voltages $|V_j^{\text{nom}}|$ of its neighboring storage devices.

To provide voltage and frequency regulation as well as active/reactive power sharing, new distributed inputs can be designed using consensus theories. These inputs include $|\dot{V}_i^{\text{nom}}| = u_i^V$, $\dot{Q}_i = u_i^Q$, and $\dot{\omega}_i^{\text{nom}} = u_i^{\omega}$. The overall dynamics of the ith energy storage device is then formulated as [14]:

$$\begin{aligned} \dot{E}_i &= \frac{-D_i^P}{3600} P_i, \\ \dot{P}_i &= u_i^P, \\ |\dot{V}_i^{\text{nom}}| &= u_i^V, \\ \dot{Q}_i &= u_i^Q, \\ \dot{\omega}_i^{\text{nom}} &= u_i^{\omega}. \end{aligned} \tag{17}$$

In the next section, control design to develop these new inputs will be elaborated in detail.

3.1 Graph theory

Some preliminary information on graph theory is needed in order to design the controllers. The multi-agent system theory is considered for designing the controller inputs so that each battery energy storage unit is considered as an agent that can communicate with neighboring agents. It is also assumed that the communication network of the system is an undirected graph $\mathcal{G}$ that has a vertex set of $\mathcal{V}$ and an edge set of $\mathcal{E}$. Each vertex represents an energy storage system and the interconnection between storage systems k and j is represented by element $(k,j) \in \mathcal{E}$.

The neighboring set of energy storage number k is expressed by $\mathcal{N}_k \triangleq \{ j \in \mathcal{V} : (k,j) \in \mathcal{E}\}$. In addition, a_{kj} is an element of the adjacency matrix $\mathcal{A}$ of $\mathcal{G}$, i.e. $a_{kj} = 1$ if $(k,j) \in \mathcal{E}$ and $a_{kj} = 0$ if $(k,j) \notin \mathcal{E}$. Finally, $\mathcal{D}$ is the degree matrix that is derived by $\mathcal{D} = \text{diag}\{d_k\}_{k=1,\dots,n}$, where $d_k \triangleq \sum_{j \in \mathcal{N}_k} a_{kj}$. It is noted that, the Laplacian matrix $\mathcal{L}$ associated to $\mathcal{G}$ can be formulated by $\mathcal{L} = \mathcal{D} - \mathcal{A}$. The leader is in charge of sending setpoints to energy storage units. The leader is represented by sub-index 0 and its neighboring storage units are denoted by $\mathcal{N}_0$. Then $a_{0i} = 1$ if $i \in \mathcal{N}_0$, while $a_{0i} = 0$ if $i \notin \mathcal{N}_0$.

3.2 Consensus control design

Let ω^{ref} and V^{ref} be the reference frequency and voltage magnitude of energy storage devices. These references serve as external commands to force the frequency and voltage magnitude of storage devices to converge precisely to the expected values. In other words, they are virtual leaders while the frequency and voltage magnitude of batteries are the followers. In this sense, the consensus design is proposed as follows,

$$\begin{aligned}
u_i^P &= \frac{-C_2}{D_i^P}\sum_{j\in\mathcal{N}_i}\left(D_i^P P_i - D_j^P P_j\right),\\
u_i^V &= -C_3\sum_{j\in\mathcal{N}_i}\left(|V_i^{\text{nom}}| - |V_j^{\text{nom}}|\right) - C_0^V a_{0i}\left(|V_i^{\text{nom}}| - D_i^Q Q_i - |V^{\text{ref}}|\right),\\
u_i^Q &= \frac{-C_3}{D_i^Q}\sum_{j\in\mathcal{N}_i}\left(D_i^Q Q_i - D_j^Q Q_j\right),\\
u_i^\omega &= -C_2\sum_{j\in\mathcal{N}_i}\left(\omega_i^{\text{nom}} - \omega_j^{\text{nom}}\right) - C_0^\omega a_{0i}\left(\omega_i^{\text{nom}} - D_i^P P_i - \omega^{\text{ref}}\right).
\end{aligned} \tag{18}$$

Note here that the controller gains for the active power and frequency are the same, and similarly the controller gains for the reactive power and voltage magnitude are also the same. The structure of the proposed distributed primary and secondary voltage/frequency controller is illustrated in **Figure 2**.

3.3 Consensus proof

Let $\tilde{P}_i \triangleq D_i^P P_i$ and $\tilde{Q}_i \triangleq D_i^Q Q_i$, then the closed-loop model of BESS with the consensus design (18) is

$$\begin{aligned}
\dot{E}_i &= \frac{-1}{3600}\tilde{P}_i,\\
\dot{\tilde{P}}_i &= -C_2\sum_{j\in\mathcal{N}_i}\left(\tilde{P}_i - \tilde{P}_j\right),\\
|\dot{V}_i^{\text{nom}}| &= -C_3\sum_{j\in\mathcal{N}_i}\left(|V_i^{\text{nom}}| - |V_j^{\text{nom}}|\right) - C_0^V a_{0i}\left(|V_i^{\text{nom}}| - \tilde{Q}_i - |V^{\text{ref}}|\right),\\
\dot{\tilde{Q}}_i &= -C_3\sum_{j\in\mathcal{N}_i}\left(\tilde{Q}_i - \tilde{Q}_j\right),\\
\dot{\omega}_i^{\text{nom}} &= -C_2\sum_{j\in\mathcal{N}_i}\left(\omega_i^{\text{nom}} - \omega_j^{\text{nom}}\right) - C_0^\omega a_{0i}\left(\omega_i^{\text{nom}} - \tilde{P}_i - \omega^{\text{ref}}\right).
\end{aligned} \tag{19}$$

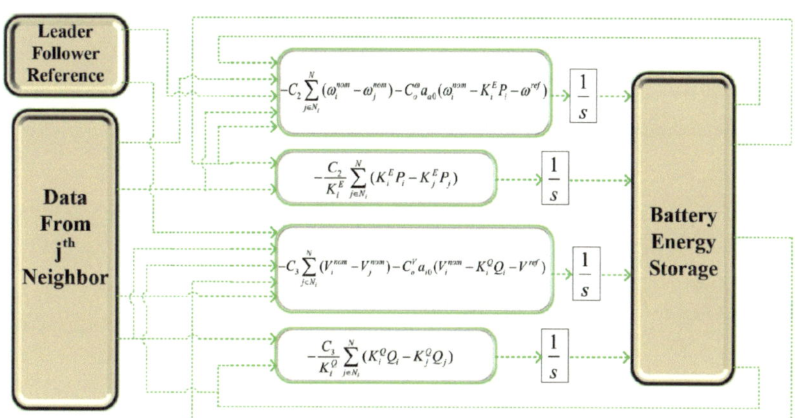

Figure 2.
Structure of the proposed distributed controller.

First, we introduce the proof for consensus of the active powers and reactive powers of storage devices. It can be easily observed from Eq. (19) that the dynamics of the proportional active powers $\tilde{P}_i$ and reactive powers $\tilde{Q}_i$ have the same form. Moreover, the consensus of the proportional active powers $\tilde{P}_i$ will lead to the consensus of batteries' energy levels E_i, if the initial state of charge of the batteries is the same. Therefore, we only present the proof for the consensus of the active powers $\tilde{P}_i$. The dynamics of $\tilde{P}_i$ is self-contained and in form of a first-order differential equation, and hence, its solution can be easily found to be

$$\tilde{P}(t) = e^{-C_2\mathcal{L}t}\tilde{P}(0), \tag{20}$$

where $\tilde{P}(0)$ is the vector of initial proportional active powers of batteries. Since $\mathcal{L}$ is a symmetric matrix, $e^{-C_2\mathcal{L}t}$ is also a symmetric matrix. Let $U \in \mathbb{R}^{N\times N}$ be an orthogonal matrix derived from diagonal matrix $\mathcal{L}$. Subsequently, it can be easily seen that

$$e^{-C_2\mathcal{L}t} = U\lim_{t\to\infty}\text{diag}\{1, e^{-C_2\lambda_2 t}, \dots, e^{-C_2\lambda_N t}\}U^T = U\text{diag}\{1, 0, \dots, 0\}U^T, \tag{21}$$

since $\lambda_2, \dots, \lambda_N$ are positive eigenvalues of $\mathcal{L}$ once $\mathcal{G}$ is connected. On the other hand, $U\text{diag}\{1, 0, \dots, 0\}U^T = [\mathbf{1}_N/\sqrt{N}, \mathbf{0}_N, \dots, \mathbf{0}_N]U^T = \mathbf{1}_N\mathbf{1}_N^T/N$ since the eigenvector associated with the zero eigenvalue of $\mathcal{L}$ is $\mathbf{1}_N$. Therefore, we obtain from (21) and (20) that

$$e^{-C_2\mathcal{L}t}\tilde{P}(0) = \frac{\mathbf{1}_N^T\tilde{P}(0)}{N}\mathbf{1}_N, \tag{22}$$

which means that the average consensus is achieved for batteries' proportional active powers. As mentioned above, similar proof can be utilized to get the consensus of batteries' proportional reactive powers.

Next, we present the proof for the consensus of the voltage magnitude and frequency of batteries to their references $|V^{\text{ref}}|$ and ω^{ref}. For brevity, only the proof for the consensus of nominal frequency ω_i^{nom} is given, while the proof for the consensus of nominal voltage magnitude $|V_i^{\text{nom}}|$ can be derived similarly because their equations are in the same form as seen in (19).

Let us denote

$$\hat{\omega}_i \triangleq \omega_i - \omega^{ref}, i = 1, \dots, N; \hat{\omega} \triangleq [\hat{\omega}_1, \dots, \hat{\omega}_N]^T.$$

Then

$$\begin{aligned}
\dot{\hat{\omega}}_i = \dot{\omega}_i &= \dot{\omega}_i^{\text{nom}} - \dot{\tilde{P}}_i \\
&= -C_2\sum_{j\in\mathcal{N}_i}\left(\omega_i^{\text{nom}} - \omega_j^{\text{nom}}\right) - C_2\sum_{j\in\mathcal{N}_i}(\tilde{P}_i - \tilde{P}_j) - C_0^{\omega}a_{0i}\hat{\omega}_i \\
&= -C_2\sum_{j\in\mathcal{N}_i}(\hat{\omega}_i - \hat{\omega}_j) - C_0^{\omega}a_{0i}\hat{\omega}_i.
\end{aligned} \tag{23}$$

Denote $\mathcal{D}_0 \triangleq \text{diag}\{a_{0i}\}_{i=1,\dots,N}$. Consequently, we obtain from (23) that

$$\dot{\hat{\omega}}^{\text{nom}} = -[C_2\mathcal{L} + C_0^{\omega}\mathcal{D}_0]\hat{\omega}. \tag{24}$$

Assume that the communication graph $\mathcal{G}$ among followers is connected and at least one follower is connected to the leader, i.e. $\mathcal{D}_0$ is not a zero matrix, then it was shown in [20] that all eigenvalues of the matrix $C_2\mathcal{L} + C_0^{\omega}\mathcal{D}_0$ have positive real parts for any $C_2 > 0$ and $C_0^{\omega} > 0$. Thus, it can be immediately concluded that the system (24) is stable, i.e. $\lim_{t\to\infty}\hat{\omega}(t) = 0$. This is equivalent to the consensus of battery frequency to the reference frequency ω^{ref}. Same analysis holds for the consensus of the nominal voltage magnitude.

4. Case studies

To validate the proposed designs, IEEE 14-bus benchmark is used. The system represents an approximation of U.S. utility system around 1962. The benchmark includes five generation units and 11 loads. Parameters of the test system were adopted from [21]. The system was modified for the current study. The generator dynamics were replaced by the battery energy storage dynamics. The system was modeled in MATLAB Simulink and a combination of MatPower and MatDyn toolboxes are used for dynamic simulation of the proposed control algorithms [22]. The MatPower toolbox was used for power flow and initial conditions of the system, where MatDyn was used for dynamic simulations and control design. The authors have extensively studied integration of battery energy storage units to IEEE benchmark cases using MatDyn toolbox in their previous publications [9, 10]. The consensus controllers then are supplemented to the model as inputs. The schematic of the modified IEEE 14-bus system used for the simulations in this study is illustrated in **Figure 3**. Parameters of the storage units are included in the Appendix section.

The communication structure of the system under investigation is illustrated in **Figure 4**. As it can be observed, the communication graph of the system is undirected and minimum number of communication links is needed to ensure the

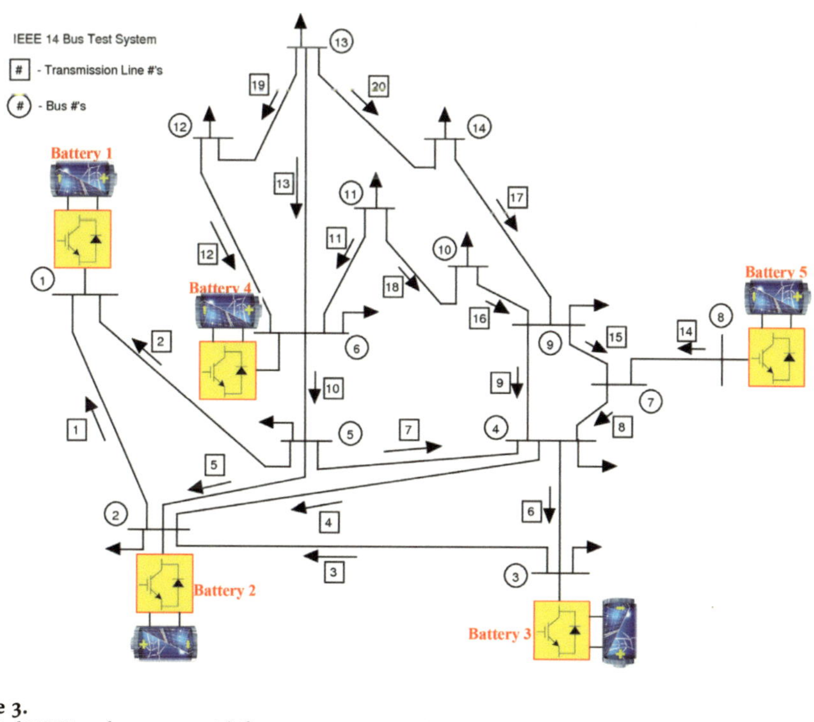

Figure 3.
Modified IEEE 14-bus system with five energy storage units.

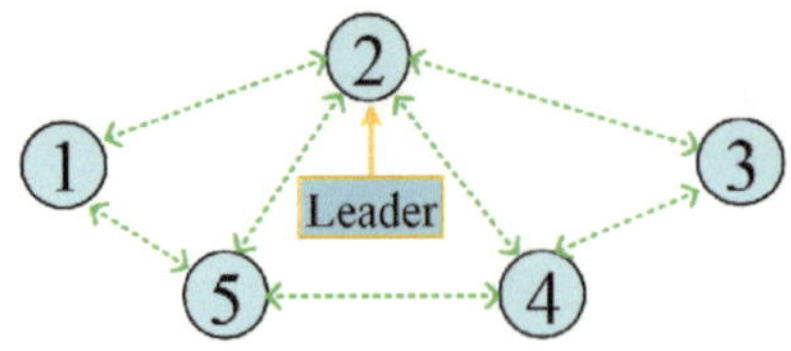

Figure 4.
Communication graph of the system.

convergence of the proposed algorithms. Furthermore, the leader is incorporated in battery energy storage number 2. The leader can be designed in any storage number as one leader is sufficient to ensure the functionality of the proposed control design.

4.1 Constant power control

In the first case, the performance of the designed voltage and frequency controller is tested when a constant load is applied to the system. The secondary controller ensures sharing the active and reactive powers between the storage units based on their capacities as well as voltage/frequency regulation. Simulation results for this case study are illustrated in **Figure 5**. For this case study, the leader is activated by setting $C_0^{\omega} = 1$ and $C_0^V = 1$. The overall load in the system (summation of loads in all busses) is $S_D = 0.5 + j0.5$ p.u. It can be shown that the active/reactive power sharing (first subplot) is achieved after 15 s, and the voltage and frequency

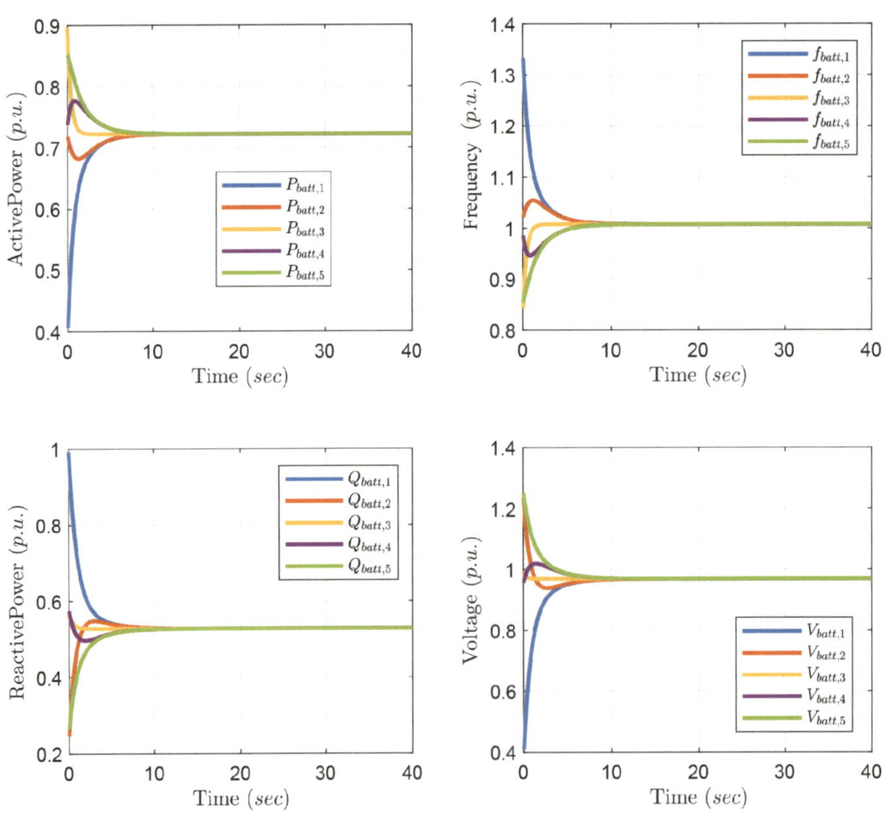

Figure 5.
Primary frequency response to a constant load.

are regulated to their reference setpoints (1 p.u.) after 15 s. The designed controller can synchronize the operation of storage units in the system very fast.

4.2 Primary frequency response

In the second case, the primary active and reactive power sharing is the main objective. The batteries should share the load power equally using the primary droop concept. The leader will be deactivated in this case enforcing consensus gains C_0^ω and C_0^V to zero. This will result in a primary voltage and frequency response, where the battery storage units will share the load active and reactive power demand, but the voltage and frequency will deviate from the nominal value. Similar to the first case study, the system starts with a $0.5 + j0.5$ p.u. load and a load change event is scheduled to increase the demand to $1 + j1$ p.u. after 20 s. Simulation results are illustrated in **Figure 6**. It is shown that the batteries can equally share the active and reactive power of the load even after the load event at 20 s. To support the active power increase in the demand, the frequency will drop and the batteries will settle in a lower frequency (0.91 p.u.). Furthermore, since the load active power has increased after 20 s, the voltage will also drop and settle to new synchronized value (0.94 p.u.).

4.3 Secondary frequency response

The third case, the performance of the secondary voltage and frequency controller during a load change event is studied. The system starts with $0.5 + j0.5$ $p.u.$

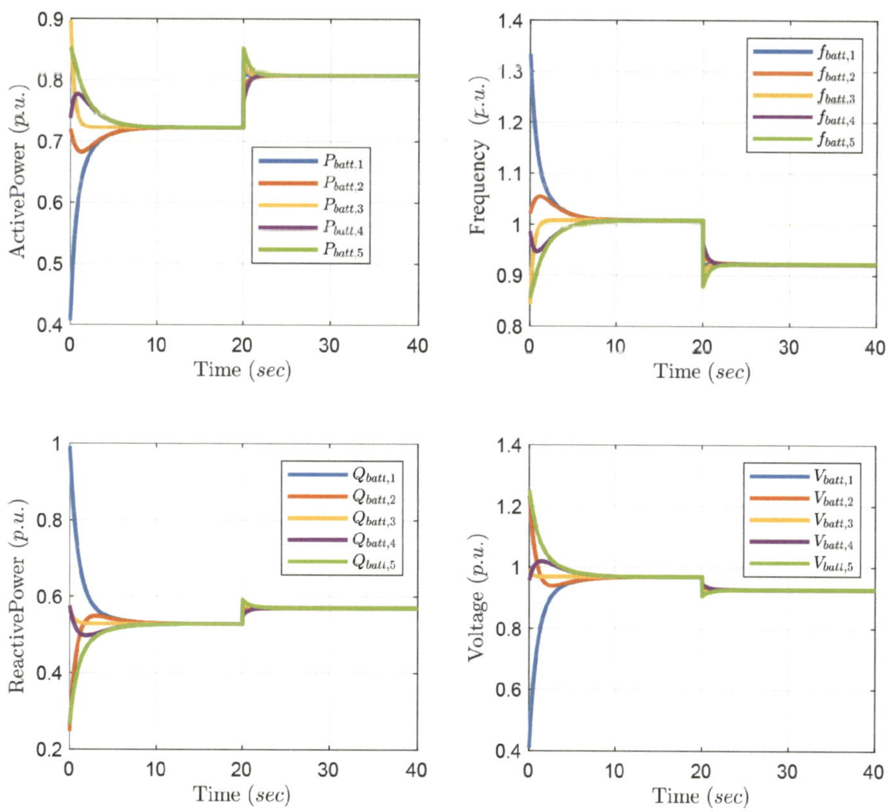

Figure 6.
Primary frequency response to a dynamic load change.

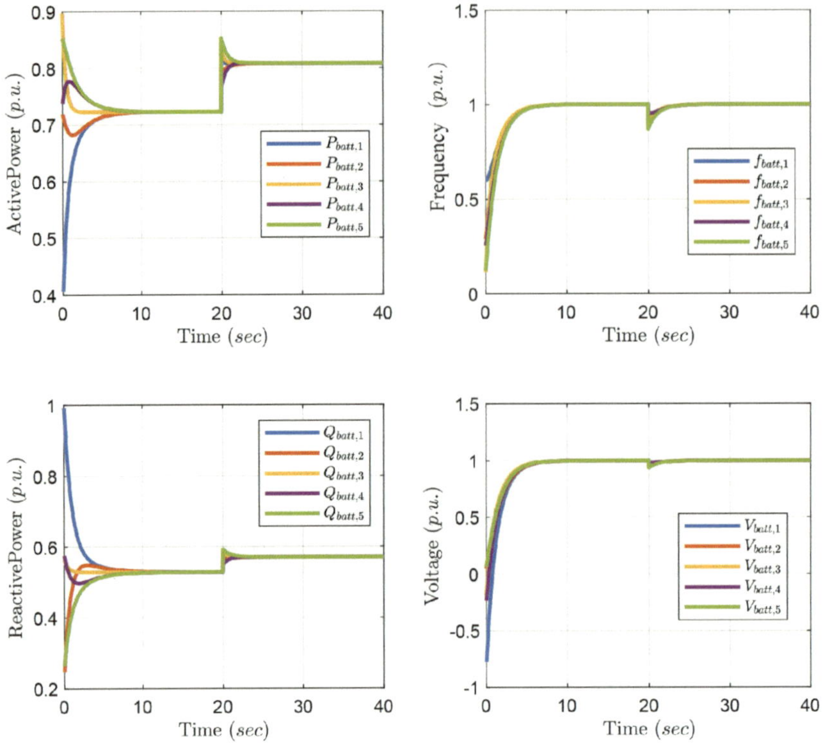

Figure 7.
Secondary frequency controller response to a dynamic load change.

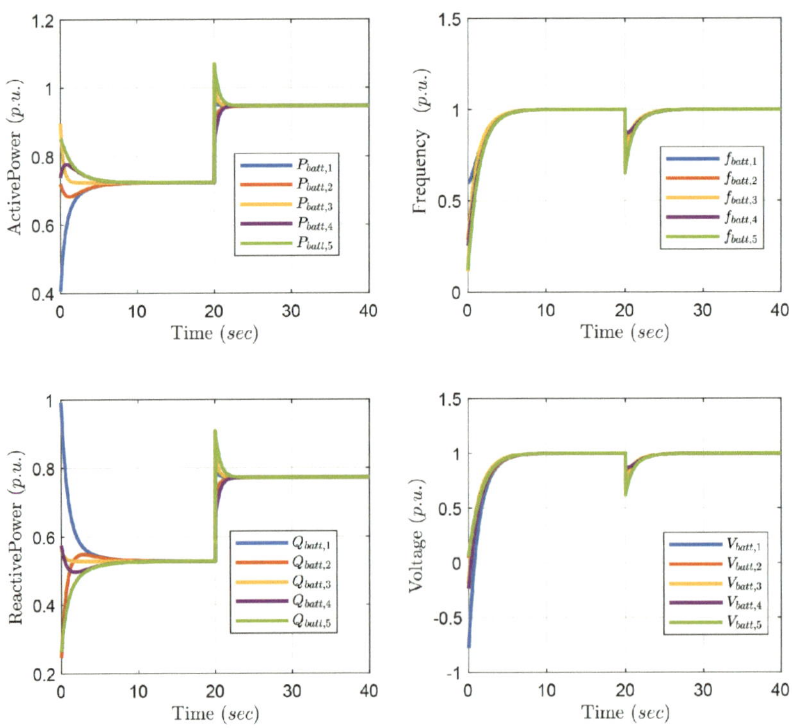

Figure 8.
Response of the system under severe dynamic load change.

load and the load increases to $1+j1\ p.u.$ after 20 s. The secondary distributed controller is reactivated by tuning C_0^{ω} and C_0^{V} to 1, as illustrated in **Figure 7**. It is observed that the load sharing is successfully achieved among the storage units. In addition, the voltage and frequency are regulated to 1 p.u. in less than 3 s after the load change.

4.4 Severe load change

In the last case, the performance of the proposed power sharing and voltage/frequency restoration algorithms in handling a severe load change is examined. The system initiates with $0.5+j0.5\ p.u.$ load and a load change of 1 p.u. is applied after 20 s. This means the total load increases to $1.5+j1.5\ p.u.$ after 20 s. As it can be observed in **Figure 8**, the proposed distributed controller can equally share the active and reactive load change while regulating the voltage and frequency. It should be noted that the frequency drops to 0.75 p.u. after the load change, but it is quickly recovered within a few seconds. Similarly, the voltage drops to 0.6 p.u. after the load change, but it is recovered within 2 s. This case study showed that the proposed controller can successfully operate under severe load changes.

5. Conclusion

This chapter proposed a novel distributed controller that can synchronize the operation of distributed energy storage units in smart grids. By minimizing the communication links between the neighboring storage units, the storage units share information (voltage/current readings) with their neighbors to share the active/reactive demand based on their capacities. Furthermore, a virtual leader is designed and supplemented to one storage unit to regulate the voltage and frequency and to provide secondary frequency response to the system. Results showed the effectiveness of the proposed algorithms in equally sharing the active power and reactive power of load during constant power load, and load change events. Furthermore, the secondary controller could successfully regulate the voltage and frequency of the system during constant load, load change, and severe load change events.

6. Future work

Future studies will focus on: (1) hardware validation of proposed approaches and (2) expansion of the developed controllers to solar and wind energy applications.

Conflict of interest

The authors declare no conflict of interest.

Appendices

Parameters of the system are shown in **Table 1**.

Parameter	**$BESS_i, i = 1, 2, \ldots, 5$**
Nominal power	100 MW
Nominal voltage	132 kV
D_i^P [p.u.]	[1 1.2 1.5 2 1.6]
D_i^Q [p.u.]	[1 1.1 1.25 1.45 1.5]
C_0	0.5 p.u.
n	5
C_1, C_2, C_3	0.1, 0.4, 0.2 p.u.

Table 1.
Parameters of the system.

Author details

Javad Khazaei[1*†] and Dinh Hoa Nguyen[2†]

1 Architectural Engineering, School of Science, Engineering, and Technology, College of Engineering, State College, Penn State Harrisburg, Pennsylvania State University, Middletown, USA

2 International Institute for Carbon-Neutral Energy Research (WPI-I²CNER) and Institute of Mathematics for Industry (IMI), Kyushu University, Fukuoka, Japan

*Address all correspondence to: jxk792@psu.edu

† These authors contributed equally.

References

[1] Alaa M, Egon O, Andreas S, Nedzad H, Danny M. Challenges in integrating distributed energy storage systems into future smart grid. In: IEEE International Symposium on Industrial Electronics. 2008. pp. 1627-1632. Available from: https://ieeexplore.ieee.org/document/4676896 [Accessed: 27 September 2019]

[2] Jones LE. Renewable Energy Integration: Practical Management of Variability, Uncertainty, and Flexibility in Power Grids. Academic Press; 2017. Available from: https://www.elsevier.com/books/renewable-energy-integration/jones/978-0-12-809592-8 [Accessed: 27 September 2019]

[3] Khanh NH, Bin SJ, Zhu H. Distributed demand side management with energy storage in smart grid. IEEE Transactions on Parallel and Distributed Systems. 2015;**26**(12):3346-3357. Available from: https://ieeexplore.ieee.org/document/6963474 [Accessed: 27 September 2019]

[4] Ehsan R, Saeed S, Roose Leon R, Marc M. Energy management at the distribution grid using a battery energy storage system (BESS). International Journal of Electrical Power & Energy Systems. 2016;77:337-344. Available from: https://www.sciencedirect.com/science/article/pii/S014206151500455X [Accessed: 27 September 2019]

[5] Antoniadou-Plytaria KE, Kouveliotis-Lysikatos IN, Georgilakis PS, Hatziargyriou ND. Distributed and decentralized voltage control of smart distribution networks: Models, methods, and future research. IEEE Transactions on Smart Grid. 2017;**8**(6): 2999-3008. Available from: https://ieeexplore.ieee.org/abstract/document/7874216 [Accessed: 27 September 2019]

[6] Lai J, Zhou H, Lu X, Yu X, Hu W. Droop-based distributed cooperative control for microgrids with time-varying delays. IEEE Transactions on Smart Grid. 2015;7(4):1775-1789. Available from: https://ieeexplore.ieee.org/abstract/document/7458197 [Accessed: 27 September 2019]

[7] Wang Y, Tan KT, Peng XY, So PL. Coordinated control of distributed energy-storage systems for voltage regulation in distribution networks. IEEE Transactions on Power Delivery. 2015;**31**(3):1132-1141. Available from: https://ieeexplore.ieee.org/abstract/document/7172533 [Accessed: 27 September 2019]

[8] Zeraati M, Golshan MEH, Guerrero JM. Distributed control of battery energy storage systems for voltage regulation in distribution networks with high PV penetration. IEEE Transactions on Smart Grid. 2016; **9**(4):3582-3593. Available from: https://ieeexplore.ieee.org/abstract/document/7775016 [Accessed: 27 September 2019]

[9] Liu W, Gu W, Yuan X, Zhang K. Fully distributed control to coordinate charging efficiencies for energy storage systems. Journal of Modern Power Systems and Clean Energy. 2018;**6**(5): 1015-1024. Available from: https://link.springer.com/article/10.1007/s40565-017-0373-1 [Accessed: 27 September 2019]

[10] Hammad E, Farraj A, Kundur D. On effective virtual inertia of storage-based distributed control for transient stability. IEEE Transactions on Smart Grid. 2017;**10**(1):327-336. Available from: https://ieeexplore.ieee.org/abstract/document/7458197 [Accessed: 27 September 2019]

[11] Xing L, Mishra Y, Tian YC, Ledwich G, Su H, Peng C, et al. Dual-Consensus-Based Distributed Frequency Control for Multiple Energy Storage Systems. IEEE Transactions on Smart

Grid: Early Access; 2019. Available from: https://ieeexplore.ieee.org/abstract/document/8664166 [Accessed: 27 September 2019]

[12] Chaghooshi AF, Pamies-Juarez L, Guyot C. Dual-consensus-based distributed frequency control for multiple energy storage systems. U.S. Patent Application No. 15/626,061. 2018. Available from: https://patents.google.com/patent/US20180364921A1/en [Accessed: 27 September 2019]

[13] Khazaei J, Miao Z. Consensus control for energy storage systems. IEEE Transactions on Smart Grid. 2016;**9**(4): 3009-3017. Available from: https://ieeexplore.ieee.org/abstract/document/7731203 [Accessed: 27 September 2019]

[14] Khazaei J, Nguyen DH. Multi-agent consensus design for heterogeneous energy storage devices with droop control in smart grids. IEEE Transactions on Smart Grid. 2017;**10**(2): 1395-1404. Available from: https://ieeexplore.ieee.org/abstract/document/8076899 [Accessed: 27 September 2019]

[15] Guan Y, Meng L, Li C, Vasquez JC, Guerrero JM. A dynamic consensus algorithm to adjust virtual impedance loops for discharge rate balancing of AC microgrid energy storage units. IEEE Transactions on Smart Grid. 2017;**9**(5): 4847-4860. Available from: https://ieeexplore.ieee.org/abstract/document/7862235 [Accessed: 27 September 2019]

[16] Khazaei J, Nguyen DH, Thao NGM. Primary and secondary voltage/frequency controller design for energy storage devices using consensus theory. In: IEEE 6th International Conference on Renewable Energy Research and Applications (ICRERA). 2017. pp. 447-453. Available from: https://ieeexplore.ieee.org/abstract/document/8191101 [Accessed: 27 September 2019]

[17] Wang D, Meng K, Gao X, Chen G, Luo F, Dong ZY. Consensus-driven distributed control of battery energy storage systems for loading management in distribution networks. In: IEEE International Conference on Smart Grid Communications. 2016. pp. 699-704. Available from: https://ieeexplore.ieee.org/abstract/document/7778843 [Accessed: 27 September 2019]

[18] Xie W, Xia X. Distributed energy dispatch of electrical energy storage systems using consensus control approach. IFAC-PapersOnLine. 2018; **51**(13):229-234. Available from: https://www.sciencedirect.com/science/article/pii/S240589631831036X [Accessed: 27 September 2019]

[19] Khazaei J, Nguyen DH. Distributed consensus for output power regulation of DFIGs with on-site energy storage. IEEE Transactions on Energy Conversion. 2018;**34**(2):1043-1051. Available from: https://ieeexplore.ieee.org/abstract/document/8469065 [Accessed: 27 September 2019]

[20] Nguyen DH, Khazaei J. Cooperative control for distributed energy storage systems with different droop schemes. In: IEEE PES GTD Grand International Conference and Exposition Asia (GTD Asia). 2019. pp. 102-107. Available from: https://ieeexplore.ieee.org/abstract/document/8715853 [Accessed: 27 September 2019]

[21] Milano F. An open source power system analysis toolbox. IEEE Transactions on Power Systems. 2005; **20**(3):1199-1206. Available from: https://ieeexplore.ieee.org/document/1490569 [Accessed: 27 September 2019]

[22] Cole S, Belmans R. MatDyn: A new matlab-based toolbox for power system dynamic simulation. IEEE Transactions on Power Systems. 2011;**26**(3): 1129-1136. Available from: https://ieeexplore.ieee.org/document/5598553 [Accessed: 27 September 2019]

Chapter 9

Power Flow Management Algorithm for a Remote Microgrid Based on Artificial Intelligence Techniques

Karim Belmokhtar and Mauricio Higuita Cano

Abstract

This paper presents a novel power flow management algorithm for remote microgrids based on artificial intelligence (AI) algorithms. The objectives of this power management system are enhancing microgrid reliability, improving renewable energy source (RES) integration, and performing active/reactive power control for remote microgrids using the fuzzy logic. This paper evaluates the performance of the proposed algorithm, which consists of both sharing diesel genset active power and regulating reactive power by using stepped and variable profiles of the load, wind speed and solar irradiation. According to the simulation results, better performance is achieved regardless of the rapid variation of different profiles. Thus, both stability and reliability of remote microgrids are demonstrated with the proposed algorithm. Indeed, the active/reactive power control algorithm responds quickly to different events on the remote microgrid, especially to rapid voltage/frequency variations on the AC-link system.

Keywords: power management system, artificial intelligence (AI), renewable energy sources (RES), remote microgrid, fuzzy logic (FL)

1. Introduction

Renewable energy sources (RES) have been positioned as an attractive solution to reduce dependence on fossil fuels while minimizing greenhouse gas emissions (GHG) [1]. RES such as wind energy and solar (PV) energy have been extensively researched in the literature [2–5]. One of the main issues with RES is that their natural intermittency can affect the stability of the microgrid. On the other hand, technical issues associated with high penetration rates of RES related to their intermittency, different dynamics and response times can be addressed by using energy storage systems (ESS) and advanced power management systems (PMS) [3].

ESS has been identified as a solution to increase energy reliability, improve the balance in energy production, and better manage load demand on the microgrid. The main purpose of ESS in the microgrid is to store excess power from an intermittent RES and return the stored energy to meet load consumption as a function of supply and demand [6]. In addition, ESS can be used in several applications in order to improve both microgrid stability and power quality. According to the literature, ESS applications include load smoothing/peak shaving, power quality improvement, increased renewable energy penetration and emergency storage systems [7–9].

Power management systems (PMS) can be divided into two categories: (i) PMS based on optimization methods and (ii) PMS based on artificial intelligence techniques [10]. Mostly, the PMS based on optimization methods involve a multi-objective function, which maximizes microgrid effectiveness, minimizes fossil fuel consumption and meets operation conditions requirements. A PMS based on optimization has recently been presented by Gao et al. [11]. In this study, optimal control is proposed using a three-level control architecture. In the third control level, a multi-objective is responsible for minimizing fuel consumption and GHGs as well as scheduling operational maintenance. The second control level is based on the discrete algorithms that regulate the frequency/voltage from active/reactive power according to the load demand. Lastly, in this study, the first control level is responsible for following the reference control between the components of the system. A multi-objective optimization in cloud platform is presented by Li et al. [12]. The optimization function is based on the particle swarm optimization (PSO) method. One of the objectives of this work is to use the cloud to perform the algorithm calculation in real time. Otherwise, hybridization between AI techniques and linear programming-based multi-objective optimization is presented in [13]. In this study, simulation results proved the effectiveness of the proposed multi-objective intelligent energy management using an FL-based expert system. A review of the advanced microgrid supervisory controllers (MGSC) and energy management systems (EMS) is presented by Meng et al. [14]. The hierarchical control, definitions and issues are presented. The centralized MGSC/EMS is usually more suitable for small-scale microgrid applications where centralized information gathering and decision-making can be performed with low communication and computing costs. Among the techniques or methods used on centralized control, a review has been presented by the authors with regard to genetic algorithms, swarm algorithms, linear and non-linear optimization, the rule-based system and machine learning systems. On the other hand, the decentralized MGSC/EMS can be more desirable when the microgrid is large or the generation, consumption and storage are widely dispersed, which makes centralized data acquisition difficult or costly. The multi-agent system (MAS) based on MGSC/EMS has become a prominent research direction, as it provides the probability to actualize decentralized management functions.

PMS based on artificial intelligence (AI) algorithms are easier to implement and are more widely used for real-time microgrid control. Furthermore, it is not necessary to know the exact mathematical model for each RES or energy storage device in the microgrid. Zahraee et al. [15] presented the applications of AI techniques for hybrid energy systems (HES). The authors present a summary of research concerning the use of AI algorithms for designing, planning and controlling problems in the fields of HES. AI algorithms are mostly used because they require less computational time, show better accuracy and better convergence in comparison to traditional methods. The research focuses on hybridization between optimization and AI techniques. These approaches have been proven to be faster, more accurate and more powerful than classical methods. Similarly, Rajesh et al. [16] presented a review of AC microgrid control. The islanded and grid-connected modes of microgrid operation are presented. The control techniques and their different hierarchy levels are identified and explained in detail. According to the review, for primary control level, droop control is used for small and large microgrids as it

DOI: http://dx.doi.org/10.5772/intechopen.93399

provides a high degree of plug-and-play capability. Centralized or decentralized control can be used on the second control level. Centralized control is more suitable for small-scale microgrid applications and decentralized control is normally used for large-scale multi-user microgrids. The tertiary control is applied when the microgrid operates in grid-connected mode. Reactive power compensation techniques for microgrids are evaluated by [17]. An overview of the challenges and power quality issues faced by microgrids are identified. Likewise, the different compensation methods, various control techniques, algorithms and devices are investigated in this study. A multi-agent system (MAS) decentralized energy management in a microgrid is presented in [18]. The microgrid system is composed of a battery/genset/FC/PV/hydro plant and variable electrical load. This study demonstrates that hybrid MAS – fuzzy Q-Learning is appropriate to solve the complex issues of energy management in a stand-alone microgrid by controlling the power flow between RES and ESS. Likewise, a control strategy for a Flywheel Energy Storage System (FESS) using the Artificial Neural Network (ANN) is presented by Daoud et al. [19]. The FESS is connected to an electric network. The charge/discharge from the electrical grid to FESS is used as grid frequency support/control, power conditioning or uninterruptible power supply (UPS) applications. The simulation and experimental tests show good results of the ANN strategy compared with classical power control strategies. In addition to its simplicity, the ANN strategy exhibits fewer tuning problems and requires less controller effort. Mallesham et al. [20] propose an automatic first-control level of microgrid using AI techniques. The difficulty in tuning a large number of parameters in complex systems can be achieved through AI techniques. This study compared various AI techniques with traditional power control strategies. The simulation results show that optimal tuning of multiple parameters in a non-linear microgrid using BFO techniques is better than PSO, GA and classical methods.

In order to address the various disadvantages of the optimal energy management systems proposed in the literature, we present in this work a novel solution of an optimal control of both active and reactive power flow for isolated microgrids based on fuzzy logic techniques. The performance of this solution that allows a better sharing of the active and reactive power flow will be presented. The stability and reliability of remote microgrids are demonstrated in this work. Active/reactive power control responds quickly to voltage/frequency variations on the AC-link system.

The rest of this paper is structured as follows: The dynamic microgrid model is presented in Section 2. The power management system based on fuzzy logic is presented in Section 3. The simulation results and discussion are presented in Section 4. Lastly, Section 5 presents the conclusions.

2. Dynamic microgrid model

2.1 PV panel model

A photovoltaic cell is basically a semiconductor diode whose p–n junction is exposed to light. Photovoltaic cells are made of several types of semiconductors using different manufacturing processes. The monocrystalline and polycrystalline silicon cells are only found at the commercial scale at the present time [21]. The basic equation from the theoretical operation of semiconductors that mathematically describes the current-voltage (I-V) characteristic of the ideal photovoltaic cell is as follows [22]:

Parameters	Value
Maximum power (*Pmax*)	220 W
Voltage at Pmax (*Vmp*)	48.3 V
Current at Pmax (*Imp*)	4.54 A
Open-circuit voltage (*Voc*)	59.26 V
Short-circuit current (*Isc*)	5.09 A
Rs	0.243 Ω
Rp	235.76 Ω

Table 1.
Electrical characteristic data from NREL system advisor model taken at STC.

$$I = I_{ph} - I\left(e^{\frac{q(V+IR_S)}{AKT}} - 1\right) - \frac{V + IR_S}{R_P} \quad (1)$$

where I_{ph} is the photo-generated current, I_O is the dark saturation current, I_{Rp} is current flowing in the shunt resistance, R_p is the cell series resistance, A is the diode quality factor, k is the Boltzmann constant at $1.38\ 10^{-23}$ J/K, and q is the electron of a charge at $1.6\ 10^{-19}$ C.

The solar photovoltaic system used in this study consists of 90 CS5P-220 panels connected in series and parallel with each other. The maximum power point tracking (MPPT) voltage and currents are 48.3 V and 4.54 A, respectively, and generate an output power of 220 W. As the output voltage is relatively low, it is necessary to increase the output voltage of the PV system to the desired value of 500 V by using a boost converter. An MPPT algorithm is used to track the MPP to control the boost converter with an appropriate duty cycle to achieve a desired continuous output voltage. The Perturb & Observe (P&O) algorithm is used to determine the desired duty cycle of the boost converter so that the MPPT is reached. The resulting power of the modeled PV generator is 19.7 kW with 90 panels, each having a maximum power of 220 W, as described in **Table 1**. Similarly, a 25 kW wind turbine was also used in this study.

2.2 Wind turbine model

A wind turbine based on the use of a Permanent Magnet Synchronous Generator (PMSG) is considered the second distributed energy resource (DER) in the configuration of our microgrid. The output power of the wind turbine [10] is given by the following relation [23]:

$$p_{mec} = \frac{1}{2}\rho \pi R^2 C_p(\lambda, \beta) v^3 \quad (2)$$

where R is the radius of the wind turbine aerodynamic rotor in meters, Ω is the rotational speed of the rotor in rad/s, and v is the wind linear velocity in m/s. ρ is the air density at the turbine in kg/m^3, Cp designates the fraction of power available in the wind that is converted into mechanical power.

Cp has a theoretical maximum value of 0.593 (Betz' limit), and basically depends on the tip speed ratio λ, and the blade pitch angle, β can be expressed as follows [23]:

$$C_p(\lambda,\beta) = 0.5\left(\frac{116}{\lambda_i} - 0.4\beta - 5\right)e^{-\frac{21}{\lambda_i}} \tag{3}$$

with

$$\lambda_i = \left(\frac{1}{\lambda + 0.08\beta} - \frac{0.035}{\beta^3+1}\right)^{-1} \tag{4}$$

Generally, wind turbines are characterized by two parameters: tip speed ratio (λ) and power coefficient (Cp). The tip speed ratio is defined as:

$$\lambda = \frac{R\Omega}{v} \tag{5}$$

Figure 1 illustrates the characteristic curves of the power coefficient obtained from (Eq. (4)). In order for the wind turbine to extract the maximum available wind power at a given wind speed, the operating point of the turbine must be kept in the λ_{opt} area. Consequently, a Maximum Power Point Tracking (MPPT) algorithm, which is detailed below, is needed to control the rotational rotor velocity of the turbine and maintain it at the maximum power.

In order to achieve the PMSG control system, its dynamic model is required. The generator model is derived from the projection of its equations on a reference coordinate system rotating synchronously with the magnet flux. In order to achieve synchronization between the dq rotating reference frame and the abc three-phase frame, a phase-locked loop (PLL) is used. Then, a dynamic model of the surface mounted PMSG is expressed as:

$$\begin{cases} v_{ds} = R_s i_{ds} + L_s \dfrac{di_{ds}}{dt} - \omega\psi_{qs} \\ v_{qs} = R_s i_{qs} + L_s \dfrac{di_{qs}}{dt} + \omega\psi_{ds} \end{cases} \tag{6}$$

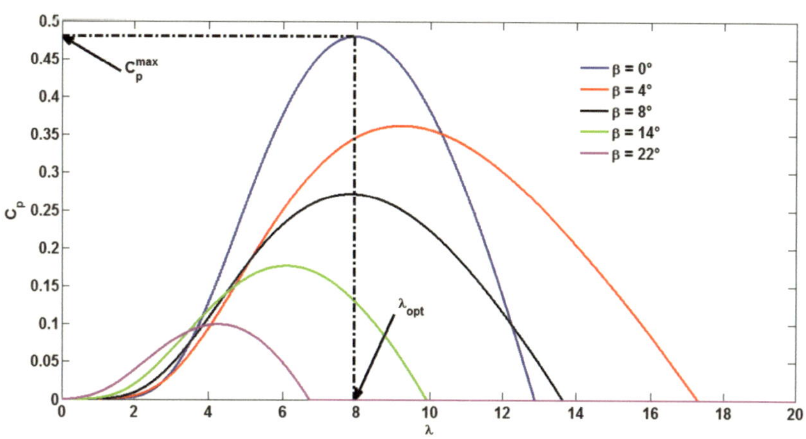

Figure 1.
Power coefficient curves versus tip speed ratio for different blade angles.

where Ls and R_s are the generator inductance and resistance, respectively, ω is the electrical generator speed, and Ψ_{ds} and Ψ_{qs} are d-axis and q-axis magnet flux, respectively, which are expressed as follows:

$$\begin{cases} \psi_{ds} = L_s i_{ds} + \varphi \\ \psi_{qs} = L_s i_{qs} \end{cases} \tag{7}$$

where Φ is the magnet flux. Then, the electrical model of PMSG in the synchronous reference frame can be expressed as:

$$\begin{cases} \dfrac{di_{ds}}{dt} = \dfrac{1}{L_s}\left(-R_s i_{ds} + \omega L_s i_{qs} + v_{ds}\right) \\ \dfrac{di_{qs}}{dt} = \dfrac{1}{L_s}\left(-R_s i_{qs} - \omega L_s i_{ds} - \omega\varphi + v_{qs}\right) \end{cases} \tag{8}$$

The electromagnetic torque of the non-salient poles PMSG is written as:

$$T_{em} = \frac{3}{2} p \ \varphi i_{qs} \tag{9}$$

where p is the number of pole pairs of the generator. Eq. (9) shows that the generator torque can be controlled directly via the q-axis current of the stator.

The mechanical dynamics model of the considered wind turbine system can be defined by the following expression:

$$J_T \frac{d\omega_r}{dt} + f\,\omega_r = T_T - T_{em} \tag{10}$$

where T_T represents the mechanical torque, J_T is the moment of inertia, F is the coefficient of friction, and ω_r is the mechanical speed, which is related to the electrical rotation as follows:

$$\omega = p\,\omega_r \tag{11}$$

2.3 Diesel genset model

The diesel generator is composed of the internal combustion engine and Wound Rotor Synchronous Generator (WRSG) [23].

2.3.1 Diesel motor model

The diesel engine model is shown in **Figure 2** [24–26]. The dynamic of the actuator is modeled by a first order model with time constant τ_1 and gain K_1 [24, 27]. The combustion block is represented with gain K_2 and delay τ_2 [26].

The actuator is modeled as follows [23]:

$$\frac{K_1}{1 + s\tau_1} \tag{12}$$

The combustion block can be expressed as follows:

DOI: http://dx.doi.org/10.5772/intechopen.93399

$$K_2 e^{s\tau_2} \tag{13}$$

As for the delay, it can be modeled as follows:

$$\tau_2 = \frac{60h}{2Nn_c} + \frac{60}{4N} \tag{14}$$

where h represents the number of strokes, n_c is the number of cylinders, N is the speed of the diesel generator (rpm), and Φ is the fuel consumption rate (kg/s) [28]. In order to maintain the grid frequency (AC-bus) constant, the speed of the diesel engine must be kept constant when the load varies.

2.3.2 Synchronous generator model

The simplified model of the Wound Rotor Synchronous Generator (WRSG) can be obtained in dq frame (conversion between abc and dq can be realized by means of the Park Transform) [29].

The stator armature windings voltages are:

$$\begin{cases} v_d = -R_s i_d + \dfrac{d\lambda_d}{dt} - \omega\lambda_q \\ v_q = -R_s i_q + \dfrac{d\lambda_q}{dt} + \omega\lambda_d \end{cases} \tag{15}$$

where R_s is the stator winding resistance. The stator fluxes are described by the following formula:

$$\begin{cases} \lambda_d = -L_d i_d + L_{md}\left(i_f + i_D\right) \\ \lambda_q = -L_d i_q + L_{mq} i_Q \end{cases} \tag{16}$$

The rotor armature winding voltage of a synchronous generator is described as:

$$v_f = -R_f i_f - L_d \frac{di_d}{dt} + L_f \frac{di_f}{dt} + L_{md} \frac{di_D}{dt} \tag{17}$$

Damper windings are expressed by the following equation:

$$\begin{cases} 0 - R_D i_D - L_{md} \dfrac{di_d}{dt} + L_{md} \dfrac{di_f}{dt} + L_D \dfrac{di_D}{dt} \\ 0 = R_Q i_Q - L_{mq} \dfrac{di_q}{dt} + L_Q \dfrac{di_Q}{dt} \end{cases} \tag{18}$$

The electromagnetic torque of the synchronous generator can be expressed as follows:

$$T_{em} = p\left(\left(L_d - L_q\right) i_d i_q + L_{md} i_f i_q + L_{md} i_q i_D - L_{mq} i_d i_Q\right) \tag{19}$$

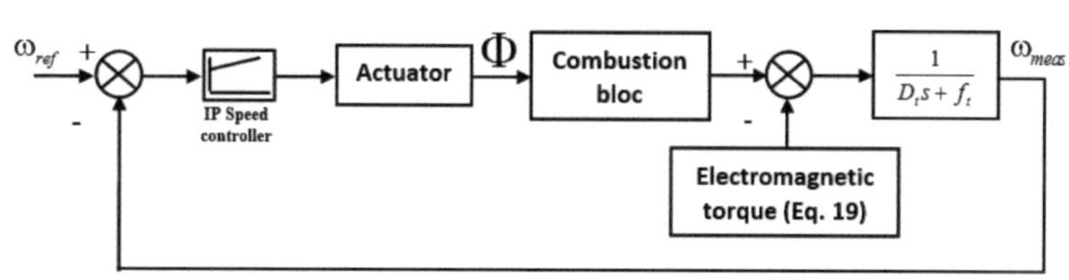

Figure 2.
Block diagram of a diesel generator model [23].

2.4 Electrical load model

In this article, we use a configuration of a dynamic three-phase load as shown in **Figure 3**, which allows us to impose any load profile. As illustrated in **Figure 3**, the gain G is used to transform the active power profile to current ones. Therefore, the three-phase active power can be expressed as follows:

$$P_{load} = \sqrt{3} V_{L-L} I \cos\phi \tag{20}$$

where $V_{L\text{-}L}$ is the line-line voltage of the microgrid. As we are using a three-phase resistor load, relation (20) can be expressed as follows:

$$I = \frac{P_{load}}{\sqrt{3} V_{L-L}} \tag{21}$$

where the gain G is calculated as follows:

$$G = \sqrt{3} V_{L-L} \tag{22}$$

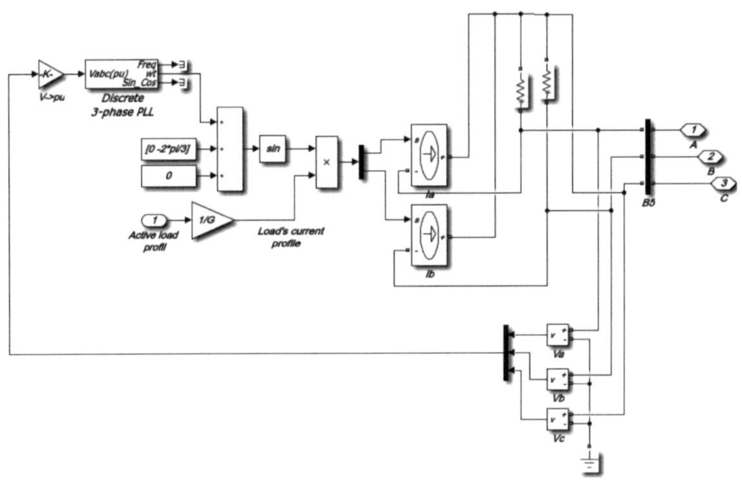

Figure 3.
Matlab block of dynamic three-phase load.

3. Methodology: power management system for remote microgrid based on fuzzy logic

The power management system (PMS) for remote microgrids is presented in this section. Two control levels have been used for the PMS. The second control level is based on the fuzzy logic algorithms, which balance the active power between two diesel generators (master and slave) and regulate the reactive power according to the load demand. The first control level is responsible for following the reference control between all components of the microgrid. The artificial intelligence (AI) algorithm maintains the stability of the remote microgrid and the supply of electricity to the load demand. The PMS based on fuzzy logic is normalized with the aim of ensuring its adaptability to different microgrid sizes.

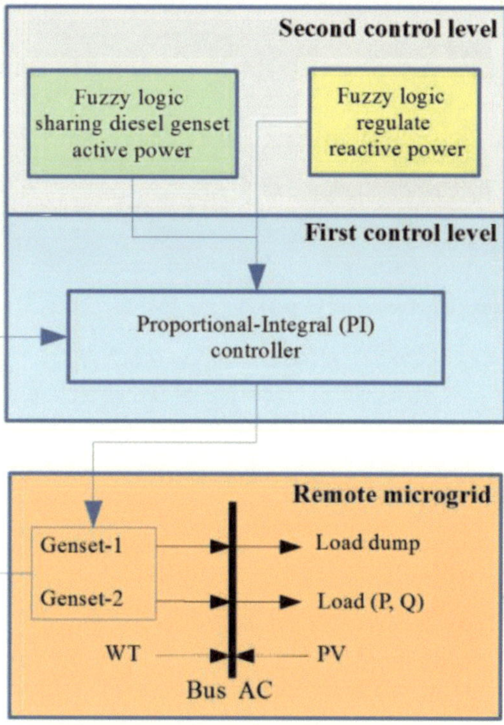

Figure 4.
Power management system for remote microgrid based on artificial intelligence (AI) algorithms.

Figure 4 presents the PMS for remote microgrids based on artificial intelligence (AI) algorithms. The remote microgrid is composed of two diesel generators (genset). The genset-1 (master) is responsible for controlling the frequency/voltage of the microgrid and to maintain its reliability. The genset-2 (slave) is used in operation when the load demand is higher than genset-1 rated power. The remote microgrid is composed of (i) an active/reactive electric load, (ii) a load dump, (iii) PV solar system and (iv) a wind turbine (WT). In this study, the energy storage system (ESS) has not been taken into consideration.

3.1 Second control level

3.1.1 Fuzzy logic controller overview: sharing diesel genset active power

An overview of the architecture of the fuzzy logic controller is presented in **Figure 5**. The fuzzy logic control system aims to balance active power between demand and generation in the remote microgrid for maintaining system reliability. Thus, the gensets (1 and 2) can run simultaneously; two outputs of fuzzy logic controllers are necessary. Pg1* and Pg2* are therefore designed as the output variables of the fuzzy logic system or set points. The centroid method is used for defuzzication.

In order to maintain the efficiency of the fuzzy control systems in terms of rules and decisions, we consider different linguistic variables. Therefore ΔP used 18 linguistic variables as shown in **Figure 6(a)**. Lastly, Pg1* and Pg2* are respectively depicted with 9 and 11 linguistic variables each, as represented in **Figure 6(b)** and **(c)**.

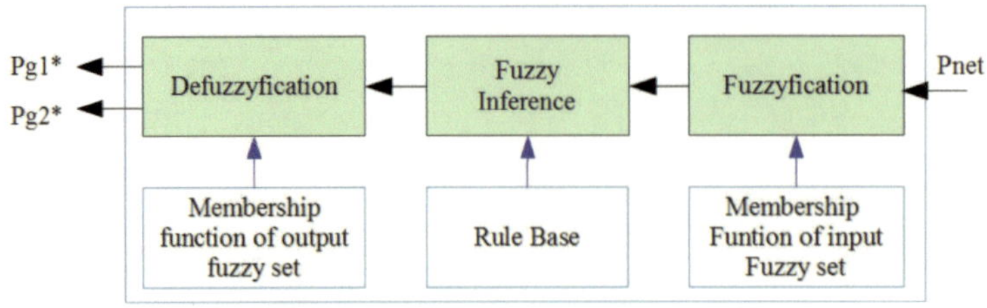

Figure 5.
Fuzzy inference system – sharing diesel genset active power.

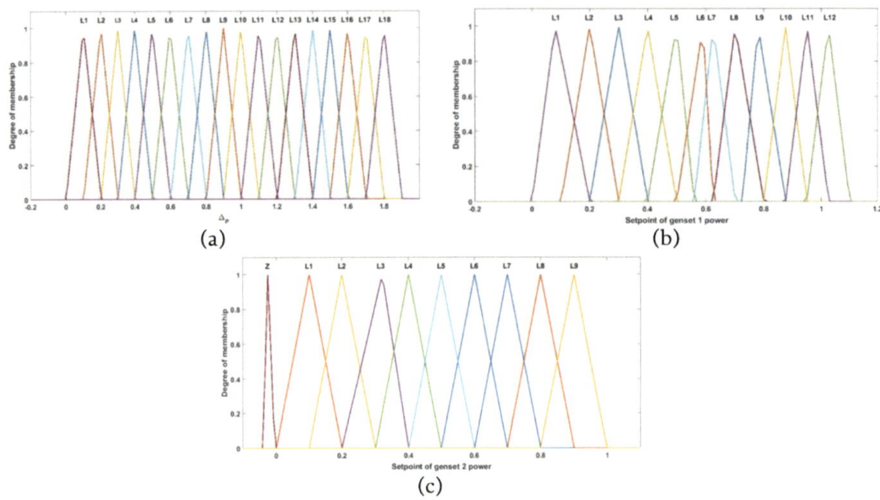

Figure 6.
FL-membership function balance active power: (a) balancing active power between demand and generation in microgrid, (b) set point of genset 1 active power (Pg1) and (c) set point of genset 2 active power Pg2*.*

The trapezoidal and triangular membership functions are used for the linguistic variables' input/output of the fuzzy control system with the aim of simplifying computer calculations on the remote microgrid control (see **Figure 6**).

3.1.2 Fuzzy logic controller overview: regulate reactive power

An overview of the architecture of the fuzzy logic controller is presented in **Figure 7**. The fuzzy logic control system aims to regulate the reactive power in

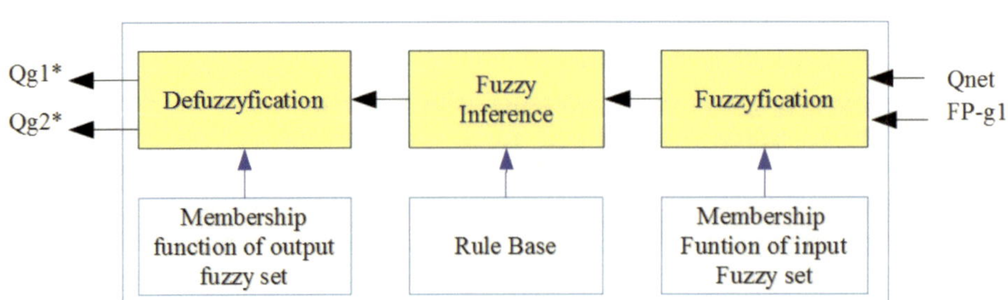

Figure 7.
Fuzzy inference system – regulate reactive power.

remote microgrids in order to maintain their reliability. Thus, regulation of reactive power across the remote microgrid can be performed simultaneously on the two-genset system; two outputs of fuzzy logic controllers are necessary. Thus, Qg1* and Qg2* are designed as the output variables of the fuzzy logic system or setpoints. The centroid method is used for defuzzication.

In order to maintain the efficiency of the fuzzy control systems in terms of rules and decisions, we consider different linguistic variables. Therefore, ΔQ used 16 linguistic variables as shown in **Figure 8(a)**. The measurement of the active power of Genset 2 used two linguistic variables each as represented in **Figure 8(b)**. The set points of Genset 1 and Genset 2 are depicted with six linguistic variables each, as represented in **Figure 8(c)** and **(d)**. Lastly, the FIS output surface of reactive power of Genset 1 is presented in **Figure 8(e)**.

The trapezoidal and triangular membership functions are used for the linguistic variables' input/output of the fuzzy control system with the aim of simplifying computer calculations on the remote microgrid control (see **Figure 8**).

3.2 First control level

The first control level is responsible for following the reference control signal for all distributed energy resources (DER). A proportional integral controller (PI) is used for controlling both the frequency and voltage of the microgrid through the

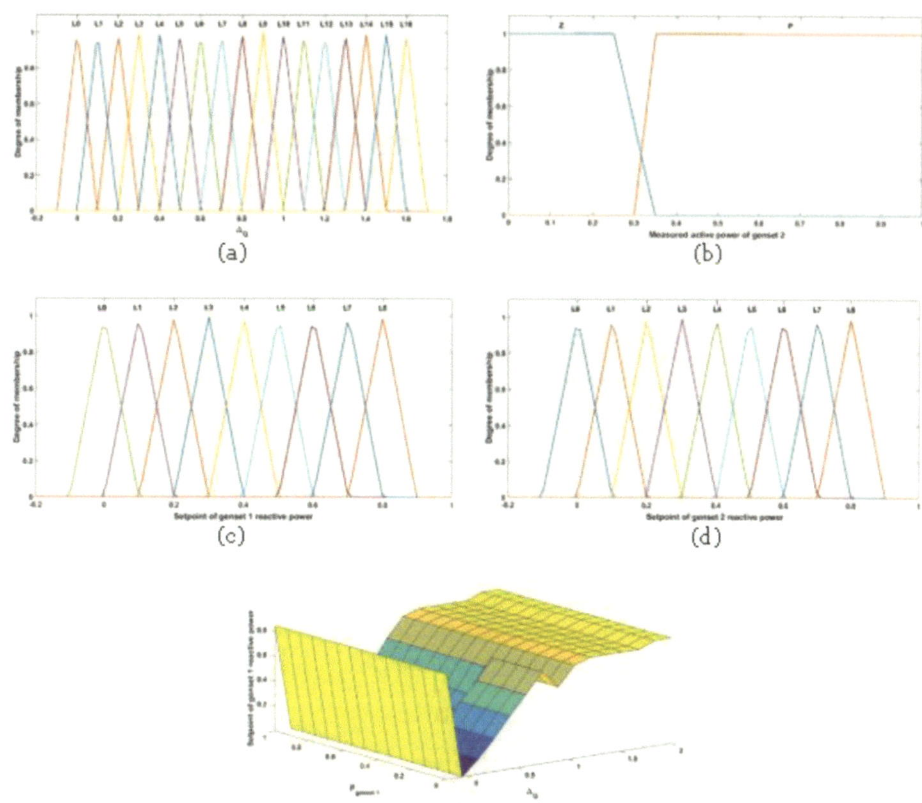

Figure 8.
FL-membership functions regulate reactive power: (a) balancing reactive power between demand and generation in microgrid, (b) measured active power of Genset 2, (c) setpoint of Genset 1 reactive power, (d) setpoint of Genset 2 reactive power and (e) FIS output surface of reactive power of Genset 1.

diesel generators. The diesel genset speed control (frequency) and the AVR control diesel genset control (voltage) are presented in this section.

Likewise, a PV solar system control using the MPPT algorithm based on the fuzzy logic technique and the PV inverter control is depicted in this section. Additionally, the wind turbine system control based on the MPPT algorithm and the grid side converter (GSC) control is presented in this section. Both MPPT algorithms are based on the perturbation and observation (P&O) technique.

3.2.1 Fuzzy logic controller overview: regulate reactive power

Figure 9 represents the speed control used for the diesel generator. The modeling of the diesel generator governor is presented by Eqs. (12)–(14). Eqs. (15)–(18) present the modeling of the synchronous generator.

Table 2 presents the PI controller parameters and the governor dynamic coefficients.

3.2.2 AVR control diesel genset

Figure 10 represents the Automatic Voltage Regulator (AVR) control diesel genset used for the diesel generator. The modeling of the synchronous generator is presented by Eqs. (15)–(18).

Table 3 presents the PI controller parameters.

3.2.3 PV solar system control

A new control algorithm for the PV system based on fuzzy logic is presented in this section. The P&O technique is used and implemented with the MPPT methodology. The power electronics converter control of the PV system is also presented in this section.

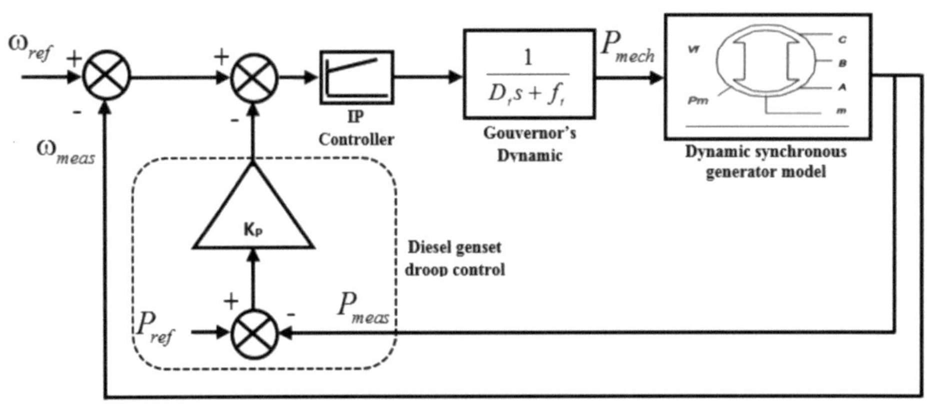

Figure 9.
Diesel genset speed control model.

PI parameters	Governor dynamic coefficients
Kp = 520	0.852
Ki = 240	2–0.15

Table 2.
PI controller parameters and governor dynamic coefficients.

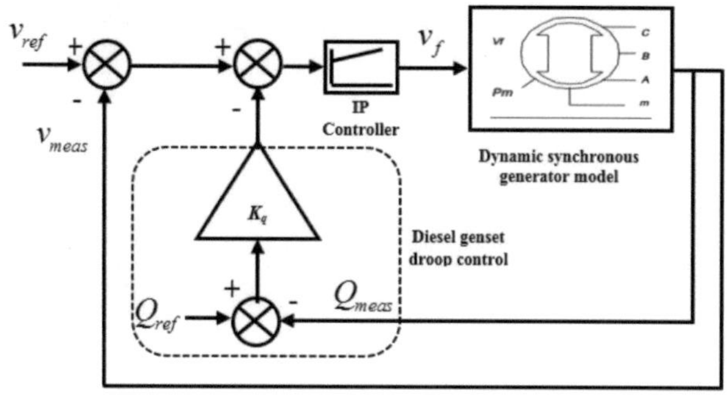

Figure 10.
Diesel genset AVR control model.

PI parameters	
Kp	12
Ki	5

Table 3.
AVR PI controller parameters.

3.2.3.1 MPPT algorithm based on fuzzy logic techniques

An overview of the architecture of MPPT algorithms based on fuzzy logic used in PV solar system control is presented in **Figure 11**. The fuzzy logic control system aims to control the PV converter system. The MPPT method is based on the perturbation & observation (P&O) technique. The maximum PV active power is extracted to every moment with the P&O technique. The variation of active power (Δp) and the variation of DC voltage (Δv) of PV systems are the inputs to the MPPT system. The duty cycle to control the dc-dc boost converter corresponds to the output of the MPPT system. The fuzzy logic used the centroid method for defuzzication.

In order to maintain the efficiency of the fuzzy control systems in terms of rules and decisions, we consider different linguistic variables. Therefore, Δp and Δv used five linguistic variables as shown in **Figure 12(a)** and **(b)**. The duty cycle used five linguistic variables each, as represented in **Figure 12(c)**. **Table 4** presents the names of the linguistic variables.

The trapezoidal and triangular membership functions are used for the linguistic variables' input/output of the fuzzy control system with the aim of simplifying computer calculations (see **Figure 12**).

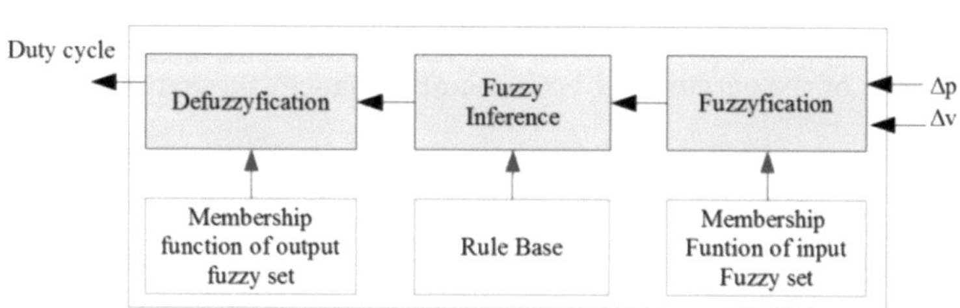

Figure 11.
Fuzzy inference system – MPPT algorithm for PV control.

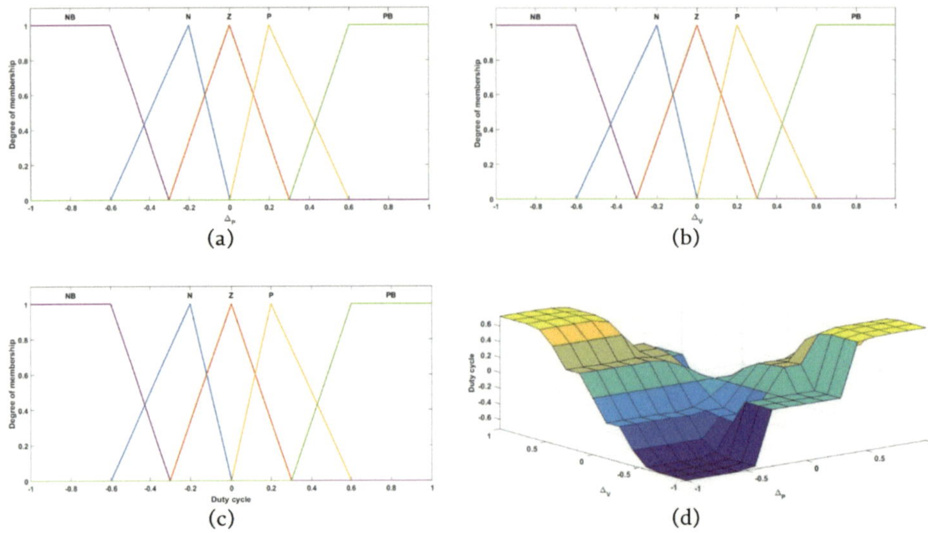

Figure 12.
FL-membership function MPPT algorithm for PV system: (a) variation of active power, (b) variation of DC voltage, (c) boost converter duty cycle and (d) FIS output surface of MPPT algorithms.

Linguistic variable	Description
NB	Negative big
N	Negative
Z	Zero
P	Positive
PB	Positive, big

Table 4.
Linguistic variables.

3.2.3.2 PV solar inverter control

The three-phase two-sage photovoltaic grid-connected system used with its overall control is illustrated in **Figure 13**. The system contains three controllers. A two-loop controller controls the three-phase PV inverter. The outer loop controls both the dc-link voltage to follow the reference value (Vdc_{ref}) and reactive power and to provide the values (Idq_{ref}) of the reference current (i_{ref}) for the inner loop. A phase-locked loop (PLL) algorithm is used to obtain this signal. The current control loop controls the inverter current (i_d and i_q) according to the reference current (i_{ref}). An MPPT algorithm is used to track the maximum power from the PV array regardless of the variation of both solar irradiance and temperature.

3.2.4 Wind turbine system control

3.2.4.1 MPPT algorithm

An overview of the architecture of MPPT algorithms for the WT system based on fuzzy logic is presented in **Figure 14**. The objective of the fuzzy logic control

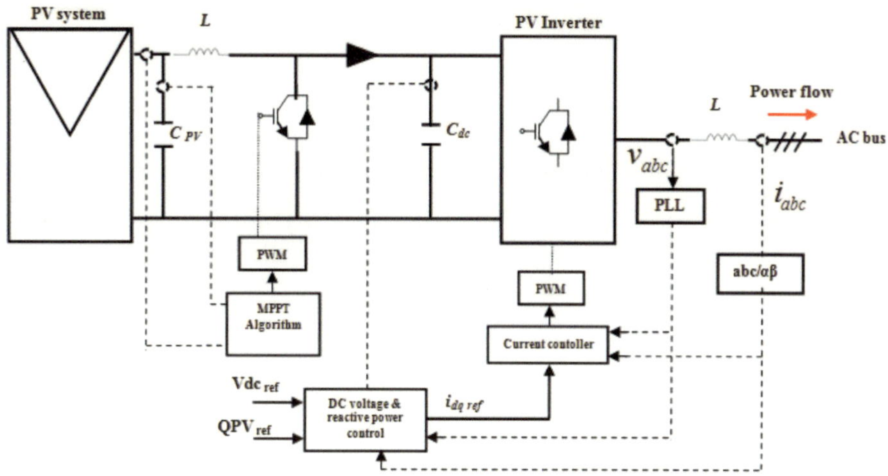

Figure 13.
Proposed three-phase, two-stage grid-connected PV system.

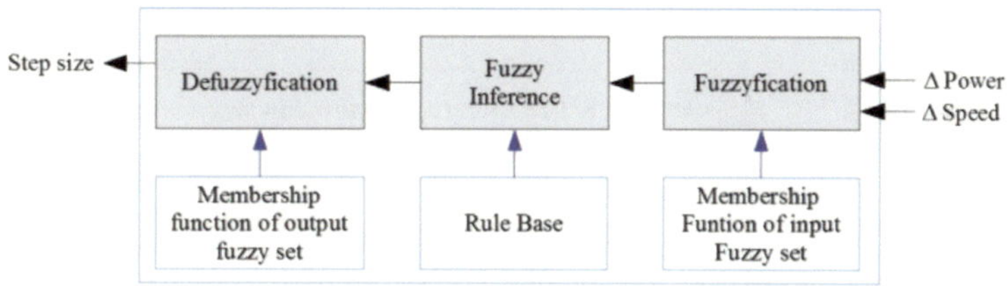

Figure 14.
Proposed wind turbine MPPT algorithm.

system is the calculation of the set point of the generator to track the maximum power point regarding the wind speed variation. The instantaneous generated wind power and its variation represent the respective inputs of the MPPT system. The setpoint of the generator is the output of the used FL MPPT algorithm.

In order to extract the maximum wind power, we have applied in this article the fuzzy logic. Thus, the set point of the speed of the generator is constantly calculated in order to follow the maximum power as a function of the variation of the wind speed. In order to avoid disturbances, the variation step of the generator speed setpoint is proportional to the difference between the maximum power and the real power. The name of the linguistic variables used is the same as represented in **Table 4**.

The control strategy of the fuzzy logic is based on an expert human operator to interpret a situation and initiate its appropriate command action [30]. Commonly, a controller based on fuzzy logic has two inputs and provides a control action. For FLC P&O, inputs are quantized into 5 levels represented by a set of linguistic variables: Negative Big (NB), Negative (N), Zero (Z), Positive (P), and Positive Big (PB). The fuzzy rule base formulation of FLCP&O is shown in **Table 5**. These rules are chosen to perform the optimization of wind generation capture as follows: (i) when the input signals are far from the optimal point, the output of the FLCP&O provides a big step size; (ii) when the inputs are close to the optimum point, the output is set to a small step size value; (iii) once the inputs are close to the optimum point, the step size is set to zero. In this article, we use the min and max operators as

Step size		ΔPower				
		NB	N	Z	P	PB
ΔSpeed	NB	NB	N	Z	P	PB
	N	NB	N	Z	P	PB
	Z	NB	N	Z	P	PB
	P	NB	N	Z	P	PB
	PB	NB	N	Z	P	PB

Table 5.
Membership functions of output variables [31].

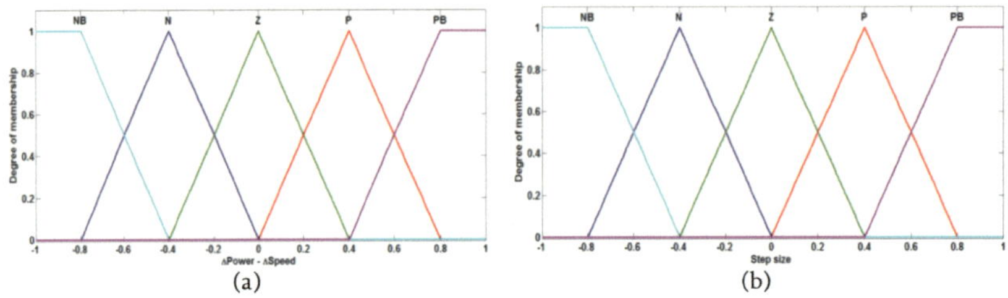

Figure 15.
FL-membership function MPPT algorithms for WT control: (a) membership functions of inputs variables and (b) membership functions of output variables.

t-norm and t-conorm, respectively. Triangular membership functions are used, principally due to their efficiency and high-performance computing. The membership adopted for both input and output variables are illustrated in **Figure 15(a)** and **(b)**, respectively [31].

3.2.4.2 Rotor side converter (RSC) and grid side converter (GSC)

Figure 16 presents the control system of a wind turbine (WT) system that uses a permanent magnet synchronous generator (PMSG). The MPPT method extracts the maximum power from the WT using the P&O technique. The MPPT output signal control or duty cycle is sent to the rotor side converter (RSC).

Figure 17 presents in detail the application of the MPPT algorithm for controlling the wind turbine. The WT rotation is measured (Ωmes) on the WT shaft. The reference value of the WT shaft rotation (Ω_{ref}) is determined by MPPT fuzzy logic.

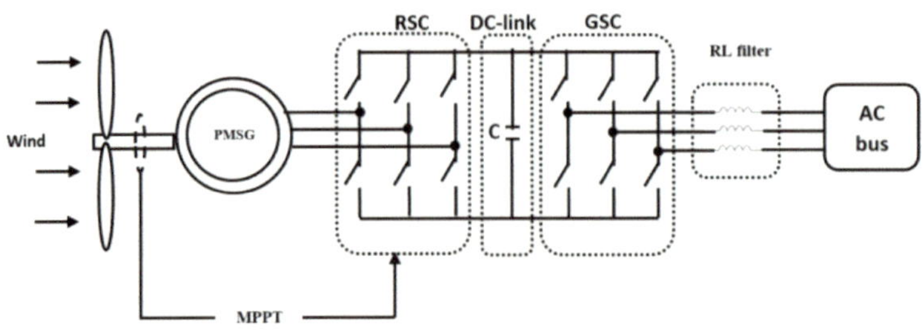

Figure 16.
Block diagram of direct-drive wind turbine system.

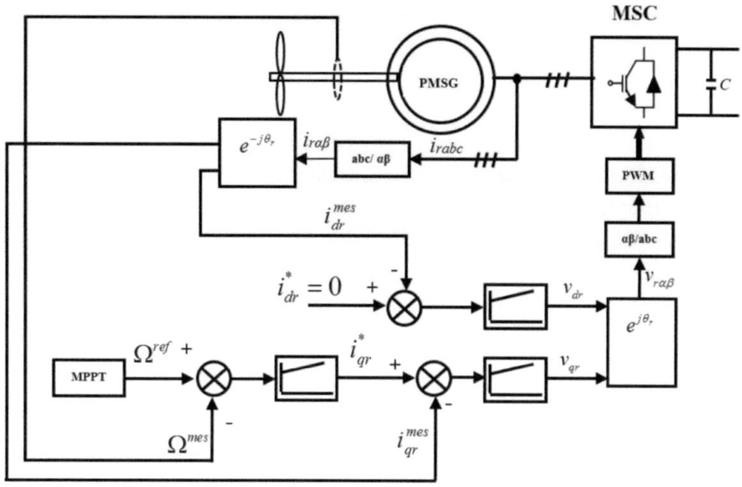

Figure 17.
MPPT control scheme for wind turbine system.

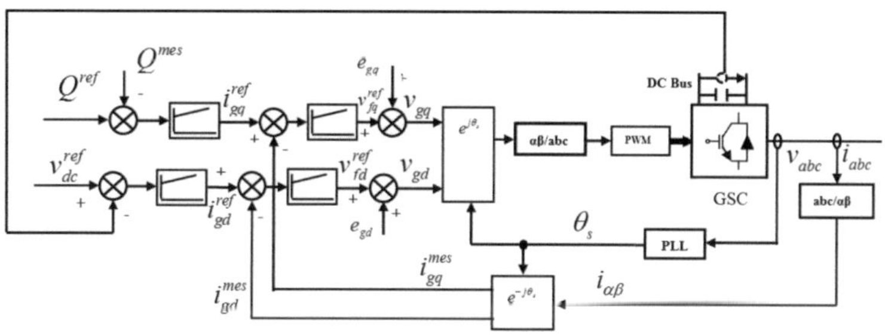

Figure 18.
GSC control scheme for wind turbine system.

The objective of the grid-side converter is to keep the DC-link voltage fixed and adjust the absorbed or provided reactive power according to grid code requirements. As shown in **Figure 18**, the active and reactive power can simply be controlled by d-axis and q-axis current using the grid voltage-oriented control. This control strategy contains two nested loops. The inner loop controls the grid current while the outer loop regulates the DC-link voltage and reactive power for the GSC. The DC-link voltage is controlled by d-axis current since it depends on the active power. On the other hand, the q-axis reference is set to zero when a unity power factor is required; otherwise, it is adjusted as a function of the reactive power needed.

4. Simulation and analysis results

The simulation results are presented below. Three simulation scenarios are proposed. The first scenario presents the results of the first control level and second control level of a remote microgrid system with wind speed and solar irradiation stepped profile. The second scenario presents the results of the remote microgrid control using a wind speed and solar irradiation fluctuating profile. Lastly, the third scenario presents the performance evaluation of MPPT-P&O based on fuzzy logic techniques.

4.1 Scenario 1: wind and solar irradiation stepped profile

4.1.1 First control level of remote microgrid

Figure 19 presents the results concerning the first control level of remote microgrids for both controllable and non-controllable energy sources such as diesel generators and renewable energy sources (WT, PV). The solar irradiance and wind speed profiles are presented in **Figure 19(a)** and **(b)**, respectively. The PV system power and the wind turbine active power are shown in **Figure 19(c)** and **(d)**. **Figure 19(e)** and **(f)** present the active power and reactive power load. Lastly, Genset 1

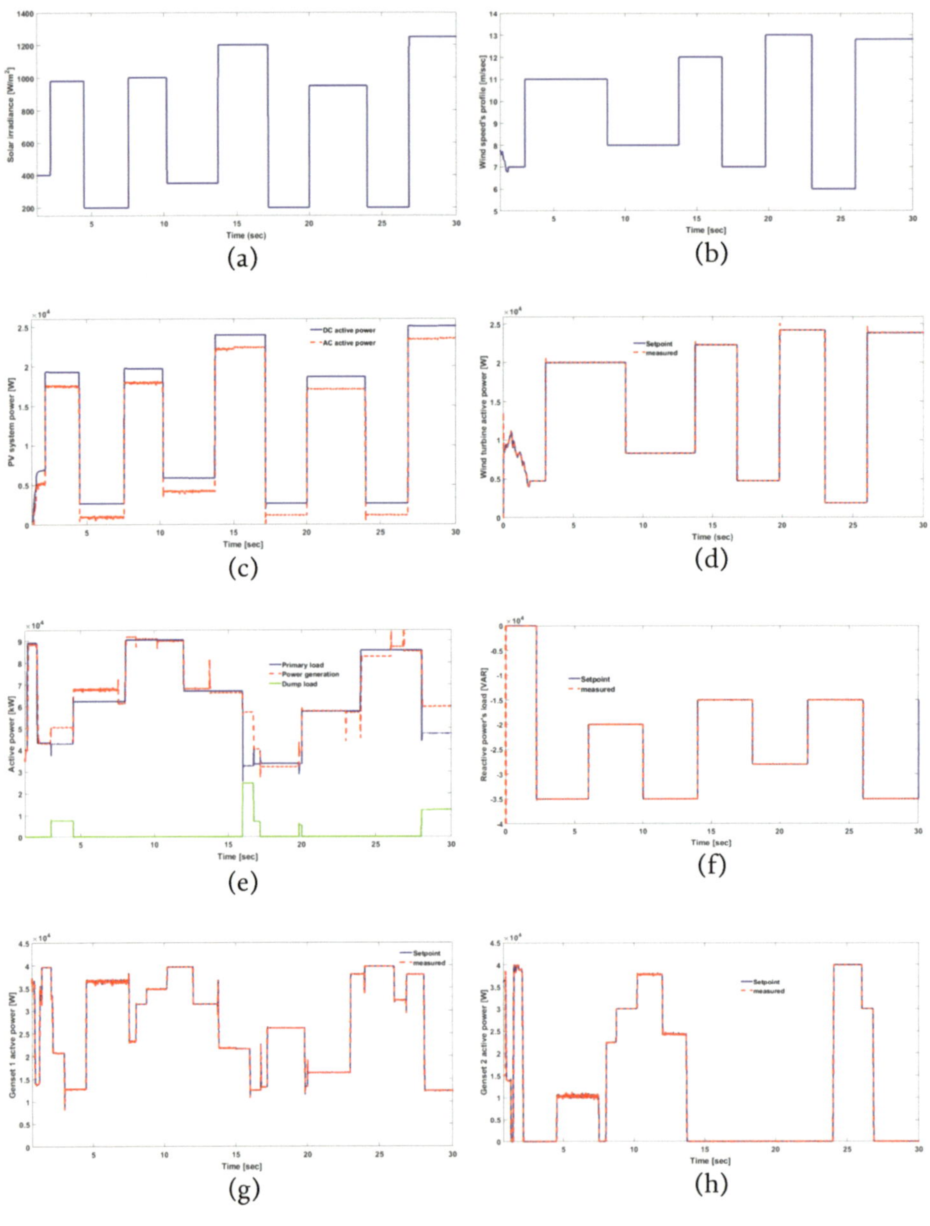

Figure 19.
First control level of remote microgrid results for wind speed and solar irradiation stepped profiles: (a) solar irradiation profile, (b) wind speed profile, (c) PV system power, (d) wind turbine active power, (e) active power, (f) reactive power load, (g) Genset 1 active power and (h) Genset 2 active power.

active power and Genset 2 active power are presented in **Figure 19(g)** and **(h)**, respectively.

4.1.2 Second control level of remote microgrid

Figure 20 presents the results concerning the second level microgrid control for both the controllable and non-controllable energy sources such as diesel generators and renewable energy sources such as PV panels. The sharing of active/reactive power is presented in **Figure 20(a)** and **(b)**, respectively. **Figure 20(c)** presents the microgrid frequency. Lastly, the DC-link voltage of the PV system is shown in **Figure 20(d)**.

4.2 Scenario 2: wind and solar irradiation continuous profile

4.2.1 First control level of remote microgrid

Figure 21 presents the results concerning the first control level of remote microgrids for both controllable and non-controllable energy sources such as diesel generators and renewable energy sources (WT, PV). The solar irradiance and wind speed profiles are presented in **Figure 21(a)** and **(b)**, respectively. The PV system power and the wind turbine active power are shown in **Figure 21(c)** and **(d)**. **Figure 21(e)** and **(f)** presents the active power with dump load and the reactive power load. Lastly, Genset 1 active power and Genset 2 active power are presented in **Figure 21(g)** and **(h)**, respectively.

4.2.2 Second control level of remote microgrid

Figure 22 presents the results concerning the second control level of remote microgrid for both controllable and non-controllable energy sources such as diesel generators and renewable energy sources such as PV panels. The sharing of active/reactive power is presented in **Figure 22(a)** and **(b)** respectively. **Figure 22(c)**

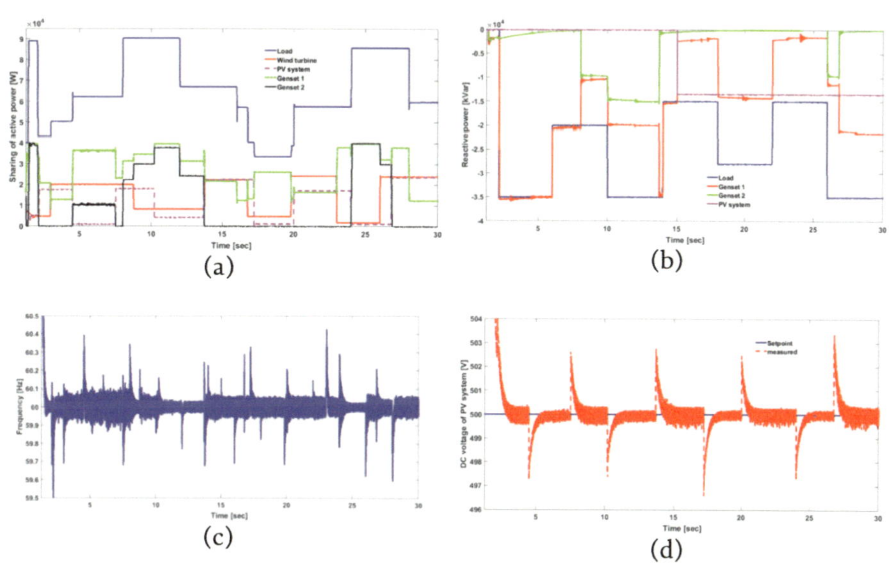

Figure 20.
Second control level remote microgrid results using fuzzy logic techniques: (a) active power sharing, (b) reactive power sharing, (c) microgrid frequency and (d) DC-link voltage of PV system.

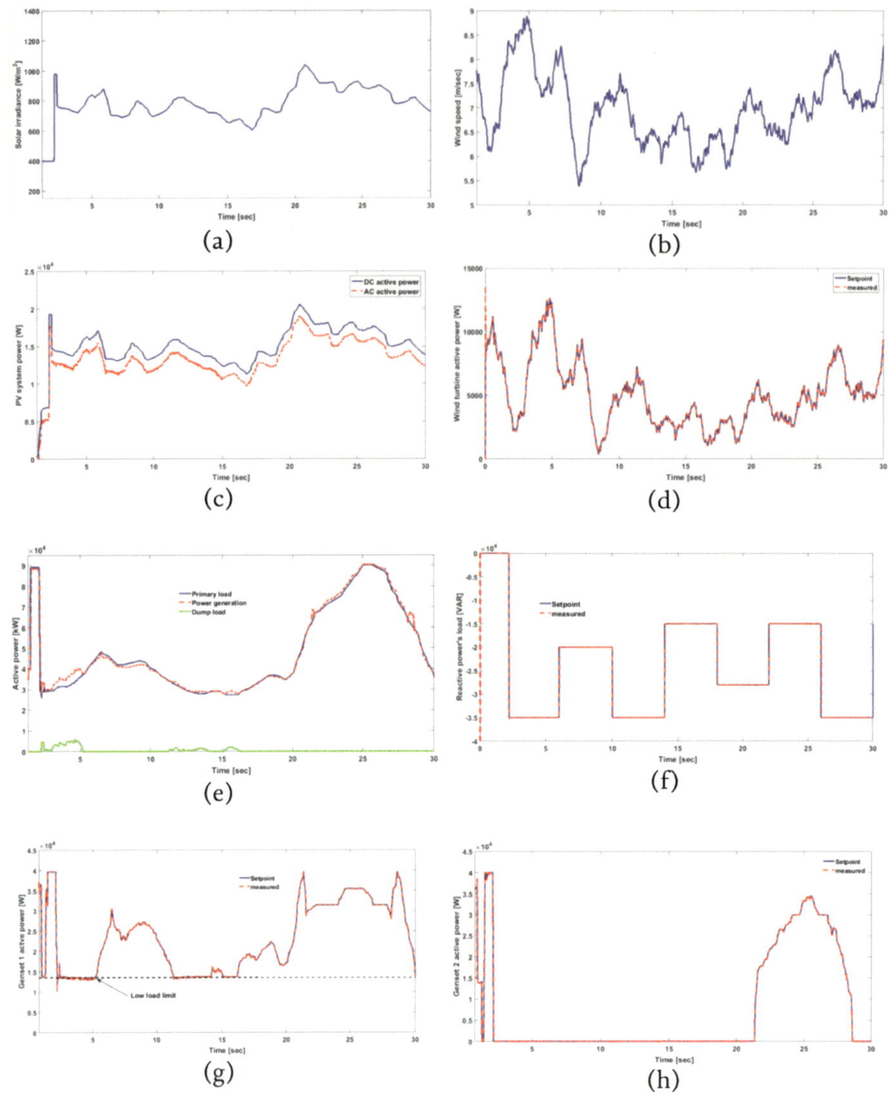

Figure 21.
First control level remote microgrid results for wind speed and solar irradiation stepped profile: (a) solar irradiation profile, (b) wind speed profile, (c) PV system power, (d) wind turbine active power, (e) active power with dump load, (f) reactive power load, (g) Genset 1 active power and (h) Genset 2 active power.

presents the microgrid frequency. Lastly, the DC-link voltage of the PV system is shown in **Figure 22(d)**.

4.3 Scenario 3: performance evaluation of PV systems control: fixed-step P&O, variable-step P&O (MPPT-FL) and inductance method

Several MPPT techniques have been proposed in the literature such as the perturbation and observation (P&O) technique, incremental conductance technique, ripple correlation technique, short-circuit current technique, and open-circuit voltage technique (OCV). These techniques vary in terms of complexity, cost, speed of convergence, required sensors, hardware implementation and efficiency [32–36].

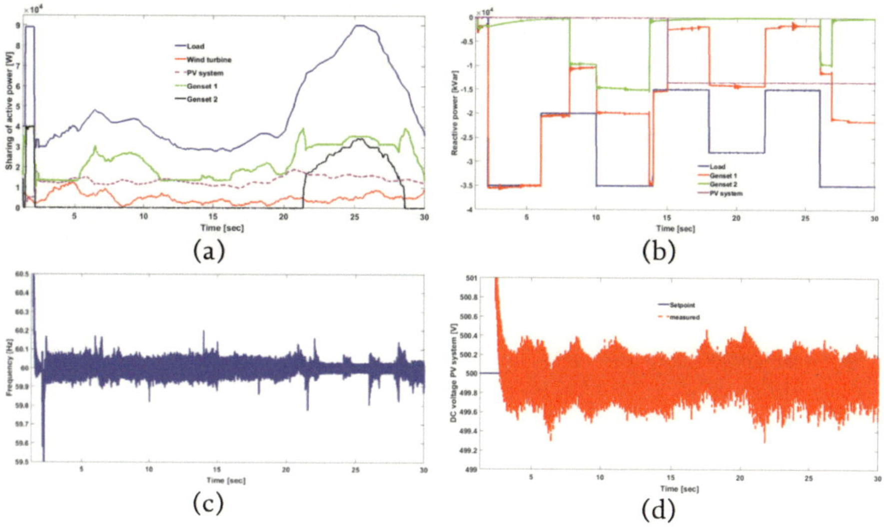

Figure 22.
Second control level remote microgrid results using fuzzy logic techniques: (a) active power sharing, (b) reactive power sharing, (c) microgrid frequency and (d) DC-link voltage of PV system.

Due to the existence of numerous MPPT methods, various researches have presented a comparative analysis of their performance. In fact, some articles present a comparative study of just a few methods while others present a comparison of multiple MPPT methods, based on simulations and from the standpoint of energy production. The MPPT techniques are evaluated while taking into account both solar irradiation and temperature variation and calculation of the total energy provided by the PV solar system.

In this work, we will focus on the simulation of two MPPT techniques (perturbation and observation, and incremental conductance) and compare their performance regarding the proposed MPPT algorithm based on the fuzzy logic techniques presented above. Both P&O and incremental conductance techniques used in this paper are widely presented in the literature [36]. It should be noted that the most significant feature among all MPPT techniques is convergence speed. Thus, any improvement in convergence speed can increase the reliability and robustness of the entire PV system. For this reason, we propose in this work to use a variable step based on fuzzy logic techniques.

In order to achieve good characterization of the different MPPT techniques, simulations were carried out using a Matlab/Simulink environment. The PV solar system was simulated under different operating conditions, especially during transient state caused by a wide variation of the solar radiation.

Table 6 illustrates both efficiency and Total Harmonic Distortion (THD) of each MPPT technique. Note that efficiency was calculated by taking into account the maximum theoretical power and the instantaneous provided power as follows [36]:

$$\eta = \frac{P_{provied}}{P_{\max}} \tag{23}$$

As can be seen in **Table 6**, all tested MPPT techniques have an acceptable THD (less than 5% regarding CEI 61727 standard) at 1000 and 800 W/m^2. However, when solar irradiation fluctuated between 600 and 400 W/m^2 the FL-P&O technique has a better THD than the other two MPPT techniques. For efficiency,

MPPT technique	Efficiency [%]				THD [%]			
	400 W/m^2	600 W/m^2	800 W/m^2	1000 W/m^2	400 W/m^2	600 W/m^2	800 W/m^2	1000 W/m^2
FL P&O	88.26	94.74	97.96	98.46	9.89	5.30	3.93	3.32
Conventional P&O	87.29	94.08	97.61	98.21	12.28	5.80	4.08	3.40
Incremental conductance	87.59	92.75	96.26	97.38	10.46	5.90	3.91	3.71

Table 6.
Performance of different MPPT techniques.

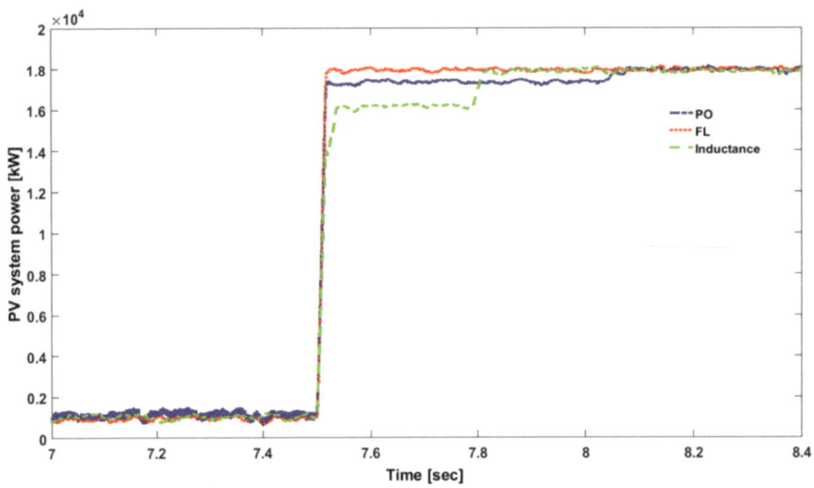

Figure 23.
Performance comparison of three PV MPPT algorithms.

notwithstanding the solar irradiance fluctuation, the FL P&O MPPT has better performance compared to P&O and inductance conductance techniques.

As shown in **Figure 23**, FL-P&O technique has the highest rise time compared to both P&O and incremental conductance techniques. In the steady state, the P&O technique has the slowest convergence speed compared to the other two MPPT techniques.

5. Conclusion

This article presented a novel power flow management algorithm for a remote microgrid based on fuzzy logic. The objectives of this power management system are improved microgrid reliability, improved renewable energy source (RES) integration and performance of active/reactive power control for remote microgrids using artificial intelligence (AI) algorithms.

Sharing of the diesel genset's active and reactive power has been based on AI algorithms such as fuzzy logic. Two simulation scenarios are proposed. The first scenario was used on a wind speed and solar irradiation stepped profile and the second scenario was used on a wind speed and solar irradiation continuous profile.

The results of both scenarios are satisfactory. The stability and reliability of the remote microgrid are demonstrated in the simulation. The active/reactive power control algorithm responds quickly to different events on the remote microgrid, especially to the voltage/frequency variations on the AC-link system. Lastly, improvement to RES integration is demonstrated with the use of a new MPPT algorithm for the PV control system. This MPPT algorithm is based on the P&O method and used fuzzy logic techniques. The simulation results demonstrate the easy adaptability, fast response and efficiency of PV control systems.

Acknowledgements

The authors would like to thank the Natural Sciences and Engineering Research Council (NSERC) for its invaluable support in making this work possible.

Author details

Karim Belmokhtar* and Mauricio Higuita Cano
Nergica, Gaspe, Quebec, Canada

*Address all correspondence to: kbelmokhtar@nergica.com

References

[1] Sen S, Ganguly S. Opportunities, barriers and issues with renewable energy development - a discussion. Renewable and Sustainable Energy Reviews. 2016;**69, 2015**:1170-1181

[2] Guney MS, Tepe Y. Classification and assessment of energy storage systems. Renewable and Sustainable Energy Reviews. 2017;**75**:1187-1197

[3] Akinyele DO, Rayudu RK. Review of energy storage technologies for sustainable power networks. Sustainable Energy Technologies and Assessments. 2014;**8**:74-91

[4] Hosseini SE, Wahid MA. Hydrogen production from renewable and sustainable energy resources: Promising green energy carrier for clean development. Renewable and Sustainable Energy Reviews. 2016;**57**:850-866

[5] Roni MS, Chowdhury S, Mamun S, Marufuzzaman M, Lein W, Johnson S. Biomass co-firing technology with policies, challenges, and opportunities: A global review. Renewable and Sustainable Energy Reviews. 2017;**78**:1089-1101

[6] Cano MH, Agbossou K, Kelouwani S, Dubé Y. Experimental evaluation of a power management system for a hybrid renewable energy system with hydrogen production. Renewable Energy. 2017; **113**:1086-1098

[7] Pegueroles-Queralt J, Bianchi FD, Gomis-Bellmunt O. Control of a lithium-ion battery storage system for microgrid applications. Journal of Power Sources. 2014;**272**:531-540

[8] Haddadi A. Modeling, Control, and Stability Analysis of an Islanded Microgrid. McGill [Theses]. McGill University Libraries. 2015

[9] Shuai Z et al. Microgrid stability: Classification and a review. Renewable and Sustainable Energy Reviews. 2016; **58**:167-179

[10] Olatomiwa L, Mekhilef S, Ismail MS, Moghavvemi M. Energy management strategies in hybrid renewable energy systems: A review. Renewable and Sustainable Energy Reviews. 2016;**62**:821-835

[11] Gao Y, Ai Q. Distributed cooperative optimal control architecture for AC microgrid with renewable generation and storage. International Journal of Electrical Power & Energy Systems. 2018;**96**(2017):324-334

[12] Li X, Li Z. Micro-grid resource allocation based on multi-objective optimization in cloud platform, In: 2017 8th IEEE International Conference on Software Engineering and Service Science (ICSESS). Beijing; 2017. pp. 509-512. DOI: 10.1109/ICSESS.2017.8342966

[13] Chaouachi A, Kamel RM, Andoulsi R, Nagasaka K. Multiobjective intelligent energy management for a microgrid – Aymen Chaouachi - Academia. IEEE Transactions on Industrial Electronics. 2013;**60**(4):1688-1699

[14] Meng L, Sanseverino ER, Luna A, Dragicevic T, Vasquez JC, Guerrero JM. Microgrid supervisory controllers and energy management systems: A literature review. Renewable and Sustainable Energy Reviews. 2016;**60**: 1263-1273

[15] Zahraee SM, Khalaji Assadi M, Saidur R. Application of artificial intelligence methods for hybrid energy system optimization. Renewable and Sustainable Energy Reviews. 2016;**66**: 617-630

[16] Rajesh KS, Dash SS, Rajagopal R, Sridhar R. A review on control of ac microgrid. Renewable and Sustainable Energy Reviews. 2017;**71**:814-819

[17] Gayatri MTL, Parimi AM, Pavan Kumar AV. A review of reactive power compensation techniques in microgrids. Renewable and Sustainable Energy Reviews. 2018;**81**:1030-1036

[18] Kofinas P, Dounis AI, Vouros GA. Fuzzy Q-learning for multi-agent decentralized energy management in microgrids. Applied Energy. 2018;**219**: 53-67

[19] Daoud MI, Abdel-Khalik AS, Elserougi A, Massoud A, Ahmed S, Abbasy NH. An artificial neural network based power control strategy of low-speed induction machine flywheel energy storage system. International Journal of Advanced Information Technology. 2013;**4**(2):61-68

[20] Mallesham G, Mishra S, Jha AN. Automatic generation control of microgrid using artificial intelligence techniques. In: IEEE Power & Energy Society General Meeting. San Diego, CA; 2012. pp. 1-8. DOI: 10.1109/PESGM.2012.6345404

[21] Banu IV, Istrate M. Modeling and simulation of photovoltaic arrays. Buletinul AGIR. 2012;**3**:161-166

[22] Tian H, Mancilla-david F, Ellis K, Jenkins P, Muljadi E. A detailed performance model for photovoltaic systems. United States. Available from: https://www.osti.gov/servlets/purl/1048979

[23] Doumbia ML, Belmokhtar K, Agbossou K. Wind diesel hybrid power system with hydrogen storage. New Developments in Renewable Energy. 2013:365

[24] Roy S, Malik OP, Hope GS. A k-step predictive scheme for speed control of diesel driven power plants. IEEE Transactions on Industry Applications. 1993;**29**(2):389-396

[25] Pena R, Cardenas R, Clare J, Asher G. Control strategy of doubly fed induction generators for a wind diesel energy system. In: IEEE 2002 28th Annu. Conf. Ind. Electron. Soc. IECON 02. Vol. 4. 2002. pp. 3297-3302

[26] Dettmer R. Revolutionary energy a wind/diesel generator with flywheel storage. IEE Review. 19 April 1990;**36**(4):149-151, DOI: 10.1049/ir:19900060

[27] PeÑa R, CÁrdenas R, Proboste J, Clare J, Asher G. Wind-diesel generation using doubly fed induction machines. In: IEEE Transactions on Energy Conversion. Vol. 23. no. 1. March 2008. pp. 202-214. DOI: 10.1109/TEC.2007.914681

[28] Jurado F, Saenz JR. Neuro-fuzzy control for autonomous wind-diesel systems using biomass. Renewable Energy. 2002;**27**(1):39-56

[29] Burton T, Sharpe D, Jenkins N, Bossanyi E. Wind Energy Handbook. Vol. 2. New York: Wiley; 2011

[30] Schneider M, Kandel A, Langholz G, Chew G. Fuzzy expert system tools. Chichester: Wiley; 1996

[31] Belmokhtar K, Ibrahim H, Doumbia ML. A maximum power point tracking control algorithms for a PMSG-based WECS for isolated applications: Critical review. IntechOpen. 2014;**2**: 283-302

[32] Elgendy MA, Zahawi B, Atkinson DJ. Evaluation of perturb and observe MPPT algorithm implementation techniques. In: 6th IET International Conference on Power Electronics, Machines and Drives (PEMD 2012). Vol. 3. 2012. p. 110

[33] Kumar H, Tripathi RK. Simulation of variable incremental conductance method with direct control method using boost converter. In: 2012 Students Conf. Eng. Syst. SCES; 2012. pp. 1-5

[34] Casadei D, Grandi G, Rossi C. Single-phase single-stage photovoltaic

generation system based on a ripple correlation control maximum power point tracking. IEEE Transactions on Energy Conversion. 2006;**21**(2):562-568

[35] Reza Reisi A, Hassan Moradi M, Jamasb S. Classification and comparison of maximum power point tracking techniques for photovoltaic system: A review. Renewable and Sustainable Energy Reviews. 2013;**19**:433-443

[36] Azar Y, DeRubertis B, Baril D, Woo K. Comparison of different MPPT algorithms with a proposed one using a power estimator for grid connected PV systems. Annals of Vascular Surgery. 2017;**43**:44

Chapter 10

Adaptive Load Frequency Control in Power Systems Using Optimization Techniques

Tarek Hassan Mohamed, Hussein Abubakr, Mahmoud M. Hussein and Gaber S. Salman

Abstract

At present, simple and classical tuned controllers are widely used in the power system load frequency control (LFC) application. Existing LFC system parameters are usually tuned based on experiences, classical methods, and trial and error approaches, and they are incapable of providing good dynamic performance over a wide range of operating conditions and various load scenarios. Therefore, the novel modeling and control approaches are strongly required, to obtain a new trade-off between efficiency and robustness. Thus, the proposed techniques in this chapter are referred to be an adaptive control technique based on new optimization methods such as Jaya, Practical Swarm Optimization Algorithm, etc., which are used to make an on-line tuning of the LFC parameters in order to face the previous challenges in LFC. The system under study is a small microgrid with a renewable energy source and variable demand load. Digital simulation results are discussed.

Keywords: load frequency control, power system, microgrid, adaptive control, optimization method

1. Introduction

A general power system is a complex electrical network that consists of generation networks, transmission networks, and distribution networks in addition to loads that are being distributed throughout the network over a large geographical area [1, 2]. In the power system, well-designed controllers are requested during the system variations to maintain the stability of the power system as well as guarantee its reliable operation.

The growth of the industries leads to increase the complexity of the power system. System frequency depends mainly on the active power, while system voltage depends on the reactive power. So, the control viewpoint of power systems can be classified into two independent issues. One is focusing on the control of the active power along with the frequency what is called load frequency control (LFC), while the other one is to deal with the reactive power along with the voltage regulation [3].

Load frequency control of an interconnected power system means the interconnection of more than one control area through tie lines. Sudden load variation in any control area of an interconnected power system will lead to both frequency change and tie line power deviation. Large frequency fluctuations may cause sometimes what is called system blackout [4].

The main objectives of load frequency control (LFC) are [5, 6] as follows: 1—regulating frequency and tracking the load demands. 2—Ensuring zero steady-state error for frequency deviations. 3—Maintaining frequency and power interchanges with neighboring control areas at the specified values. 4—Controlling the change in tie line power between control areas. 5—Maintaining acceptable overshoot and settling time on the frequency and tie line power deviations.

1.1 Reasons for limiting frequency deviations

Frequency deviations should be within restricted limits for some reasons [7]:

i. To keep the three phase AC machines, in which its running speed relates proportionally to the supply frequency.

ii. To keep the turbine's blades that are designed to operate at a particular speed, but the change of supply frequency will cause variation in this speed. This speed change may cause damage of the turbine blades.

iii. When frequency goes below rated frequency at the case of constant system voltage, then the flux in the core increases, and then the transformer core goes into the saturation region.

iv. The frequency error may affect negatively on the digital storage and retrieval process.

1.2 Load frequency control (LFC) problems

If it's not required to maintain the frequency constant, then the system frequency and speed change with the characteristics of the governor as the load changes, and the operator is not required to change the setting of the generator. On the other hand, if constant frequency is required, the operator can adjust the velocity of the turbine by changing the characteristics of the governor.

Most published research in this field neglects uncertainties [8] and practical constraints [9] and furthermore, suggest complex control structures with impractical frameworks, which may have some difficulties when implementing in real-time applications [10].

As a result of considerable degree of interconnection, the presence of technical and economic constraints, and the traditional requirements of system reliability and security, operating the power system in the new environment will certainly be more complex than in the past. At present, simple and classical tuned controllers are widely used in the power system LFC task. Existing LFC system parameters are usually tuned based on experiences, classical methods, and trial and error approaches, and they are incapable of providing good dynamical performance over a wide range of operating conditions and various load scenarios. Therefore, the novel modeling and control approaches are strongly required, to obtain a new trade-off between efficiency and robustness.

Thus, this chapter presents an adaptive control technique that uses new optimization methods to make an on-line tuning of the LFC parameters to deal with both load demand changes and fluctuations resulted from renewable energy sources.

2. System under study

Recently, remote off-grid MGs have been widely developed especially for rural and distant areas, in which providing electrical energy from the main utility grid is costly and has destructive environmental effects. There are many real MGs installed for providing the electrical energy for distant areas [11–14].

Low inertia and dynamic complexity are the most important challenges in the MGs. Therefore, if a mismatch between the load and power generation occurs, the MG frequency deviation is inevitable. Therefore, it seems that the robust control design strategies can be considered as powerful solutions to achieve robust performance and stability [15–16]. Several optimization techniques have been proposed by researchers to tune the control parameters using simulation of the entire system and to damp the frequency fluctuation [17] such as optimization of controller parameters [18–19].

The system used in this chapter is a microgrid power system shown in **Figure 1** consisting of a 20 MW diesel generator and 17 MW load. The nominal parameters of the system are listed in **Table 1**. The simulation results have two scenarios.

The dynamic model of the proposed microgrid power system can be described in the following equations [5]:

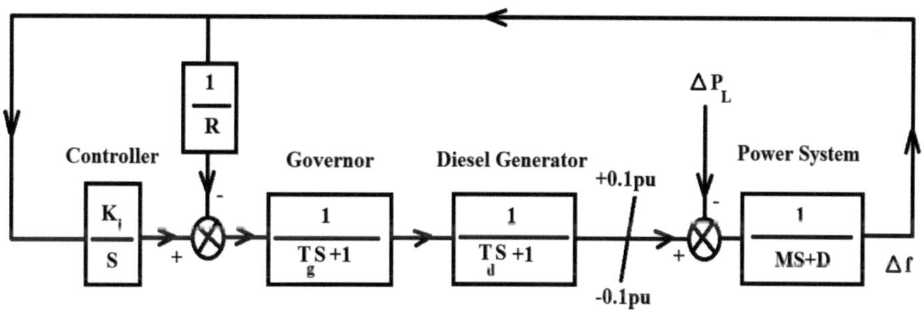

Figure 1.
Block diagram of the system under study.

K_i	D (pu/Hz)	M (pu.sec)	R (Hz/pu)	T_g (sec)	T_d (sec)
-0.3	0.015	0.08335	3	0.08	0.4

Table 1.
Data of the microgrid power system [5].

$$\Delta\dot{f} = \left(\frac{1}{M}\right).\Delta P_d - \left(\frac{1}{M}\right).\Delta P_L - \left(\frac{D}{M}\right).\Delta f \tag{1}$$

$$\Delta\dot{P}_d = \left(\frac{1}{T_t}\right).\Delta P_g - \left(\frac{1}{T_t}\right).\Delta P_d \tag{2}$$

$$\Delta\dot{P}_g = \left(\frac{1}{T_g}\right).\Delta P_c - \left(\frac{1}{R.T_g}\right).\Delta f - \left(\frac{1}{T_g}\right).\Delta P_g \tag{3}$$

and ($\Delta\dot{f}$, $\Delta\dot{P}_d$, $\Delta\dot{P}_g$) equal to $\left(\frac{df}{dt}, \frac{dP_d}{dt}, \frac{dP_g}{dt}\right)$, respectively.

3. Optimization techniques

3.1 Particle swarm optimization (PSO)

PSO is one of the famous optimization techniques. It has been derived from the social-psychological theory. PSO has some features such as:

PSO is basically developed through simulation of the bird flocking in two-dimensional space. The position of each particle is represented by XY axis position, and the velocity is represented by Vx and Vy. The position and velocity information will guide the modification of the particle position. Bird flocking optimizes a certain objective function. Each particle knows both of its XY position and its best value (P_{best}). Each particle knows the best value so far in the group (g_{best}) among (P_{best}).

This information is analogy of knowledge of how the other particles around them have performed. Namely, each particle tries to modify its position using the following information [20–22]:

- The distance between the current position and P_{best} and the distance between the current position and g_{best}.
- The current positions (x, y) and the current velocities (Vx, Vy).

Velocity of each particle can be modified by the following equation:

$$V_i^{k+1} = wV_i^k + c_1 * rand_1\left(P_{best,i}^k - s_i^k\right) - c_2 * rand_2\left(g_{best,i}^k - S_i^k\right) \quad (4)$$

where

S_i^k	current position of particle i at iteration k
V_i^k	velocity of particle i at iteration k
$p_{best,i}^k$	personal best of ith particle at iteration k
$g_{best,i}^k$	global best of ith particle at iteration k.
c1, c2	social parameters
w	the inertia weight used to accelerate the obtaining of the global best solution in the search space.
rand1, rand2	positive random numbers drawn form a uniform distribution between [0,1].

The inertia weighting function is utilized as follows:

$$w = \frac{w_{\max} - w_{\min}}{iter_{\max}} \times Iter \quad (5)$$

where:

$w_{\max}$	Initial velocity
$w_{\min}$	Final velocity
$iter_{\max}$	Maximum iteration number

Using Eq. (4) and Eq. (5), a certain velocity (which gradually gets close to (P_{best} and g_{best})) can be calculated, and the current position can be modified by the following equation:

$$S_i^{k+1} = S_i^k + V_i^{k+1} \tag{6}$$

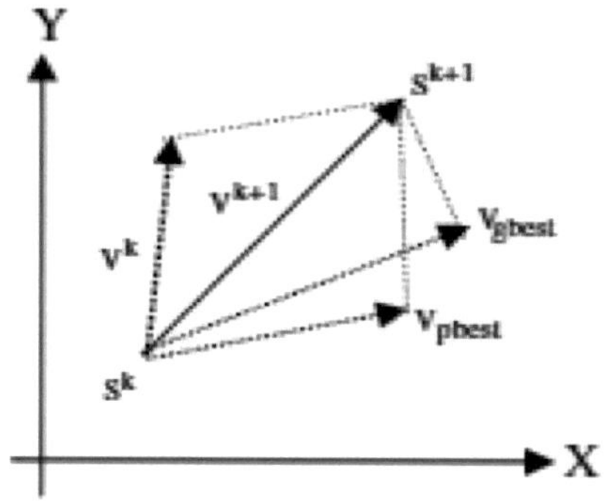

Figure 2.
PSO's concept of searching point.

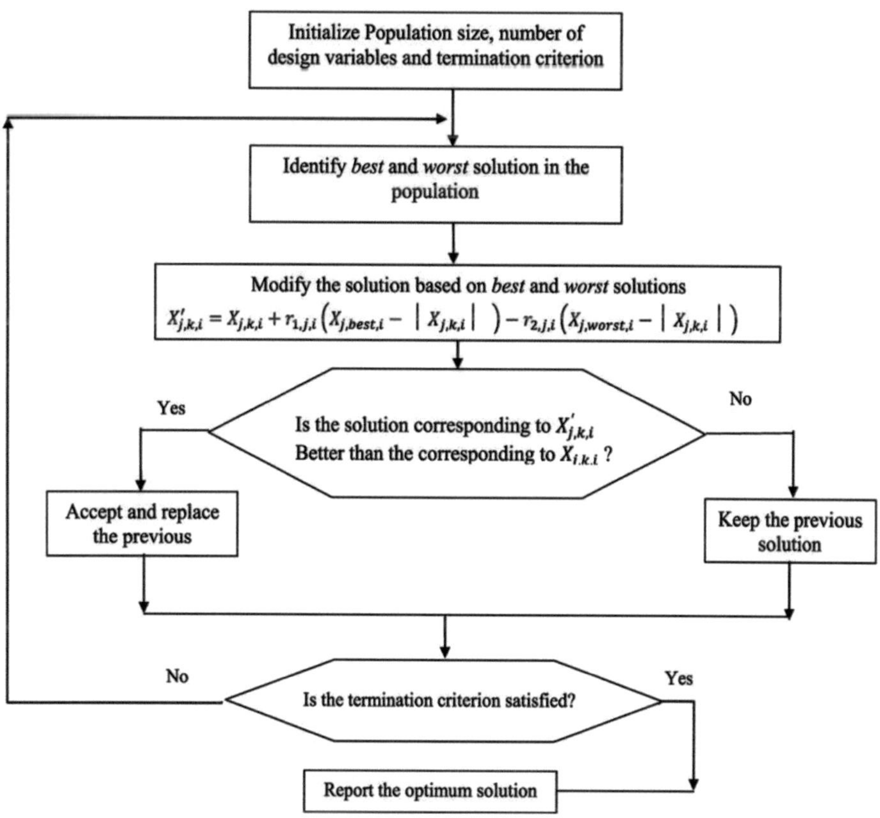

Figure 3.
Flow chart of Jaya method [10].

where

S_i^{k+1}: modified position of particle i at iteration k.

V_i^{k+1}: modified velocity of particle i at iteration k.

Figure 2 illustrates the modification concept of searching point by PSO.

3.2 Jaya optimization method

In2015, Venkata Rao has presented as a new optimization algorithm. One of the main advantages of Jaya is that there is no need to tune of its parameters. There are similarities between Jaya and the Teaching-Learning-Based Optimization (TLBO) [22].

The idea of Jaya is pushing the problem to move towards the best solution and avoid moving towards the worst solution. The flowchart illustrated in **Figure 3** shows the work procedures of the Jaya algorithm.

The advantages of Jaya algorithm can be concluded as follows [23, 24]:

1. It does not contain the problem of the selection of algorithm-specific control parameters.

2. It can solve unconstraint and constraint problems.

3. It is suitable for discrete optimization problems.

4. Ease of solving.

5. Jaya algorithm has a victorious nature, and this leads it to be more powerful.

4. Simplified microgrid model for optimization methods

It will be more effective to build the objective function of the optimization algorithm, if the total transfer function of the controlled system is in standard second order form, so it will be easy to use the standard parameters such as natural frequency, settling time, and rise time $\omega_n, T_s, and\, T_r$.

A simplified microgrid model shown in **Figure 4** is applied to drive the standard second order transfer function

$$T.F = \frac{\omega_n^2}{S^{2+2\eta\omega_n S+\omega_n^2}} = \frac{\left(k_i/M\right)}{S^{2+\left(R/M.(1+R)\right)S+\left(k_i/M\right)}} \tag{7}$$

From this transfer function, the parameters $\omega_n, T_s, and\ T_r$ can be calculated. These parameters can be applied in the objective function of the optimization methods.

$$J = \omega_n + T_s + T_r \tag{8}$$

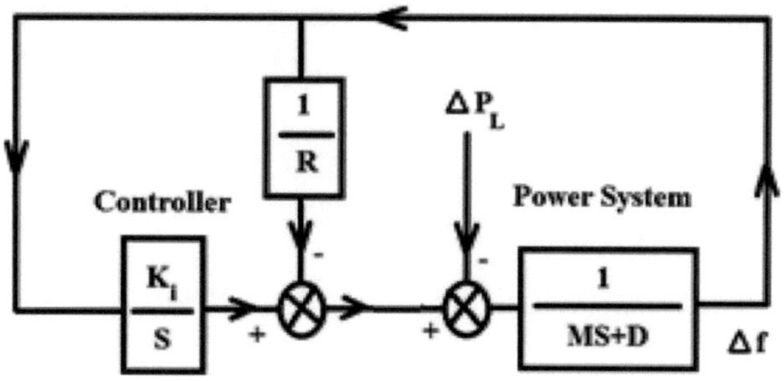

Figure 4.
Block diagram of simplified microgrid model.

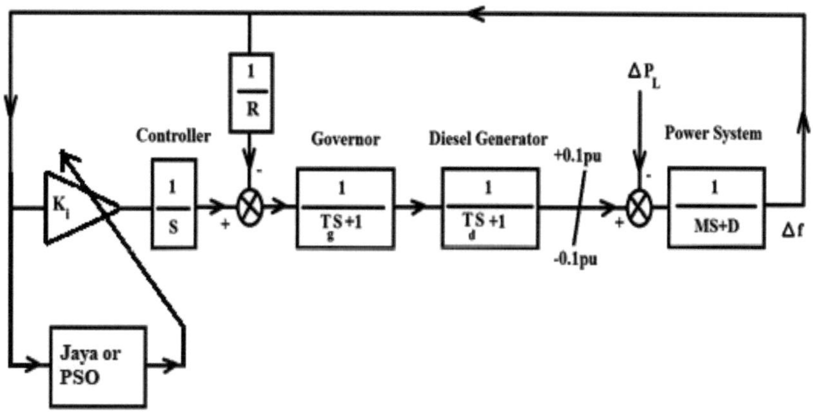

Figure 5.
Block diagram of the adaptive LFC system.

5. Adaptive load frequency controller-based optimization techniques

Figure 5 illustrates the block diagram of an adaptive load frequency controller of micro-grid power system. In this technique, optimization methods such as Jaya and PSO are used to make on-line tuning of the gain of the area controller.

6. Simulation results

6.1 Adaptive LFC-based PSO

Figure 6 shows the system response of the system with adaptive LFC controller-based PSO in case of step load demand ($\ddot{A}P_L = 0.02\ pu$. At t = 3 s). Both frequency and diesel power are illustrated in the figure. It can be noted clearly that the adaptive controller could improve the system responses compared with fixed parameters controller. **Figure 7** shows the output of PSO.

6.2 Adaptive LFC-based Jaya algorithm

Figure 8 illustrates the system response of the system with adaptive LFC controller-based Jaya algorithm in case of the same step load demand. The figure supported that the system with the adaptive controller gives more robust performance compared with the system with fixed parameters controller. Also, **Figure 9** shows the output of Jaya.

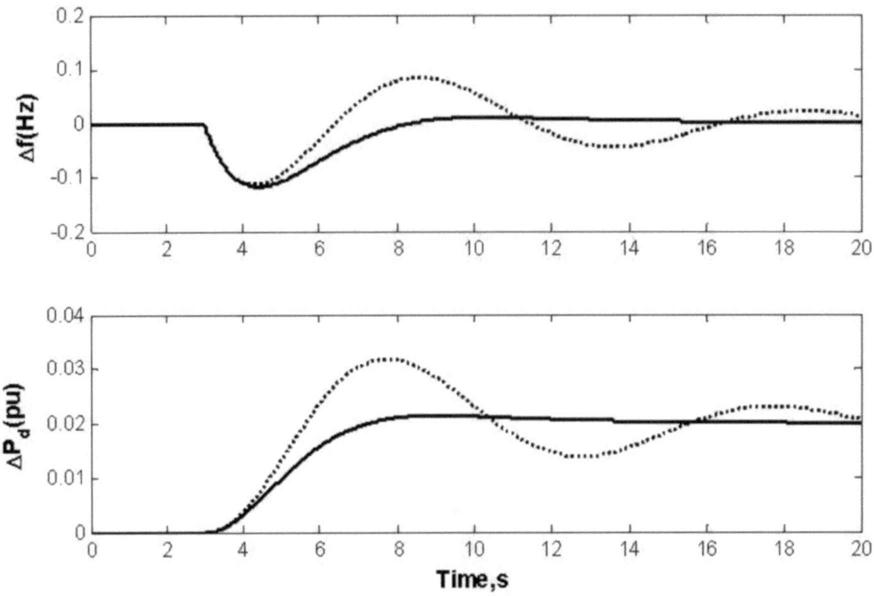

Figure 6.
System response using adaptive LFC-based PSO (...... fixed parameters controller —— adaptive controller-based PSO).

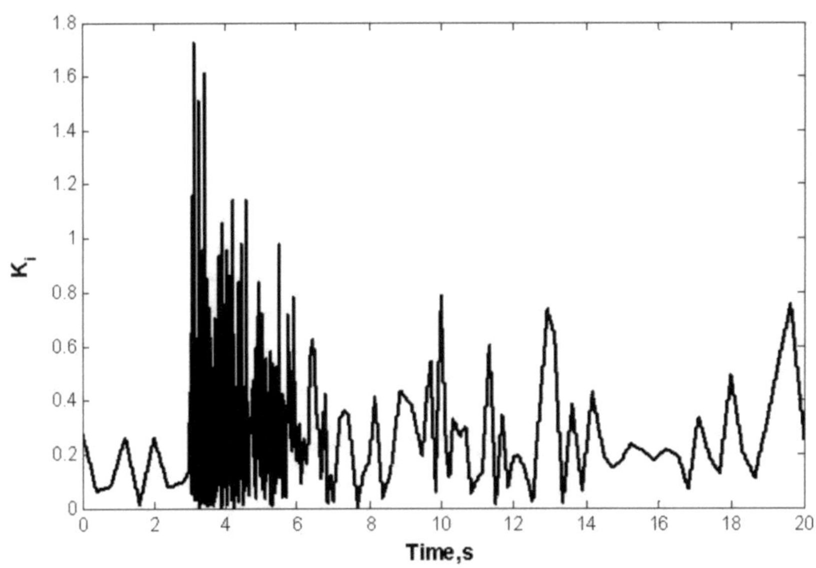

Figure 7.
Tuned controller gain using PSO.

6.3 Comparing Jaya algorithm with PSO

Table 2 presents a comparative performance analysis of Jaya algorithm with PSO. Each test function used the same number of iterations and population. PSO parameters are stated in **Table 3**. **Table 2** shows that the Jaya technique can give good speed convergence characteristics as compared to PSO.

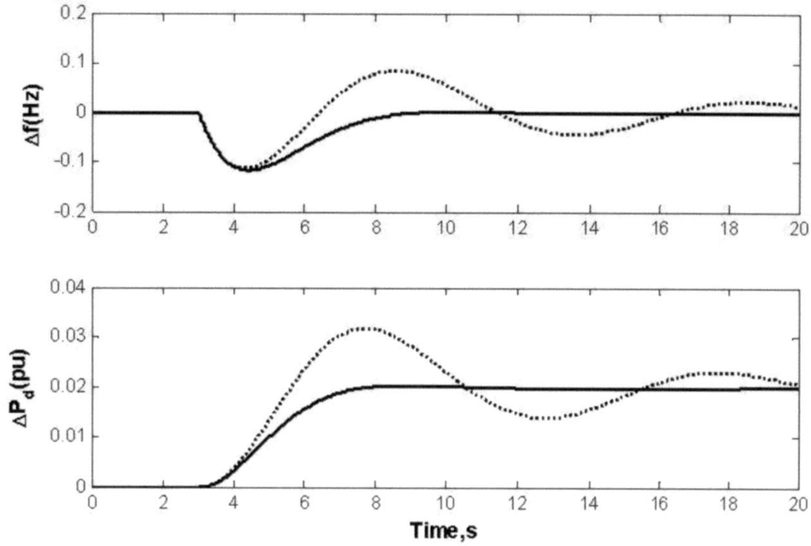

Figure 8.
System response using adaptive LFC-based PSO (...... fixed parameters controller —— adaptive controller-based Jaya).

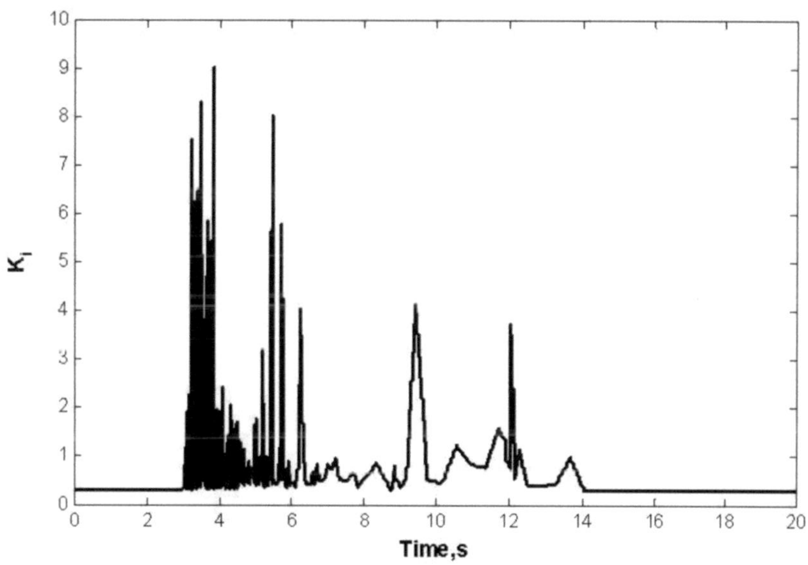

Figure 9.
Tuned controller gain using Jaya.

Functions	D[a]	Search space	Statistical values	PSO	Jaya
Ackley	5	[-10,10]	Best	3.85e-05	2.70e-16
			Worst	3.88e-05	2.70e-16
Sphere	5	[-10,10]	Best	1.79e-08	4.77e-14
			Worst	1.82e-08	4.77e-14

Table 2.
Comparative performance indexes of different test functions.

Parameters	Values
Swarm size	50
Inertia weight (w)	1
Inertia weight damping ratio (w_{damp})	0.99
Personal learning coefficient (C1)	1.5
Global learning coefficient (C2)	2.0

Table 3.
Algorithm-specific parameters values for PSO.

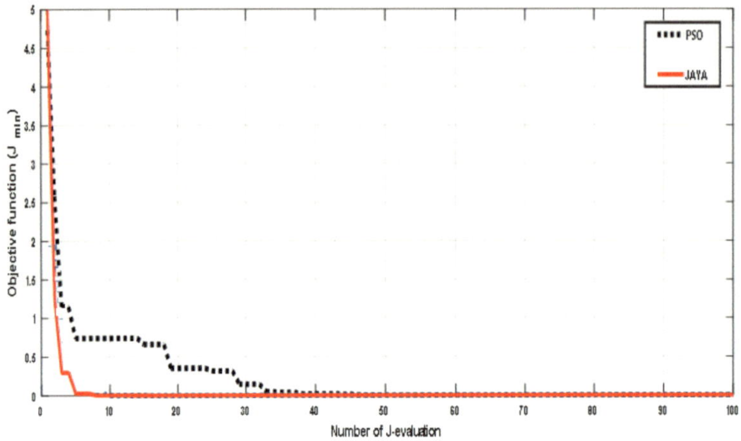

Figure 10.
Convergence characteristics for Matyas function.

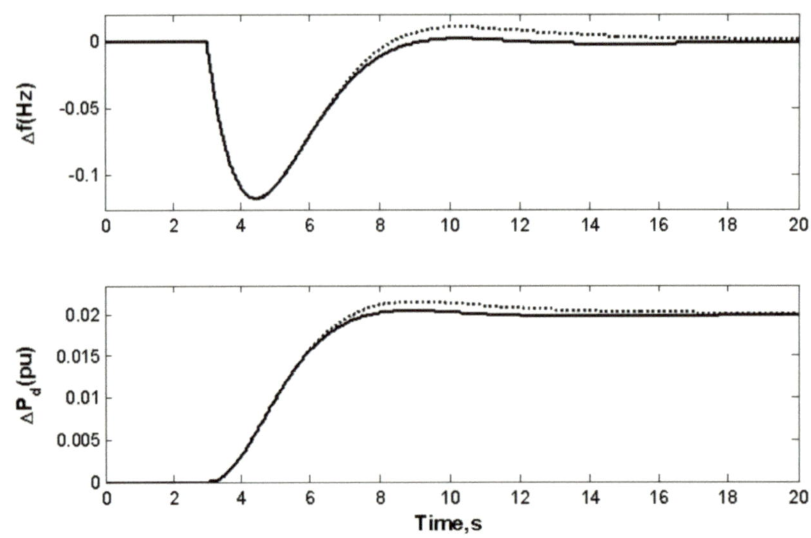

Figure 11.
System response using adaptive LFC-based PSO (...... adaptive controller-based PSO —— adaptive controller-based Jaya).

Figure 10 shows the value of the objective function (J_{min}) with the number of J-evaluation in case of using Matyas function for 50 population size. It can be noted that the Jaya technique converges relatively faster than PSO.

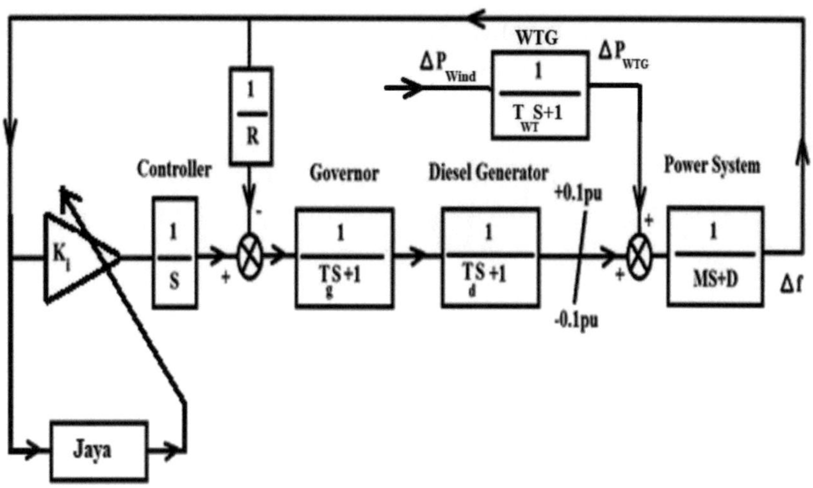

Figure 12.
Block diagram of the adaptive LFC system in the presence of a wind energy source.

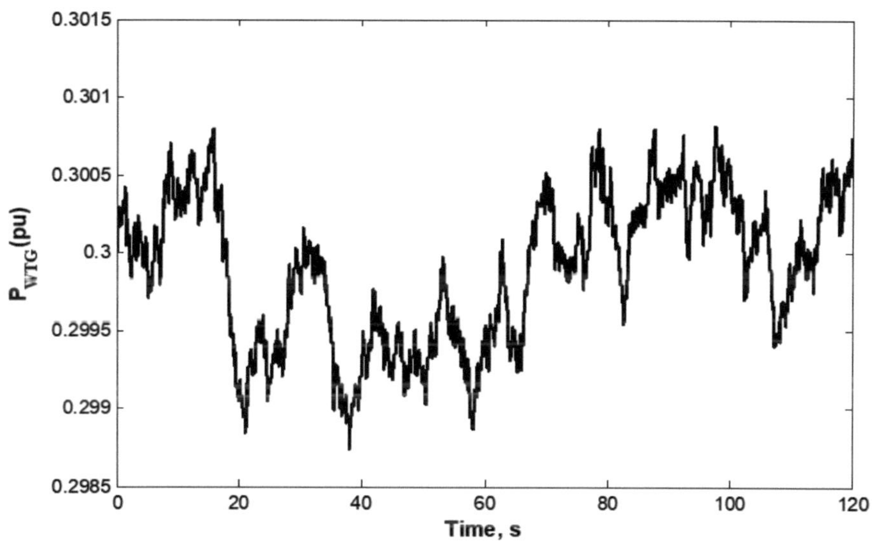

Figure 13.
Output power of the wind turbine.

Figure 11 illustrates a comparison between Jaya and PSO using the proposed objective function presented in Eq. (8). The figure indicates that system with adaptive controller tuned by the Jaya optimization method gives good response compared with the system with adaptive controller tuned by PSO. It could minimize the overshoot and the settling time.

6.4 Case of presence of renewable energy source

In this case of study, the system with proposed controller has been tested under fluctuation resulted from renewable power generation such as power generated from wind turbine as shown in **Figure 12**. The simplified dynamic model of the wind turbine is presented in the following transfer function:

$$\Delta P_{WTG} = \left(\frac{1}{T_{WT}S + 1}\right)\Delta P_{wind} \tag{9}$$

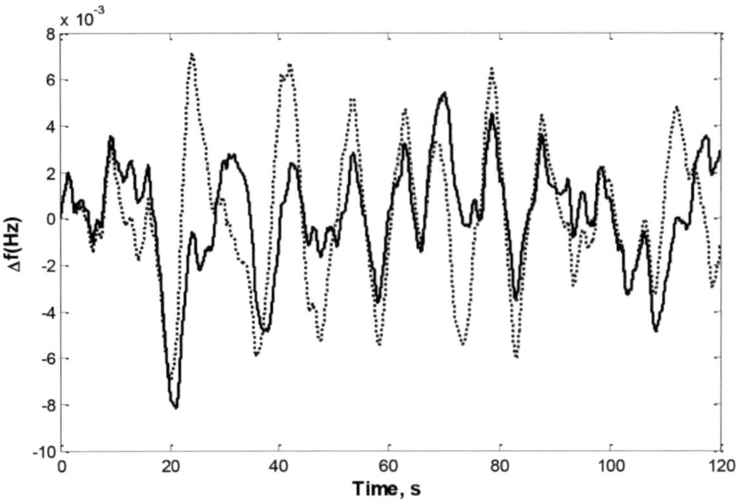

Figure 14.
Frequency response in case of presence of the wind energy source (...... fixed parameters controller —— adaptive controller-based Jaya).

Figure 13 illustrates the power of the wind turbine.

The system with proposed controller tuned by Jaya algorithm has been compared with the system with conventional fixed parameter controller, and the result is shown in **Figure 14**. this result supports the efficiency of the controller with Jaya optimization in dealing with frequency variation that resulted from the wind energy source.

7. Conclusions

This chapter presents an adaptive load frequency controller in a microgrid power system. The gain of the proposed controller is tuned by optimization techniques. The system under study consists of a microgrid with a 20 MW diesel generator and 17 MW demand load. PSO and Jaya optimization algorithms have been used to tune the gain of the system controller. The system with Jaya has been compared to the system with PSO and the system with fixed controller parameters in the case of step load change. Simulation results indicated that the system with Jaya optimization can give the best performance at the moment of step load demand. In addition, the system with Jaya algorithm has been compared to the system with a conventional controller in case of frequency fluctuations resulting from a wind energy source. Digital simulations supported the superiority of the Jaya optimization method.

Acknowledgements

The authors want to thank both Professor Gaber Shabib and Dr. Mahmoud M. Hussein for their efforts to finish this chapter.

Author details

Tarek Hassan Mohamed*, Hussein Abubakr, Mahmoud M. Hussein and Gaber S. Salman
Faculty of Energy Engineering, Aswan University, Aswan, Egypt

*Address all correspondence to: tarekhie@aswu.edu.eg

References

[1] Kundur P. Power System Stability and Control. New York: McGraw-Hill Education; 1994

[2] Zhang Y, Dong L, Gao Z. Load frequency control for multiple-area power systems. In: American Control Conference, Louis, MO, USA. 2009

[3] Kothari DP, Nagrath IJ. Modern Power System Analysis. 3rd ed. New Delhi: Tata McGraw-Hill; 2003

[4] Kirby BJ, Dyer J, Shoureshi DRA, Guttromson R, Dagle J. Frequency Control Concerns In the North American Electric Power System. USA: SciTech Connect; 2003

[5] Bevrani H. Robust Power System Frequency Control. New York: Springer; 2009. pp. 15-61. DOI: 10.1007/978-0-387-84878-5

[6] Mukta, Surjan BS. Load frequency control of interconnected power system in deregulated environment: A literature review. International Journal of Engineering and Advanced Technology (IJEAT). 2013;**2**(3):435-441

[7] Wadhwa CL. Electrical Power System. New Delhi: New Age International Publisher; 2010

[8] Wen T. Load frequency control: Problems and solutions. In: The 30th Chinese IEEE Control Conference (CCC), Yantai. 2011

[9] Ohba S, Ohnishi H, Iwamoto, Shinichi. An advanced LFC design considering parameter uncertainties in power systems. In: The 39th North American IEEE conference on Power Symposium, Las Cruces, NM. 2007. DOI: 10.1109/NAPS.2007.4402376

[10] Jaleeli N, Ewart DN, Fink LH, et al. Understanding automatic generation control. IEEE Transactions on Power Systems. 1992;**7**(3):1106-1122. DOI: 10.1109/59.207324

[11] Al-Saedi W, Lachowicz SW, Habibi D, Bass O. Power quality enhancement in autonomous microgrid operation using particle swarm optimization. International Journal of Electrical Power & Energy Systems. 2012;**42**(1):139-149. DOI: 10.1155/2014/342019

[12] Dekker J, Nthontho M, Chowdhury S, Chowdhury SP. Economic analysis of PV/diesel hybrid power systems in different climatic zones of South Africa. International Journal of Electrical Power & Energy Systems. 2012;**40**(1):104-112. DOI: 10.1155/2014/626251

[13] Kamalapur GD, Udaykumar RY. Rural electrification in India and feasibility of photovoltaic solar home systems. International Journal of Electrical Power & Energy Systems. 2011;**33**(3):594-599. DOI: 10.1016/j.ijepes.2010.12.014

[14] Alzola JA. Microgrids project, Part 2: Design of an electrification kit with high content of renewable energy sources in Senegal. Renewable Energy. 2009;**34**(10):2151-2159. DOI: 10.1016/j.renene.2009.01.013

[15] Gu W, Liu W, Wu Z, Zhao B, Chen W. Cooperative control to enhance the frequency stability of islanded microgrids with DFIG-SMES. Energies. 2013;**6**(8):3951-3971. DOI: 10.3390/en6083951

[16] Serban I, Marinescu C. Aggregate load-frequency control of a wind-hydro autonomous microgrid. Renewable Energy. 2011;**36**(12):3345-3354. DOI: 10.1016/j.renene.2011.05.012

[17] Lei Y, Lin X, Zhu Y. Passivity-based control strategy for SMES under an

unbalanced voltage condition. IEEE Access. 2018;**6**:28768-28776. DOI: 10.1109/ACCESS.2018.2831251

[18] Wang Y, Zhou R, Wen C. Robust load-frequency controller design for power systems. In: IEE Proceedings C (Generation, Transmission and Distribution). 1993. pp. 11-16

[19] Bohn E, Miniesy SM. Optimum load-frequency sampled-data control with randomly varying system disturbances. IEEE Transactions on Power Apparatus and Systems. 1972:1916-1923. DOI: 10.1109/TPAS.1972.293519

[20] Kennedy J, Eberhart R. Particle swarm optimization. In: Proceedings of the IEEE International Conference on Neural Networks, Perth: Australia. vol. 4. 1995. pp. 1942-1948

[21] Shi Y, Eberhart R. A modified particle swarm optimizer. In: Proceedings of the IEEE International Conference on Evolutionary Computation, Anchorage, UK. 1998. pp. 69-73

[22] Yoshida H, Kawata K, Fukuyama Y. A particle swarm optimization for reactive power and voltage control considering voltage security assessment. IEEE Transactions on Power Systems. 2000;**15**:1232-1239. DOI: 10.1109/59.898095

[23] Venkata Rao R. Jaya: A simple and new optimization algorithm for solving constrained and unconstrained optimization problems. International Journal of Industrial Engineering Computations. 2016;7:19-34. DOI: 10.5267/j.ijiec.2015.8.004

[24] Rao R, Rai D, Ramkumar J, Balic J. A new multi-objective Jaya algorithm for optimization of modern machining processes. Advances in Production Engineering & Management. December 2016;**1**(4):271-286. DOI: 10.14743/apem2016.4.226

Chapter 11

Modeling the Hidden Risk of Polyethylene Contaminants within the Supply Chain

Gladys Bonilla-Enríquez, Patricia Cano-Olivos,
José-Luis Martínez-Flores, Diana Sánchez-Partida
and Santiago-Omar Caballero-Morales

Abstract

Inventory management is very important to support the supply chain of the manufacturing and service industries. All inventories involve warehousing; however, most of the products and packages are associated to plastic which is the main generator of polyethylene (phthalate) pollution in the air and water resources. In fact, phthalate has been identified as the cause of serious health conditions and its impact within the operation of logistic processes has not been studied. In this work, we perform research on the generation of phthalate as the control on these emissions is important to adjust the supply strategy to reduce the human risk exposure and contamination of the environment. For this purpose, generation of phthalate is modeled through the use of artificial neural networks (ANNs) and its impact on the supply strategy is assessed through its integration within a stochastic inventory control model. As presented, it is possible to adjust the supply strategy to reduce the cumulative generation of phthalate within the warehouse and thus reduce its impact on human health and environment sustainability.

Keywords: sustainability, phthalate contamination, inventory control, supply strategy, artificial neural networks

1. Introduction

An important aspect to consider for sustainable proposals is the growth of the world population, which is projected to increase from 7 to 9 billion people by the year 2050 [1, 2]. This is a challenge for companies to comply with economic, environmental, and social needs.

As presented in **Figure 1**, to address economic, environmental, and social issues, companies must address multidisciplinary issues such as follows:

- sustainable development (SD) which emphasizes the balance between economic well-being, natural resources, and society without compromising the quality of life of the human population [3];

- supply chain (SC) management, which is focused on optimizing the flow of goods and services through the supply chain, considering the procurement of

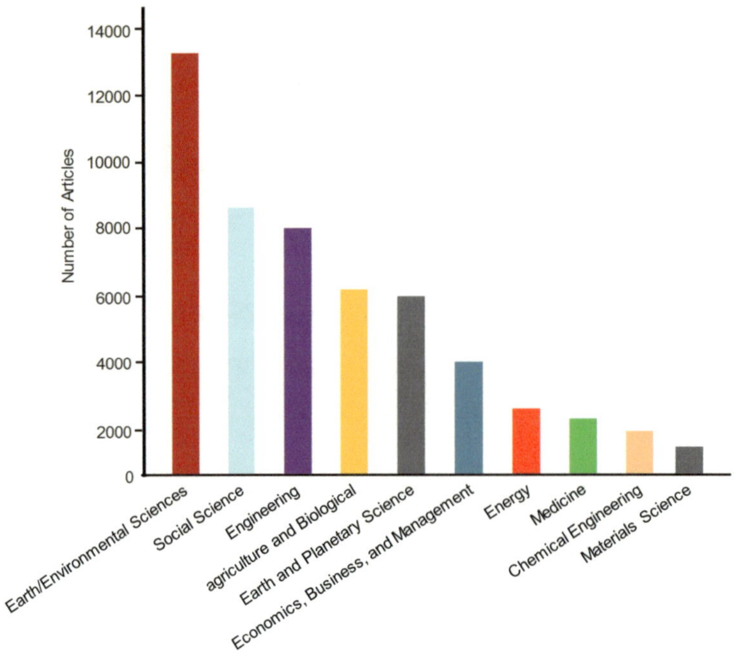

Figure 1.
Interdisciplinary studies in sustainability (adapted and edited from Ref. [4]).

raw materials, the distribution from suppliers to manufacturers, the transformation into final products and warehousing operations, and the distribution of final products from manufacturers to retailers. This to actively streamline the company's supply-side activities to maximize customer value and gain a competitive advantage in the marketplace.

- inventory control and management, which is focused on determining the appropriate inventory replenishment strategies to ensure efficient supply and distribution of products when needed at the minimum cost.

Here, it is important to observe that different storage/warehousing is performed through all the SC. Just in the last decade, health risk was identified for people who work at storage facilities due to the presence of semi-volatile organic compound pollutants (SVOC) and plastic contaminants which are generated by the stored inventory [5–7]. In this regard, the most abundant SVOCs found among the 58 classified SVOCs are phthalates.

In general, there are six types of phthalate that have been found in outdoor and indoor air and surfaces [6, 8]. Phthalates are distributed worldwide, having a global presence ranging from the most remote regions in the Arctic to isolated rainforests of the Amazon [8, 9]. On March 2019, the United States Environmental Protection Agency (EPA) issued a priority list of 40 chemicals to determine if they are of high or low risk for human health. Phthalates were considered within the "high priority" list [6].

There are studies that have concluded that the pollution of phthalate in the air is harmful to human health [8–12]. Also, it has been determined that humans are exposed through ingestion, inhalation, and dermal exposure, even since intrauterine development [13, 14].

Because phthalates, which are organic lipophilic compounds, are mainly used to increase the flexibility of plastic polymers, they are frequently used in printing inks

Country	Place	Work
Japan	Home	[19]
China	Home/office	[20]
Sweden	Pre-school	[21]
Canada	Home	[22]
USA (California)	Child care facility	[23]

Table 1.
Research works performed on indoor phthalate (taken from Ref. [8]).

and food packages [14, 15]. Thus, exposure to phthalates mainly occurs via food ingestion [10, 14, 16–18].

Inhalation is the second route of exposure as phthalates degrade into particles that diffuse through the air [10]. This can be the main route of exposure for individuals who work in plasticizing processes [10] or in closed places where products with phthalate are stored for long periods of time such as in warehouses. Because phthalates are used as plasticizers in numerous consumer products, commodities, and building materials, this compound has been found in offices, work places, homes, bathrooms, gardens, and food containers. **Table 1** presents an overview of the places where people are more exposed to indoor phthalates.

As presented in **Table 1**, phthalates are found in human residential and occupational environment in high concentrations, both in air and in dust [24]. Thus, we can consider the facilities of industries such as warehouses, productions areas, and scrap areas, to be frequently contaminated with this compound.

In the case of products which are stored during long periods of time, there are economic, environmental, and health implications on SD. If the product is not used or moved (e.g., low inventory rotation), it may deteriorate and/or become obsolete, leading to economic losses. Also, a deteriorating product may release other harmful chemicals. Finally, in the social aspect, there is the health risk for employees who are exposed to harmful chemicals generated by the stored products. If the management fails to determine the optimal inventory levels and lots, the environmental, social, and economic risks can affect all entities through the SC.

To extend on these findings, we perform an updated review of the presence of phthalates and their effect on human health. Also, we extend on the adaptation of supply strategies to reduce these effects through the SC and on the environment. This is performed through the modeling of phthalate generation and integration within a stochastic inventory control strategy.

2. Pollution related to phthalates

Phthalates are chemicals which are produced in high volumes, accounting for 70% of the world consumption of plasticizers in 2014. In this context, Asia, Western Europe, and the USA accounted for 59, 14, and 16%, respectively, of the world plasticizer consumption in 2014 [25].

More recently, phthalates accounted for 65% of the world consumption of plasticizers in 2017. **Figure 2** presents the main consumers of plasticizers in 2017. However, in 2005, this amount was higher (approximately 88%), and it was forecasted to decrease to 60% by 2022. This decrease was defined to be caused by [25]:

- rapid consumption growth of non-phthalate plasticizers, mainly terephthalates, epoxy, aliphatics, and benzoates, as replacements for DEHP and other phthalates such as DINP and BBP;
- continued growth of non-phthalates in different applications and markets; and
- ongoing pressure from retailers and consumers to limit the use of phthalates, especially in developed regions.

However, consumption of phthalate plasticizers has been also forecasted to grow at an average annual rate of 1.3% during 2017–2022. Today, phthalates are known pollutants, which can affect human health. Human bio-monitoring studies from 2000 to 2015 have determined that exposure to phthalates can cause adverse health outcomes like fertility problems, respiratory diseases, childhood obesity, and neuropsychological disorders [9, 12, 15, 26]. Other studies found that it may disrupt fetal testicular testosterone production [27, 28].

Although many studies have researched on the impact of outdoor pollution on human health, few studies have investigated the impact of indoor pollution on the human health. Because people spend most of their time indoors, it is crucial to understand how an indoor pollutant, including household dust, affects human health [19, 29].

Indoor pollution is largely influenced by outdoor sources, but indoor activities (e.g., cooking, cleaning, and the use of consumer products and building materials) are also sources of indoor pollution [5, 11, 20, 23, 30, 31]. Phthalate levels build up over time in indoor environments where their main sources like children's toys, cosmetics, flexible PVC flooring, and cable insulation among others are found [8, 31].

Consequently, six phthalates, namely, dimethyl phthalate (DMP), diethyl phthalate (DEP), di-n-butyl phthalate (DBP), butyl-benzyl phthalate (BBP), di(2-ethylhexyl) phthalate (DEHP), and di-n-octyl phthalate (DnOP) have been identified as priority pollutants by the United States Environmental Protection Agency

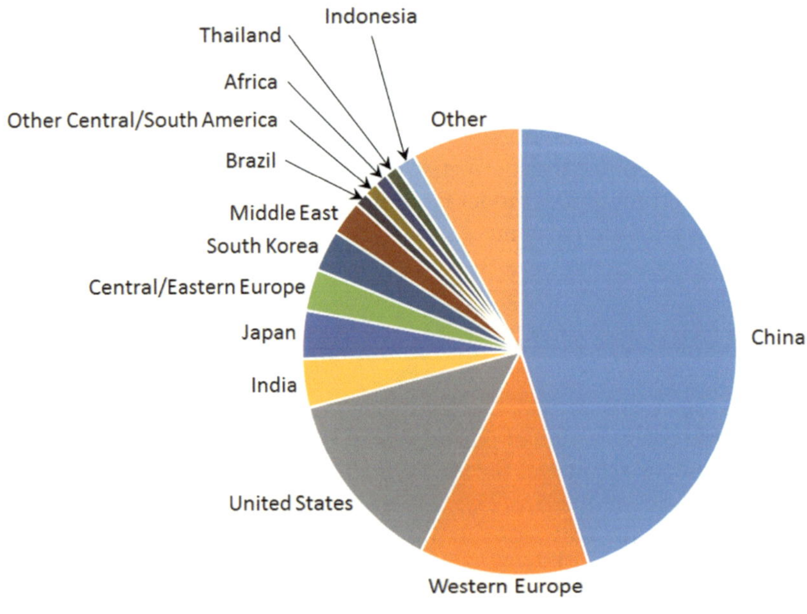

Figure 2.
World consumption of plasticizers 2017 (adapted and edited from Ref. [25]).

(US EPA) and the European Union (EU). Usage of DEHP, DBP, BBP and DnOP has been limited to ≤0.1% in toys and childcare articles by EU (Directive 2005/84/EC), US (CPSIA—Consumer Product Safety Improvement Act of 2008), China (China National Standard GB 6675, 2014), India (BIS, 2011), and Japan (Japan Toy Safety Standard ST-2002 Part 3, 2011). Recently in 2015, DEHP, DBP, and BBP were classified as reproductive toxicant category 1B and completely banned from any application without prior approval in the EU.

To compare two exposure scenarios, different dust particle fractions were analyzed: inhaled (<5 μm) and ingested (<75 μm) fraction sizes. Results showed that the daily intake of dust-contaminated phthalate was 2 to 12 times (inhalation and ingestion, respectively) higher for 2-year-old children than for adults [11].

However, phthalate exposure (phthalate metabolite levels in urine) among countries indicates the highest exposure for people living in Europe (2.1×10^2 μg/l) closely followed by USA (2.0×10^2 μg/l) and least in Asia (1.3×10^2 μg/l) [26]. In this context, there are reported discrepancies between trends of industrial consumption and human exposure [8, 14, 18].

The highest concentrations of phthalates in different items have been found in the range of 300–461 g/kg for DEHP, 283–345 g/kg for DBP, 150 g/kg for DnOP, and 20–33 g/kg for BBP in floorings, shower curtains, gloves, plastic sandals, plastic balls, and soap packaging [8, 18].

Particularly in Latin America, a phthalate presence study was carried out on beverages, where the results brought that the bottles contained average 2.62 g/kg of diethylhexyl (FDEH) [10, 32].

Increased phthalate levels have been found in the presence of temperature changes (i.e., bottled water exposed to higher than 35°C or sunlight) [31, 33–35]. Elevated temperatures considered in various studies do not represent ambient temperature but become important in case of heating of for example food in products containing phthalates. However, the influence of temperature on phthalate emissions in dust requires further investigation.

Also, higher relative humidity has been reported to increase hydrolysis of phthalates, which results in a gradual decrease in concentration of phthalates in the source and a sink [8]. An increase in temperature increases the emission rates of non-covalently bound phthalates from their polymer matrices resulting in a higher concentration in warmer months [30]. In good agreement, studies have found higher phthalate levels in summer in indoor as well as outdoor surfaces [8, 9].

These studies have been carried out in houses, kinder gardens, and offices in places where cleaning is regular [8, 15]. Hence, in closed places where products with this chemical are stored, the toxicity risk is even higher. Thus, there is a necessity to reduce the effect of this chemical by storing just the optimal lots.

3. Where the phthalates are within logistic facilities and what can we do?

Within all industries, management is focused on solving the essential decision-making associated to product design/placement, organization, picking operations, facility layout, and distribution. These operations take place within spaces or logistic facilities such as offices, workshops, and warehouses, where workers spend most of their time (40–60 hours per week). In these closed places, airflow distributes many impurities, including phthalates which can be ingested or inhaled [9, 14, 15, 32].

The importance of airflow has been studied when designing the warehouses because phthalates are not chemically bound to the plastics and they can leach into

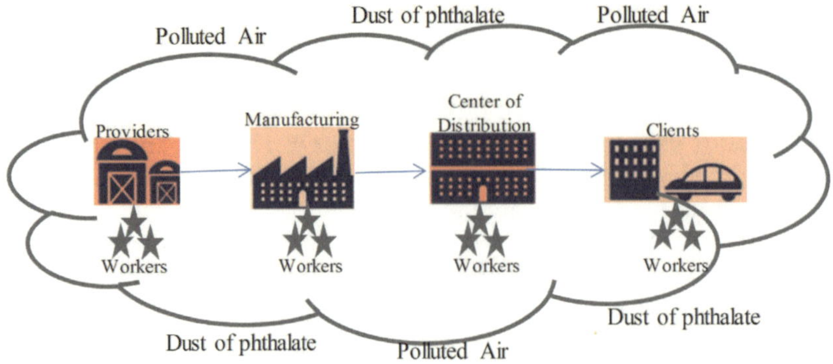

Figure 3.
Theoretical exposure of phthalates through the air (own work).

water, indoor dust, and air, resulting in cyclic exposure [14, 36]. In these closed spaces, the stored products can spread out this pollutant which may be increased based on temperature conditions. In this situation, co-workers inhale phthalates during labor time [14, 15] and contamination by phthalates (more than 100 mg/l) has been reported to be common in heavily industrialized areas.

Additional to airflow, inventory movement is another aspect to contribute to phthalate accumulation. As inventory commonly stores packaged products (bottles, bags, paper and board packaging, etc.) if it remains static, it can generate more dust and particles [9–11, 32]. Based on this information, **Figure 3** presents the theoretical cycle of emission of phthalate within the facilities present in the SC.

From this cycle of emission, we can identify that inventory movement through the facilities is dependent of the inventory turnover, which can be optimized through the use of proper supply strategies. This task requires consideration of real market conditions which are characterized by demand of products with large variability.

Thus, stochastic demand patterns are a main aspect to consider within the strategies to reduce phthalate accumulation/emission and improve inventory turnover.

4. Modeling the phthalate emission within the supply strategy

Within logistic management of inventories, there are strategies to improve inventory turnover and reduce static inventory and inventory levels. As an example, consider the cost equation of the continuous review strategy to determine the optimal lot size Q of inventory to reduce operational costs [37]:

$$E(C) = \frac{DC_o}{Q} + \frac{C_h Q}{2} + C_h[R - \mu_{LT} + \sigma_{LT} L(z)] + \frac{pAD}{Q}, \tag{1}$$

where C_o is the order cost per lot, C_h is the holding cost per unit within Q, p is the unit stock-out cost, D is the cumulative demand through a planning horizon, μ_{LT} and σ_{LT} are the mean and standard deviation of the demand during the lead time, $L(z)$ is the standard loss function with $z = \Phi^{-1}(1 - (QC_h)/(pD))$, and A is the expected stock-out units per inventory cycle ($=\sigma_{LT} L(z)$).

As emission of phthalates is associated to the size of the warehoused lot (i.e., Q), this aspect can be integrated into this strategy to determine a more appropriate lot size. For this purpose, consider that $f(t)$ is the general emission function of phthalate through time t per stored unit and H is the maximum safety level of phthalate within a closed space. In this case, the cumulative generation of phthalate through time associated to the average stored lot Q then can be estimated as follows:

$$\frac{Q}{2}\int_0^t f(t). \tag{2}$$

To determine the optimal lot size considering the minimization of phthalate, the following mathematical formulation can be defined:

$$Minimize\ E(C) = \frac{DC_o}{Q} + \frac{C_h Q}{2} + C_h[R - \mu_{LT} + \sigma_{LT} L(z)] + \frac{pAD}{Q} \tag{3}$$

Subject to:

$$\frac{Q}{2}\int_0^t f(t) \le H, \tag{4}$$

$$Q \in \Re^+, \tag{5}$$

where (3) is the objective function, (4) is the restriction to ensure that the lot size does not lead to increase the cumulative phthalate over a permissible limit H, and (5) is the restriction over Q to consist of real positive values.

Additionally, periodic surface cleaning and handwashing has been identified as appropriate measures to reduce accumulation and exposure to phthalate [38]. Particularly for co-workers, the use of protective wear within warehouses is highly recommended.

Research performed to model $f(t)$ has led to different proposals and values. This is understandable due to the different considered environments and contexts (i.e., home, office, plants, etc.). As we are concerned regarding the applicability of the model in the supply strategy, we propose a general emission model that can be adapted to different contexts by the use of artificial neural networks (ANNs).

The advantage of ANNs to model data when compared to standard regression approaches is that regression only performs well if the regression equation fits very closely the considered data. By contrast, the use of hidden neuron layers provides ANNs with more flexibility to fit any data pattern.

As input data for modeling, we considered the estimations presented by Afshari et al. and Liang et al. [39, 40] regarding phthalate concentrations (μg/m^3) generated by PVC and different materials in indoor spaces. **Figure 4** presents a review of the approximate concentration values reported in Refs. [39, 40].

As presented in **Figure 4**, significant differences are found depending on the considered material and environment. For assessment purposes of the proposed model, we consider an average concentration which is also presented in **Figure 4**.

For modeling through ANNs, it was important to match the concentration data based on m^3 to stored units in the warehouse. To accomplish this task, the following variables are considered:

V_Q = volume (m^3) associated to each product unit and E_t = cumulative emission per m^3 associated to a product unit at a time t.

Thus, (4) can be represented as:

$$\frac{Q}{2}\int_0^t f(t) \le H \rightarrow \frac{Q}{2} \times V_Q \times E_t \le H, \tag{6}$$

where E_t is obtained through the regression achieved with ANNs.

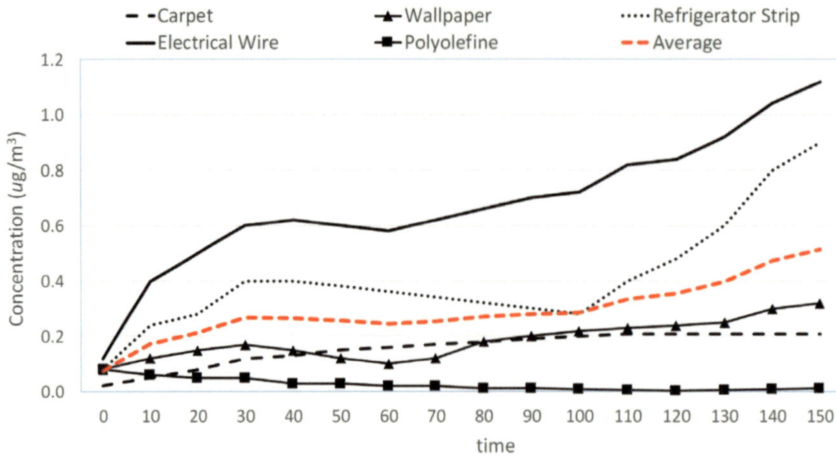

Figure 4.
Review of concentration patterns through time on different indoor spaces and materials (own work based on data reported in Refs. [39, 40]).

For this purpose, we considered the nonlinear autoregressive with external (exogenous) input (NARX) [41] time series ANN. For time series modeling, the NARX ANN can associate the current value of a time series to (a) past values of the same series and (b) current and past values of an external series that influences the series of interest [42]. Thus, it can predict a series $y(t)$ given n past values of $y(t)$ and another series $z(t)$. This is leads to more accurate modeling when compared to nonlinear input-output ANNs.

In this case, $z(t)$ is the time series (in days) and $y(t)$ is the cumulative average concentration (μg/m^3). Implementation of the NARX ANN was performed with the MATLAB R2016a software on a DELL laptop computer with Intel i7 CPU at 2.06 GHz and 8 MB RAM. **Table 2** and **Figure 5** present the details of the ANN and the training algorithm.

Figure 6 presents the comparison of the performance of the ANN for t = 0:360 days and the original average data with t = 0:150 (as presented in

Hidden layers	2
Neurons	10
Training method	Levenberg-Marquardt

Table 2.
Training details of the NARX ANN.

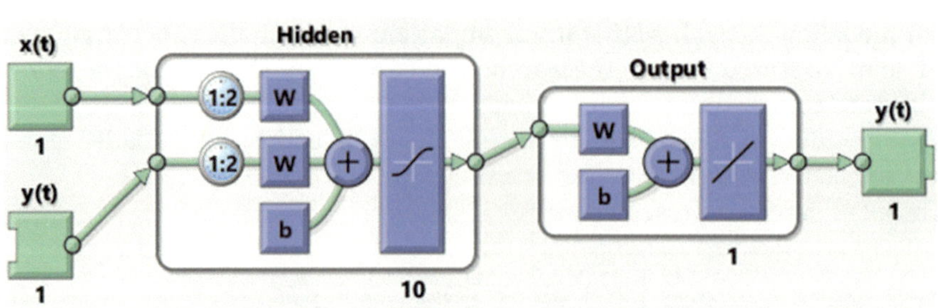

Figure 5.
Structure details of the NARX ANN.

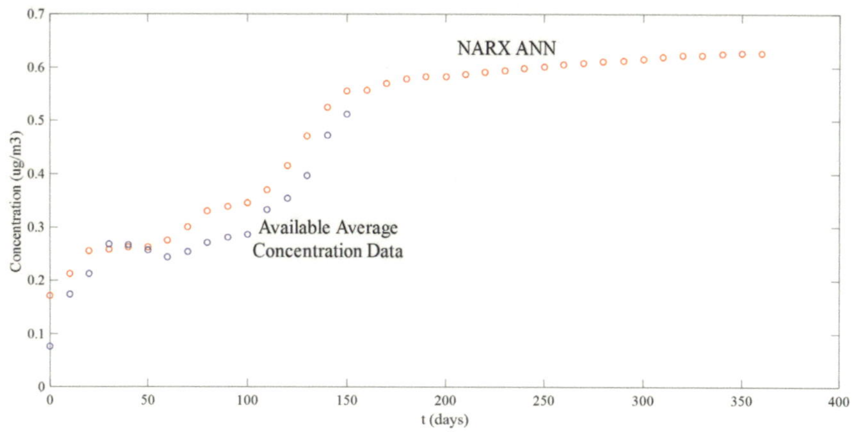

Figure 6.
Performance of the NARX ANN for extended time periods.

Figure 4). It can be observed that the prediction of the ANN closely resembles the available data used for training within the period from t = 0:150 days.

With the estimated concentrations, we can proceed to assess the model described by (3)–(6) with a supply example. This assessment considers two scenarios: (a) scenario with no control on the phthalate concentrations and (b) scenario with restriction H on the phthalate concentration per inventory cycle. These details of this assessment are presented in the following section.

5. The economic and environmental impacts of phthalate emission control within the supply strategy

In the previous section, we provided the means for phthalate concentration modeling through ANNs and its integration within an inventory supply strategy. To address the impact of this strategy, we proceed to quantitatively evaluate the economic and environmental results on an inventory supply case.

Table 3 presents the numerical data of the considered supply case (this is, the set of values for (3)–(6)). This case considers the supply strategy over a 5-year period (1800 days with a cumulative demand of 60,000 units). Each unit of product is assumed to have a standard size of 0.50 m × 0.50 m × 0.50 m = 0.125 m^3. At t = 250 days, the cumulative concentration is expected to be at 0.61 μg/m^3 and an estimate of H = 50.0 μg/m^3 is considered as a safety limit.

Planning horizon = 1800 days	D = 60,000 units
C_o = 340 USD	Daily demand = D/1800 = 60,000/1800 = 34 units
C_h = 12 USD	Daily standard deviation = 8 units
Lead time = 180 days	p = 45 USD
V_Q = 0.125 m^3	t = 250 days
H = 50.0 μg/m^3	$E250$ = 0.61 μg/m^3

Table 3.
Assessment data for the integrated model with phthalate emission.

Scenario	1	2
Objective function	$Minimize\ E(C) = \frac{DC_o}{Q} + \frac{C_h Q}{2} + C_h[R - \mu_{LT} + \sigma_{LT}L(z)] + \frac{pAD}{Q}$	
Restrictions	Q > 0	Q > 0 $\frac{Q}{2} \times V_Q \times E_t \leq H$
Q	1880	1312
R	6257	6271
$E(C)$	25,646	27,088
$\frac{Q}{2} \times V_Q \times E_t$	71.66 μg/m^3	50.00 μg/m^3

Table 4.
Results of the integrated model with the assessment data.

With these data, we proceed to evaluate two scenarios:

a. Scenario 1: determination of the supply lot size Q is performed based only on the economic aspects of (3) without the phthalate emission factor.

b. Scenario 2: determination of the supply lot size Q is performed based on the economic aspects of (3) and considering the phthalate emission factor defined by (4) and (6).

Both scenarios were solved with the Solver Tool ® of MS Excel. The results which were obtained are presented in **Table 4**.

If no restriction on the concentration of phthalate is considered, then large lots can be ordered (Q = 1880 units). This minimizes the overall operating costs (E(C) = 25,646). Also, these large lots can lead to cumulative phthalate concentration up to 71.66 μg/m^3.

If the restriction on the cumulative phthalate is considered, a reduction of 30.22% can be obtained (50.00 μg/m^3). However, as this is dependent of the lot size, smaller lot sizes are required (Q = 1312 units). As consequence, this can lead to an increase in operational costs up to 5.62% (E(C) = 27,088).

These findings are very important to establish strategies to balance economic and environmental/health benefits. Particularly within the supply chain, suppliers, manufacturers, and distributors are continuously exposed to phthalates and thus represent health risks in the long term.

6. Conclusions

Minimizing the exposure to phthalate is an important task within all contexts in our society. These chemicals are present in office buildings, schools, homes, vehicles, food packaging, and warehouses, among others. The sources of phthalates which are used in building materials are more permanent in nature and their removal requires regulatory intervention, while other sources such as plastic materials and foam mattresses are easier to be replaced or removed [8].

In manufacturing, where inventories are the main resource for production, supply, and distribution, phthalates are continuously present. However, determining the possible risk based on phthalate concentration through time in warehousing facilities is not widely studied.

In this work, we explored on this aspect and proposed an integrated inventory control model with phthalate emission factor. Also, we addressed phthalate emission through the use of ANNs to estimate concentrations for different time periods where are commonly considered during inventory control strategies.

As presented, if phthalate concentrations are not considered, these can be increased in the presence of large lots, which frequently decrease operational costs associated to inventory ordering/re-supply.

If considering the phthalate concentration restriction to a certain permissible level, this can lead to reduce the ordering lots and, thus, to increase the operational costs. Nevertheless, the cost increase may be minimal in comparison to the reduction in phthalate concentration.

Thus, the proposed model can be used to support measures to control the presence of phthalate while keeping also under control the operational costs. Also, the model can be used as a basis for extended or alternative models considering the costs of cleaning tasks and the risk of specific health complications in certain environments/contexts.

Author details

Gladys Bonilla-Enríquez*, Patricia Cano-Olivos, José-Luis Martínez-Flores, Diana Sánchez-Partida and Santiago-Omar Caballero-Morales
People's Autonomous University of the State of Puebla (UPAEP), Puebla, México

*Address all correspondence to: gladys.bonilla@upaep.mx; gbonillae1973@gmail.com

References

[1] Godfray H, Beddington J, Crute I, Haddad L, Lawrence D, Muir J, et al. Food security: The challenge of feeding 9 billion people. Science. 2010;**327** (5967):812-818. DOI: 10.1126/science.1185383

[2] van der Vorst J, Joost S. Developments and Needs for Sustainable Agro-Logistics in Developing Countries [Internet]. 2014. Available from: https://openknowledge.worldbank.org/handle/10986/17834 [Accessed: 06 June 2018]

[3] Kahle L, Gurel-Atay E. Communicating Sustainability for the Green Economy. New York: M.E. Sharpe; 2014

[4] Linton J, Klassen R, Jayaraman V. Sustainable supply chains: An introduction. Journal of Operations Management. 2007;**25**(6):1075-1082. DOI: 10.1016/j.jom.2007.01.012

[5] Xu Y, Zhang J. Understanding SVOCs. ASHRAE Journal. 2011;**53**(12): 121-125

[6] EPA. United States Environmental Protection Agency [Internet]. 2019. Available from: https://www.epa.gov/history [Accessed: 01 April 2019]

[7] ECHA. European Chemicals Agency (ECHA) [Internet]. 2019. Available from: https://echa.europa.eu/legislation [Accessed: 01 April 2019]

[8] Kashyap D, Agarwal T. Concentration and factors affecting the distribution of phthalates in the air and dust: A global scenario. Science of the Total Environment. 2018;**635**:817-827. DOI: 10.1016/j.scitotenv.2018.04.158

[9] Young-Min L, Jung-Eun L, Wooseok C, Taeyeon K, Ji-Young L, Younglim K, et al. Distribution of phthalate esters in air, water, sediments, and fish in the Asan Lake of Korea. Environment International. 2019;**126**: 635-643. DOI: 10.1016/j.envint.2019.02.059

[10] García F, Bustamante L, García M. Presencia de ftalatos en bebidas en el estado de México. Iberoamericana para la Investigación y el Desarrollo Educativo. 2013;**11**:1-17

[11] Weiss J, Gustafsson A, Gerde P, Bergman A, Lindh C, Krais A. Daily intake of phthalates, MEHP, and DINCH by ingestion and inhalation. Chemosphere. 2018;**208**:1-43. DOI: 10.1016/j.chemosphere.2018.05.094

[12] Sun Y, Zhang Q, Hou J, Wang P, Sundell J. Exposure of phthalates in residential buildings and its health effects. Procedia Engineering. 2017;**205**: 1901-1904. DOI: 10.1016/j.proeng.2017.10.286

[13] Heudorf U, Mersch-Sundermann V, Angerer J. Phthalates: Toxicology and exposure. International Journal of Hygiene and Environmental Health. 2007;**210**(5):623-634. DOI: 10.1016/j.ijheh.2007.07.011

[14] Xiyan M, Ying H, Jia L, Ke Y, Wenbo Y, Gongming S, et al. New insights into the mechanism of phthalate-induced new insights into the mechanism of phthalate-induced. Environmental Pollution. 2018;**241**: 674-683. DOI: 10.1016/j.envpol.2018.05.095

[15] Saravanabhavan G, Murray J. Human biological monitoring of diisononyl phthalate and diisodecyl phthalate: A review. Journal of Environmental and Public Health. 2012; **810501**:1-12. DOI: 10.1155/2012/810501

[16] Fierens T, Servaes K, Van Holderbeke M, Geerts L, De Henauw S, Sioen I, et al. Analysis of phthalates in

food products and packaging materials sold on the Belgian market. Food and Chemical Toxicology. 2012;**50**(7): 2575-2583. DOI: 10.1016/j.fct.2012.04.029

[17] Aurela B, Kulmala H, Soderhjelm L. Phthalates in paper and board packing and their migration into Tenax and sugar. Food Additives & Contaminants. 1999;**16**(12):571-577. DOI: 10.1080/026520399283713

[18] Mérida-Ortega A, Hernández-Alcaraz C, Hernández-Ramírez R, García-Martínez A, Trejo-Valdivia B, Salinas-Rodríguez A, et al. Phthalate exposure, flavonoid consumption and breast cancer risk among Mexican women. Environment International. 2016;**16**:167-172. DOI: 10.1016/j.envint.2016.08.023

[19] Takeuchi S, Tanaka-Kagawa T, Saito I, Kojima H, Jin K, Satoh M, et al. Differential determination of plasticizers and organophosphorus flame retardants in residential indoor air in Japan. Environmental Science and Pollution Research. 2015;**25**(8): 7113-7120. DOI: 10.1007/s11356-015-4858-z

[20] Wang X, Tao W, Xu Y, Feng J, Wang F. Indoor phthalate concentration and exposure in residential and office buildings in Xi'an, China. Atmospheric Environment. 2014;**87**:146-152. DOI: 10.1016/j.atmosenv.2014.01.018

[21] Larsson K, Lindh CH, Jönsson BA, Giovanoulis G, Bibi M, Bottai M, et al. Phthalates, non-phthalate plasticizers and bisphenols in Swedish preschool dust in relation to children's exposure. Environment International. 2017;**102**: 114-124. DOI: 10.1016/j.envint.2017.02.006

[22] Kubwabo C, Rasmussen P, Fan X, Kosarac I, Wu F, Zidek A, et al. Analysis of selected phthalates in Canadian indoor dust collected using household vacuum and standardized sampling techniques. Indoor Air. 2013;**23**(6): 506-514. DOI: 10.1111/ina.12048

[23] Gaspar F, Castorina R, Maddalena R, Nishioka M, McKone T, Bradman A. Phthalate exposure and risk assessment in California child care facilities. Environmental Science and Technology. 2014;**48**(13):7593-7601. DOI: 10.1021/es501189t

[24] Wormuth M, Scheringer M, Vollenweider M, Hungerbuhler K. What are the sources of exposure to eight frequently used phthalic acid esters in Europeans? Risk Analysis. 2006;**26**(3): 803-824. DOI: 10.1111/j.1539-6924.2006.00770.x

[25] IHS Markit. Chemical Economics Handbook: Plasticizers [Internet]. 2018. Available from: https://ihsmarkit.com/products/plasticizers-chemical-economics-handbook.html [Accessed: 06 June 2018]

[26] Katsikantami I, Sifakis S, Tzatzarakis M, Tzatzarakis E, Kalantzi O, Tsatsakis A, et al. A global assessment of phthalates burden and related links to health effects. Environment International. 2016;**97**:212-236. DOI: 10.1016/j.envint.2016.09.013

[27] Drake A, Van den Driesche S, Scott H, Hutchison G, Seckl J, Sharpe R. Glucocorticoids amplify dibutyl phthalate-induced disruption of testosterone production and male reproductive development. Endocrinology. 2009;**150**(11): 5055-5064. DOI: 10.1210/en.2009-0700

[28] Chang W-S, Tsai Y-S, Wang J-Y, Chen H-L, Yang W-H, Lee C-C. Sex hormones and oxidative stress mediated phthalate-induced effects in prostatic enlargement. Environment International. 2019;**126**:184-192. DOI: 10.1016/j.envint.2019.02.006

[29] Weschler C, Nazaroff W. SVOC partitioning between the gas phase and

settled dust indoors. Atmospheric Environment. 2010;**44**:3609-3620. DOI: 10.1016/j.atmosenv.2010.06.029

[30] Yirui L, Ying X. Emission of phthalates and phthalate alternatives from vinyl flooring and crib mattress covers: The influence of temperature. Environmental Science & Technology. 2014;**48**(24):14228-14237. DOI: 10.1021/es504801x

[31] Yirui L, Ying X. Improved method for measuring and characterizing phthalate emissions from building materials and its application to exposure assessment. Environmental Science & Technology. 2014;**48**(8):4475-4484. DOI: 10.1021/es405809r

[32] Farhoodi M, Emam-Djomeh Z, Reza M, Oromiehie A. Effect of environmental conditions on the migration of di(2-ethylhexyl)phthalate from pet bottles into yogurt drinks: Influence of time, temperature, and food simulant. The Arabian Journal for Science and Engineering. 2008;**33**(2B): 279-287

[33] Muhamad S, Esmail L, Hasan S. Effect of storage temperature and sunlight exposure on the physicochemical properties of bottled water in Kurdistan region - Iraq. Journal of Applied Sciences and Environmental Management. 2011;**15, 147**(1):154

[34] Garg A, Gupta N, Tyagi S. Levels of benzene, toluene, ethylbenzene, and xylene near a traffic-congested area of East Delhi. Environmental Claims Journal. 2018;**30**(4):1-11. DOI: 10.1080/10406026.2018.1525025

[35] Fujii M, Shinohara N, Lim A, Otake O, Kumagai K, Yanagisawa Y. A study on emission of phthalate esters from plastic materials using a passive flux sampler. Atmospheric Environment. 2003;**37**:5495-5504. DOI: 10.1016/j.atmosenv.2003.09.026

[36] Bassok A, Hurvitz PM, Bae CC, Larson C. Measuring neighbourhood air pollution: The case of Seattle's international district. Journal of Environmental Planning and Management. 2010;**53**(1):23-29. DOI: 10.1080/09640560903399640

[37] Gel E, Keskinocak P. ISYE3104/Lot Size/Reorder Level (Q-R) Models. Atlanta, United States of America: Georgia Tech; 2013

[38] Chen C, Chou Y, Lin S, Lee C. Developing an intervention strategy to reduce phthalate exposure in Taiwanese girls. Science of the Total Environment. 2015;**517**:125-131. DOI: 10.1016/j.scitotenv.2015.02.021

[39] Afshari A, Gunnarsen L, Clausen PA, Hansen V. Emission of phthalates from PVC and other materials. Indoor Air. 2004;**14**:120-128. DOI: 10.1046/j.1600-0668.2003.00220.x

[40] Liang Y, Bi C, Wang X, Xu Y. A general mechanistic model for predicting the fate and transport of phthalates in indoor environments. Indoor Air. 2019;**29**:55-69. DOI: 10.1111/ina.12514

[41] Billings SA. Nonlinear System Identification: NARMAX Methods in the Time, Frequency, and Spatio-Temporal Domains. New York: Wiley; 2013

[42] Boussaada Z, Curea O, Remaci A, Camblong H, Bellaaj NM. A nonlinear autoregressive exogenous (NARX) neural network model for the prediction of the daily direct solar radiation. Energies. 2018;**11**:1-21. DOI: 10.3390/en11030620

Chapter 12

Sustainable Energy Management of Institutional Buildings through Load Prediction Models: Review and Case Study

Antonio Santos Sánchez, Maria João Regufe, Ana Mafalda Ribeiro and Idelfonso B.R. Nogueira

Abstract

Institutional buildings need smart techniques to predict the energy consumption in a smart grids' framework. Here, the importance of dynamic load forecasting as a tool to support the decision in smart grids is addressed. In addition, it is reviewed the energy consumption patterns of institutional buildings and the state-of-the-art of load forecast modeling using artificial neural networks. The discussion is supported by historical data from energy consumption in a university building. These data are used to develop a reliable model for the prediction of the electric load in a campus. A neural network model was developed, which can forecast the load with an average error of 6.5%, and this model can also be used as a decision tool to assess the convenience of supplying this load with a set of renewable energy sources. Statistical data that measure the availability of the local renewable sources can be compared with a load model in order to assess how well these energy sources match the energy needs of buildings. This novel application of load models was applied to the campus where a good correlation (Pearson coefficient of 0.803) was found between energy demand and the availability of the solar resource in the campus.

Keywords: sustainable energy management, renewable energy, load prediction, artificial intelligence, smart systems, smart grid

1. Introduction

Institutional buildings present similar patterns in their occupancy level and therefore in their energy consumption. Examples of this type of buildings are museums, hospitals, libraries, schools (secondary and University), non-profit foundations, governmental administrative offices, and prisons. Sometimes, as in the case of administrative and hospital complexes or University campuses, a set of buildings are grouped within a vast area reaching the energy consumption level of a small city. They all offer opportunities for energy improvement [1] which reflect in the saving of public money. Moreover, due to their similar characteristics, these buildings can share a similar energy-efficiency approach [2, 3].

There is a growing interest in technologies to perform effective management of these buildings, leading them to the transition into energy efficient smart buildings.

Among the research trends, two are assessed in this paper. The first one refers to smart techniques to predict the energy consumption in a smart grids' framework. In particular, it will be discussed the importance of dynamic load forecasting as a decision support system for a smart grid. The smart grid concept can be defined as an electrical grid that utilizes advanced control and telecommunication in order to optimize the energy generation, distribution, and consumption. This concept will be discussed and applied to the small electric network of a University Campus. After a review of load forecast models using artificial neural networks, a case-study using real data from a University building is presented. The main objectives of this work are:

- Offer an insight about the importance of load forecasting in smart grids;
- Apply the smart grid concept to a complex of institutional buildings;
- Review the state-of-the-art of load forecast modeling using artificial neural networks;
- To detail and develop an accurate model for the prediction of the load demand in a University campus.

In addition, a second research trend will be assessed in this paper. Future institutional buildings and smart campuses will also have an increasing level of self-supply through renewable energy sources. Therefore, it is presented a new approach that, to our knowledge, has not been done previously: To use the load forecast model for studying the correlation between the energy demand and the availability of renewable energy sources in the campus (solar and wind power).

We hope readers will appreciate this novelty. Overall, this work aims to contribute to the interesting topic that is the development of smart grids in institutional buildings.

2. The smart grid concept and the importance of load forecasting

The graphical representation of the demand of energy in a power system is called a load curve or load profile. Therefore, a load curve is a graph that illustrates the variation in demand/electrical load over a specific time, typically cycles of 24 h (daily load curve), 7 days, and 12 months (yearly load curve).

Load curves are determined based on the historical records of energy consumption of the system. Available data can be obtained from direct metering or other means: transformers' readings, utility meter load profilers and smart-grid automatic meters, or even customer billing [4]. Other influential parameters can be added to these energy consumption data in order to develop an energy demand model capable of forecasting the variation of the electric load. These models consider the weight of each type of consumer (residential, commercial, and industrial) in the system, their behavior and variables such as temperature variation or seasonal holydays.

Reliable and dynamic energy demand models are crucial elements of any smart grid [5–7]. They allow a better management of an electric system, so power supply can match demand in a more efficient way. The energy demand of a region is constituted by the sum of the effect of residential, commercial, and industrial loads and can vary greatly within a short period of time (hours). Power generation must fit this demand in an effective way or otherwise imports/exports of energy should be needed, if available. Nuclear or coal thermal plants lack the flexibility of varying their output and thus constitute the baseline of power generation. Based on load forecasts, the power output of the most flexible generation units (such as gas thermal plants) can be scheduled according to daily and seasonal cycles. Typically, gas power plants work at their maximum to supply daily peaks of load and have their output reduced during low demand hours. Hydroelectric power plants have also some capacity of power regulation and, in the case of pumped-storage hydroelectricity, can absorb the excess of power generated during night time and return it during peak times. Renewable energy, in particular wind power, arises as a destabilizing source of the system due to its intermittent and unpredictable characteristics. Its effective integration in the electric system is one of the main technical challenges for smart grids. Also, in demand-side management (demand response), daily load curves are used to set up electric tariffs in order to influence demand. Better prices of energy during low-demand hours encourage some consumers to move their activity to those hours and thus reduce the intensity of load peaks.

When talking about a much smaller system, such as a University campus or a small village, the situation is quite different, but knowing the local load profile can also lead to optimum operation as well as important energy savings.

In such a small system, the generation capacity would be represented by local distributed generation systems, such as roof-top solar systems or small wind turbines. Biomass boilers could also make use of neighboring agricultural residues, woods, or pruning waste. The latter resource should not be neglected as several institutional buildings such as University campuses, administrative and hospital complexes or prisons count with vast green areas in their surroundings. Diesel-fueled generators are present in many on-grid electric systems. In the case of commercial buildings, depending on the energy tariffs, it could be economic to switch off the building from the grid during peak hours and supply its own power demand burning diesel or other fuel. In the case of some institutional buildings such as hospitals and prisons, or some administrative buildings with data-centers, emergency generators are generally mandatory. Besides the use of diesel generators to supply power during peak times, some big commercial buildings resort to co-generation. In those buildings where HVAC systems are responsible for most of the power demand, it may be profitable the use of gas engines for the combined generation of electric power and heat. The latter can be transformed into refrigeration through thermal-chemical or other absorption system.

In addition, diesel generators can be coupled with energy systems that make use of local renewable resources conforming hybrid systems (mixture of PV solar, wind turbines, and biomass). Hybrid systems are a convenient option to gain reliability and diminish the intermittency problem of renewable sources, especially when coupled with batteries and are widely used in small isolated off-grid systems [8]. For small-scale systems, batteries are practically the only available form of energy storage. They can be big battery packs made from sodium-sulfur, vanadium-redox flow batteries, or other materials, grouped in "battery farms," or the smaller lithium-ion batteries from electric cars plugged to the system. Gónzalez et al. assessed the infrastructure needed for enabling the transition to a smart grid in a University campus, and in particular peak shaving of load with battery storage, concluding that for such case it is only economically feasible with limited battery sizes, and only when there are renewable energy sources available on-site [9]. Besides batteries for electricity storage, a building complex could also have thermal storage for its HVAC needs. In such case, thermal storage would influence the load profile and should be included in the load forecasting models [10]. Whatever the case, energy storage is one of the main components to be considered in a smart grid, as shown in **Figure 1**.

As can be observed in the previous figure, distributed or embedded generation (either from intermittent renewable sources or from diesel/gas generators)

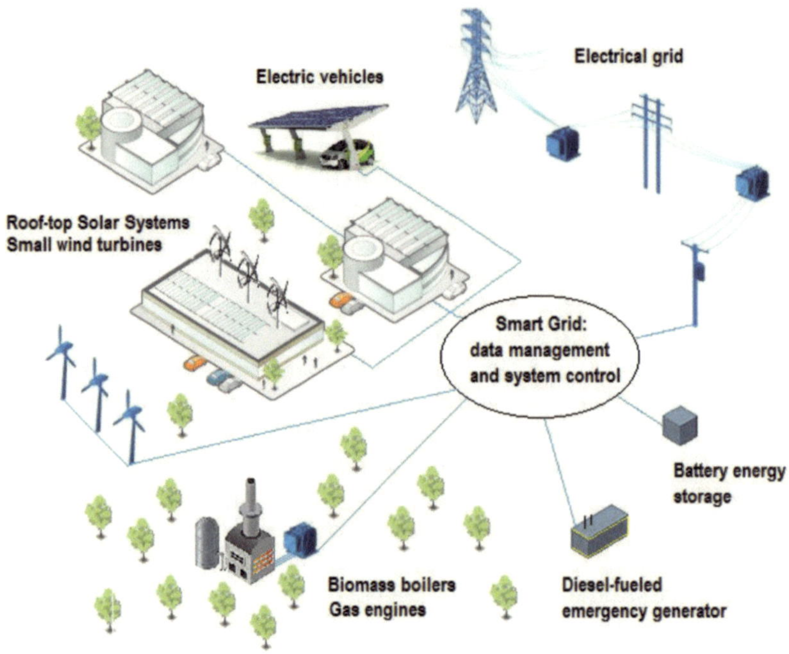

Figure 1.
Concept of a smart energy grid for a set of institutional buildings.

plays an important role in the design and operation of smart grids. The generation capacity could temporarily excess the local demand and then it would be necessary to either sell the excess power to the main grid or shut down the system if this option is not feasible (if local wind turbines are the ones to be turned off then it is called wind curtailment). When talking about the smart grid concept, a third option must be considered: to store that temporary surplus of energy. This can be done through the use of battery banks, as above-mentioned, or by increasing the energy consumption of a few selected utilities. Some examples: the HVAC system (cooling chillers, electric heaters, and heat pumps) could ramp its refrigeration/heat production and store the excess in a tank insulation system. Similarly, the local water/wastewater system could increase the consumption of pumps (switching them on or increasing their rotation through variable-frequency drives) to absorb a part of the excess of energy. The concept is similar to that of a load balancer in smart telecommunication grids, which distributes workloads across multiple computing resources [11, 12]. Another option usually considered in smart grids is the use of electric vehicles. In the case of institutional buildings with charging/discharging infrastructure for electric vehicles, those are more prone to act as a load to supply than as a source that can return the stored energy if needed. The reason is that in this type of buildings, the majority of the vehicles remain parked within the facilities only during workday while the charging time for electric vehicles currently requires periods of some hours. Therefore, the use of the vehicle's batteries by the local grid could leave them inoperative during some hours that could be coincident with the time that those vehicles are required.

There must be a system controller (an automated controller supervised by humans) that decides what to do, in each moment, to overcome a temporary surplus or deficit of energy forecasted for a close period of time. This controller has to deal with a number of input variables such as the state of the batteries (available storage capacity) or the number of electric vehicles plugged, as well as with short-term forecasts: predictions of weather (including solar and wind power), water and HVAC demand, and of course the forecasted electric load [13]. Therefore, the operation of a smart grid consists of an iterative process that considers the dynamic modeling of the load using a series of variables, with the aim of anticipating a situation through short-term predictions. Then, it uses this load forecast for the control process of the smart grid system and obtains feed-back through smart meters in the buildings facilities. Finally, it recalculates the load model and elaborates a new load prediction starting the control process again. **Figure 2** shows a diagram that schematizes the control process of a smart grid.

As shown in **Figure 2**, the advanced dynamic load model uses a historical database that is constantly refreshed with real-time measurements of energy demands [6]. Smart energy meters, deployed over the set of buildings and facilities, are thus a central part of the system. Those smart meters and sensors must transmit data to the control system through radio frequencies, Ethernet, Bluetooth, Wi-Fi, 6LoWPAN, Z-Wave or other technologies [14]. ZigBee wireless technology is the option chosen for the smart grid in the Illinois Institute of Technology main campus, which aims to reduce 20% of energy and 10% of gas consumption each year during a 5 years' period of time [15]. Other examples of smart grid design and concept applied in University campuses can be found in [9, 16].

Besides the smart grid concept, the use of data-driven analytical insights is widely used for a better energy management in buildings and in the power systems that supply them. Overall, the forecasting of energy demand in a building can lead to the following benefits:

1. To choose the most suitable tariff (contract power purchases);

2. Utilities and power system operators can respond quickly and confidently to forecasts and can improve performance for planning horizons that range from very short-term to very long-term. Forecasting peaks of energy demand is crucial to avoid black-outs, outages, and system failures;

3. Provides solid background to optimize the calculation of the power system components of the building. The most useful information is the maximum

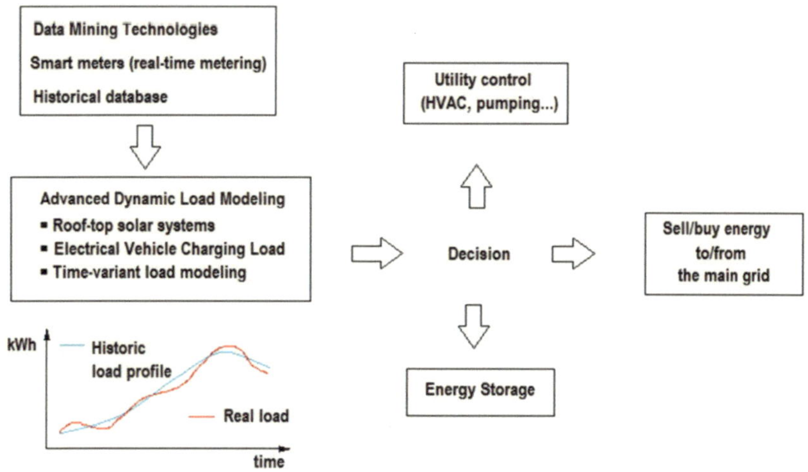

Figure 2.
Use of the dynamic load modeling for the control of a smart grid.

daily peak. Knowing the maximum expected current under normal conditions is crucial to calculate the transformers capacity and the size of conductors, as well as the power system protections. The hourly forecast of load is used in the calculation of either thermal or energy storage capacity;

4. Allows to define normal values of daily consumption and to compare different buildings of the same type that should present similar load profile. This is of particular interest for energy conservation programs in public, institutional buildings;

5. As highlighted by Dong et al. [17], the prediction of building energy consumption is increasingly important for building energy baseline model development and for performance Measurement and Verification Protocol (MVP). Having a computational model that models the energy consumption of a building along time is useful to verify savings after implementing energy conservation measures. Through calibrated simulation, any energy demand model can be tested and refined until it matches the actual energy performance measured in the facility with a high accuracy. Such a model may be valid for similar buildings of the same type and reliable in determining the savings of an energy efficiency project or calculating the energy consumption during the building life-time;

6. Energy consumption prediction for Building Energy Management systems (BEMS) allows building owners to optimize energy usage. In a similar way as the one described for smart grids, a smart building can vary its operation issues to respond to the demand signals from its sensors. Some authors agree that BEMS can be considered as one of the key factors in the success of energy saving measures in modern building operation [18].

3. State-of-the-art of load forecasting in buildings

Several computational models are used to forecast the demand of energy of different electric systems, ranging from small buildings and households [19] to big markets composed of several interconnected regions [20]. Multiple regression models are used, in which combinations of variables are tested sequentially for model improvement. Examples of these models are genetic algorithms [21], particle swarm optimization [22, 23], ant colony optimization [24], Fourier series [25], Support Vector Regression (SVR) [26–30], Support Vector Machine (SVM) [31], Autoregressive Integrated Moving Average (ARIMA) [20, 27, 28, 32–35], multiple linear regression [20, 26, 36, 37], Fuzzy logic [20, 38, 39], case-based reasoning [40], decision trees [41], and other data-driven forecasting algorithms [42–49], with special highlights to artificial neural networks [50]. For short-term load forecasting (daily demand profiles), exponential smoothing [51], least-square regression [52], and other methods may be more suitable while for a very short-term prediction, such as the prediction period of 1 hour, some authors have proposed a simple adaptive time-series model that considers the measurement history together with weather data [53]. Some complete reviews of buildings energy prediction techniques may be viewed at [54, 55].

This manuscript has the focus on load demand forecasting using artificial neural networks (ANN). Many readers are already familiar with these machine learning models that mimic a human neural system.

Among the Artificial Intelligence techniques, the ANN can be highlighted by its ability to track relationships between data groups. Their capacity to extract important information from data makes the ANNs an important tool in several fields. The overall structure of a ANN is composed by an input layer (where the data are presented to the model), hidden layers (where the extracted information is stored), and output layer where the response is given, as shown in **Figure 3**.

ANN can be used for forecasting water [56], gas [57–59], steam [60], and electricity demand in a set of buildings. They have also been proposed as a tool for evaluating energy performance of buildings and grant the correspondent energy performance certificates [61]. ANNs can model parameters that greatly influence the energy consumption of buildings such as HVAC performance [62, 63] or solar radiation [64, 65] and can also be used to accurately control and predict the performance of wind and solar energy systems [66–69].

Generally, the number of input variables would determine the complexity of the model. The three shown in **Figure 3** are the most common among the models found in the available literature. The "calendar" group of variables considers working days, holydays, and working hours. This type of variables has a great impact on office, administrative or University buildings as it determines the occupation level of the building, which is linked to its energy demand. The number of light hours per day, which affect the lighting needs of the building, can be modeled for each day of the year and therefore can be considered as a "calendar" variable. Sometimes there may be strikes or unexpected events, but their effect in the load prediction can be minimized with the use of the second group of variables: the load from the previous hours. The "weather conditions" type of variables directly influences the consumption of the HVAC systems. Some authors propose to develop an indicator of whether a building is likely to be weather sensitive (which measures the degree to which building loads are driven directly by local weather), for instance by using a Spearman Rank Order Correlation function [70]. Examples of this type of variables are dry bulb outdoor/indoor temperature and humidity. Ideally, these variables are measured in real time by wireless sensors and their variation trend is taken as an input for the model. If real-time

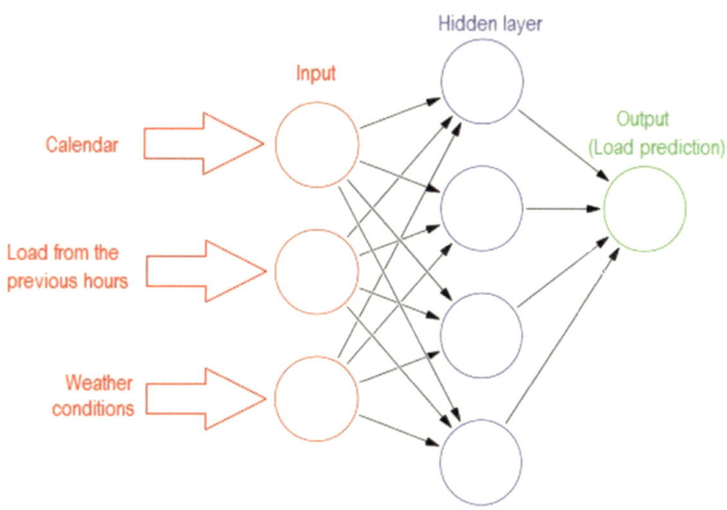

Figure 3.
Example of the architecture of an ANN that forecasts load in a building using three inputs.

measurement is not available, the input can be approximated with annual profiles from local historical data. Let us remember that, in addition to energy demand, "weather conditions" would have a great impact in solar and wind power production (the first one more predictable than the latter) so the monitoring of variables such as solar irradiation or wind speed/intensity would also be valuable for the forecast of the renewable energy generation of the building that aims to supply a part of the load.

The end-use approach aims to forecast separately the load demand of each of the main sub-systems that conform the building. In that approach, there is an ANN model for the HVAC system, another one for the water pumps, another one for the lighting needs, and so on. The final forecasted load will be the sum of the outputs of the set of models.

Other models may consider as inputs the state of the batteries or thermal tanks (available energy storage capacity) or the number of electric vehicles plugged.

The inputs presented to an ANN are weighted by parameters known as "weights." Moreover, each neuron will have a bias, which is another structure parameter. The product between the weights and inputs plus the bias will form the input argument of the so-called activation function. The output of the activation function will be the input of the subsequent layer and the final output of the model. In order to estimate the structure parameters, a train group is necessary, which will contain known inputs and outputs that is wanted to be tracked. Thus, the ANN prediction is compared to the known output for a given input. This "comparison" constitutes the objective function of the model training. Mean absolute percentage deviation (MAPE) and the coefficient of variation (CV) are usually used to evaluate the model performance during the training. In the present case, this error is function of consumption and the ANN prediction, given by:

$$MAPE = \frac{1}{n} \cdot \sum_{t=1}^{n} \left| \frac{C_t - F_t}{C_t} \right| \cdot 100\,(\%) \qquad (1)$$

where C_t is the actual value (the measured consumption in the instant t) and F_t is the forecast value for that instant. The difference between C_t and F_t is divided by the actual value C_t again and the absolute value of the resulting division is summed for every forecasted point and divided by the number of fitted points n.

Meanwhile, the coefficient of variation (CV), also known as relative standard deviation (RSD), is a standardized measure of dispersion of a probability (frequency) distribution. As in the case of MAPE, it is often expressed as a percentage. It is defined as the ratio of the standard deviation to the mean or to the absolute value of the mean (Eq. (3)):

$$CV = \frac{\sigma}{\mu} \cdot 100\,(\%) \qquad (2)$$

where σ is the standard deviation and μ is the mean.

A comprehensive review of applications of ANNs in the predictions of building's energy demand can be found in [71]. Following, in **Table 1**, a selected literature review is offered with the aim to offer a wide insight of the strategies and architectures used for load prediction using ANNs.

Type of system	Type of ANN (artificial neural network)	Accuracy (MAPE)	Accuracy (CV)	Year	Ref.
Main electric network	Cascaded neural network (CANN); short-term load forecasting	2.7%	—	1997	[72]
Main electric network	The annual growth rate is extracted from the data used for the ANN model	2.0%	—	2007	[73]
Main electric network	Nonlinear autoregressive with exogenous (NARX)	1.67%	3.60%	2015	[74]
Main electric network	Inputs: temperature and weather. Generalized regression neural network with decreasing step fruit fly optimization algorithm	0.024% (RMSE)	—	2017	[75]
Main electric network	Boosted neural network	1.42%	—	2017	[76]
Main electric network	Nonlinear autoregressive with exogenous (NARX)	1.0%	—	2017	[77]
Low-voltage smart electricity microgrid	Feed forward neural networks	4.0%	—	2016	[78]
Households (residential)	Elman ANN trained with the "back-propagation with momentum" algorithm. Multi-layer perception (MLP) with two inputs: weather data and electricity demand; short-term load forecasting	3.1%	0.36%	2008	[79]
Households (residential)	Feed-forward ANN and the Levenberg-Marquardt algorithm	10.0–23.5%	1.06%	2014	[80]
Households (residential) powered with wind and solar sources	Empirical mode decomposition, cascade-forward neural network (for solar and wind forecast) and a fuzzy logic-based controller (for load demand)	0.47% (wind) 19.2% (solar)	—	2014	[81]
Residential and commercial buildings with different wall types and insulation thickness	Backpropagation neural network	1.5%	3.43%	2009	[82]
Residential and commercial	Gated ensemble method (ordinary least squares and k-nearest neighbors)	55.8% (residential) and 7.5% (commercial)	—	2015	[83]
Residential and commercial	Nonlinear autoregressive with exogenous (NARX)	11.7%	55.89%	2017	[84]
Commercial building	Adaptive ANN models: accumulative training (AT) and sliding window training (SW)	13.3% (AT) and 12.9% (SW)	2.53% (AT) and 0.26% (SW)	2005	[6]
Commercial building	Adaptive ANN: accumulative training and sliding window training	—	2.50% and 0.36%	2005	[6]

Type of system	Type of ANN (artificial neural network)	Accuracy (MAPE)	Accuracy (CV)	Year	Ref.
Commercial building	Feed forward neural networks with hypothesis testing, information criteria and cross validation; 24 h forecast	1.5%	2.39%	2006	[85]
Commercial building	ANN model with Bayesian regularization algorithm; short-term load forecasting	5.0%	10.00%	2015	[86]
Commercial building	Three-layered perceptron with the logistic activation function and BFGS algorithm	1.3% (cooling energy consumption) and 2.4% (heating)	—	2017	[87]
Generic commercial building: ASHRAE contest	Input: relationship between load/temperature. Feedback ANN trained by hybrid algorithm	0.0033%	1.40%	2005	[88]
Commercial and industrial buildings	Seasonal ANN. Multi-layer perception (MLP) with two inputs: weather data and electricity demand	2.0–9.0%	—	2014	[89]
University campus	Input: temperature. ANN prediction method based on building end-uses	6.5%	—	2011	[90]
University campus (Library building)	Feed forward neural network with a single hidden layer of tansig neurons	—	0.03–0.10%	2011	[40]
University campus	Input: time temperature curve (TTC) forecast model. ANN prediction method based on building end-uses	6.3%	—	2013	[91]
University campus	Feed-forward with “Bayesian regularization” training algorithm	2.06%	—	2016	[92]
Institutional solar-powered building	17 inputs: weather data (indoor/outdoor sensors) and electricity demand; short-term load forecasting	11.5%	1.00–1.50%	2014	[93]
Institutional building	Feed forward neural network; short-term load forecasting	7.3–8.5%	—	2015	[41]
University campus	Feed-forward ANN trained with the Levenberg-Marquardt (LM) back-propagation algorithms	—	—	2018	[94]
Shopping mall	Optimized backpropagation and Levenberg-Marquardt back-propagation	4.267%	—	2018	[95]
Building energy consumption	Conditional restricted Boltzmann machine (CRBM) and factored conditional restricted Boltzmann machine (FCRBM)	—	—	2016	[96]

Table 1.
Literature review of load prediction using ANNs.

4. Energy profile and characteristics of the studied campus

This section presents an analysis of the characteristics that influence the load profile of the studied institutional building. The behavior of this building can be taken as representative for the set of buildings that compose the whole University campus in which it is inserted. Not surprisingly, all the buildings present the same occupation profile concentrated during working hours and workdays. In addition, almost all the buildings are of the same age and materials. The campus is located in the coast of Northeast Brazil, within a humid tropical region at 12° 58′ 16″ Latitude. In these conditions, the thermal comfort zone can be achieved through natural ventilation and several buildings were designed in that way, but as the University expanded the buildings ended up closing their indoor spaces in detriment of natural ventilation. Nowadays they are characterized by bad thermal insulation and by the massive use of small-size air-conditioning units instead of more efficient centralized units composed by chillers and cooling towers. This peculiarity, common in the majority of the Brazilian campuses and institutional buildings, is reflected in high energy consumption for cooling needs as well as a high dependence of the load curve with temperature. In other words, the building's load presents high weather sensitivity. Typically, the maximum load demand of the year occurs during the central hours of hot summer days.

The region is characterized by abundant renewable energy resources [97] but with water and energy supply problems [98]. Energy and water conservation are of crucial importance for both the region and the University institution. A great part of the budget of the campus is dedicated to water and energy. In this context, campus managers and researchers are considering options such as rainwater harvesting [99], water and energy conservation programs [100], and the transition into a smart grid [101, 102].

This campus has 15 university units within an area of almost 50 ha, providing services for approximately 15,000 students. Among these units, the Polytechnic School is composed of a main building and ancillary laboratories. Daily, almost 6,000 students as well as the correspondent University staff work and study at this particular facility.

The Polytechnic School presents mixed occupancies, which means that it may have multiple occupancies mainly educational, administrative, laboratory, and storage uses, as well as areas intended for food and drink consumption. The average energy consumption on a high-occupancy day is 462 kWh. The main end uses for energy are air conditioning (46.1%), lighting (30.9%), and electronic equipment (18.2%) as shown in **Figure 4**.

The rest of uses speak for almost 5% of the energy consumption of the building. Elevator and escalators typically represent from 3–8% of the energy used in most buildings [101]. However, during the period studied (years 2013 and 2014), the four elevators of the building were removed due to a reform. Besides the removal of the elevators, the reform did not have any other significant impact on the energy consumption.

The two following graphs illustrate very well the two main afore-mentioned variables that drive the load of the building. **Figure 5** shows the typical behavior of a daily load (period of 24 consecutive hours) for a working and a non-working day.

As can be observed in **Figure 5**, the daily profile of the load is directly dependent on the occupancy level of the building. Between 23 and 5 h, the energy demand remains at its minimum as the only load is outdoor lighting. On a working day (blue line), the load curve starts to ramp abruptly at 6 h and reaches a maximum at 9 h 30. There is a slight decrease in the load at lunch time, between 12 and 13 h, and then the load continues at its highest level until 18 h when it starts to decrease. Differently, on a non-working day (red line), the building remains unoccupied and the consumption continues at its lowest level, even with a slight decrease during the day as the outdoor lighting is automatically switched off.

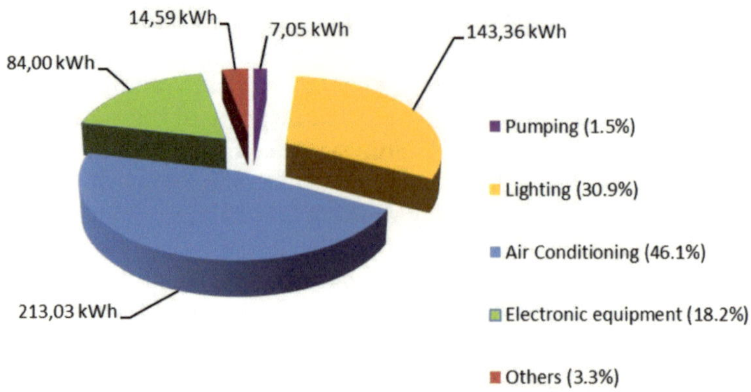

Figure 4.
Final uses of electric energy in the building (kWh/day).

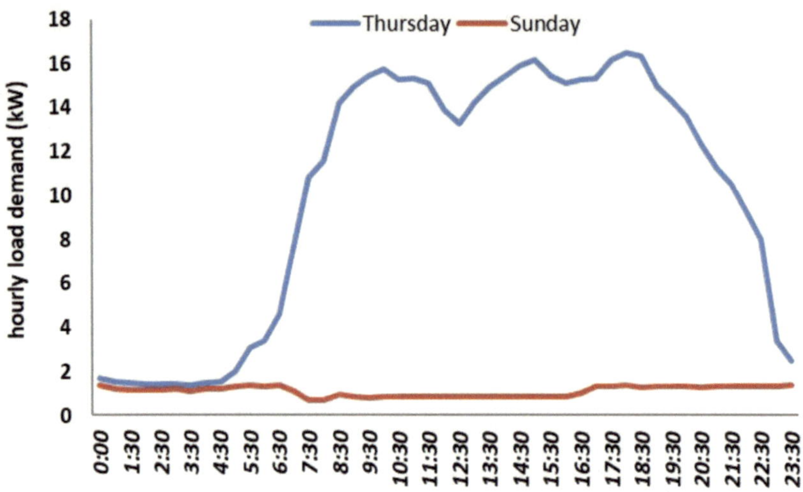

Figure 5.
Average daily load profiles of the building in both a working and a non-working day.

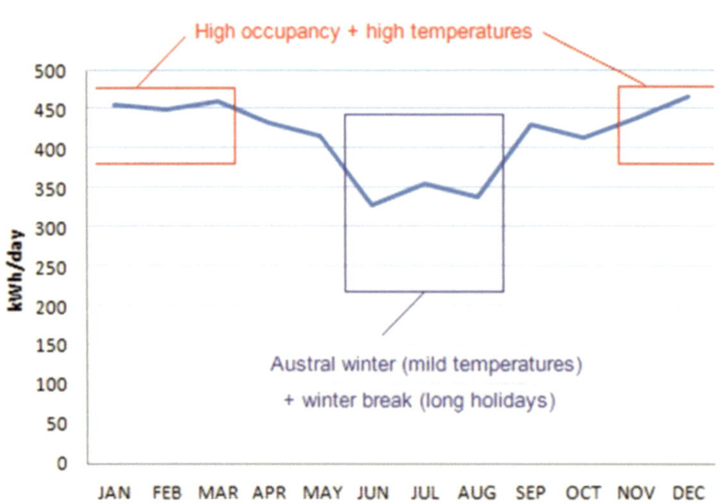

Figure 6.
Average curve of energy consumption in the building during years 2013/2014.

Figure 6 shows that the average daily consumption of energy in the building can vary ±30% because of the combined effects of temperature and calendar. The local temperature ranges from a minimum of 21.2°C in August to a maximum of 37.1°C in December.

5. Methodology

As historical data, it was used a database [102] containing energy consumption records from the building during more than 300 consecutive days. These data will serve as the foundation for a model that has to reflect as accurately as possible the effect of occupancy and temperature patterns in the load of any building in the campus, disregarding other effects in which the energy demand does not depend on.

When considering historical series of electric energy demand, especially in big electric networks, we must take into account that there is a rising tendency due to the influence of economic and population growth. This tendency must be extracted and modeled separately, typically as a constant rate related to the annual economic growth rate. It can also be modeled using ANN and regression models [103]. What remains is the fluctuation caused by the difference in demand from month to month, which depends among other factors on the seasonal variation of temperature. This fluctuation generates the annual load curve and must be modeled separately. After doing so, both effects can be summed up to obtain the series forecasting for upcoming months or years. The result is a more accurate model, achieving in some cases (with neural networks) values of the mean absolute percentage error (MAPE) of around 2% [73].

University buildings and campuses are within a much smaller scale. The only possible ways they can present the aforementioned growing trend in their energy consumption is due to:

- the use of new technologies and equipment, the implementation of new activities or the increase of existing ones, all of the above having a significant (and constant) impact on the energy consumption.

- an increase in the number of building occupants (alumni and workers).

Conversely, the energy consumption can present a constant decreasing trend, due to a decrease in the number of building occupants and – more frequently – due to the effects of energy conservation measures. In both cases, it is important to quantify and separate these rising/decreasing trends from the consumption pattern that it is intended to model.

However, this is not the case of the studied campus. During the one-year period of historical data, the energy consumption per capita has been constant. No major breakthroughs have occurred during that year, as was the case in some previous years thanks to, for example, the replacing of incandescent light bulbs with energy-efficient light bulbs, which produced a significant decrease in the load demand for the same occupation pattern. Moreover, the number of occupants in the building during that period (students and workers) also remained constant.

In addition, as pointed out by [90], the load in institutional buildings is also subjected to unpredictable factors: there are factors that may affect the consumption such as a failure of the HVAC system, strikes, etc. These events should be detected, and data must be filtered from the historical records in order to build a more reliable model. Those outliers were identified and removed prior to the development of the ANN model that is detailed from this point on.

The daily consumption is directly related to the period of the year and the day of the week. For this reason, the model structure may be a simple feed-forward as the one that was shown in **Figure 3**. However, the demand at any day may present some correlation with the one from the previous day. In order to take into consideration possible correlations between the daily demands, it is proposed a more evolved structure: a non-linear autoregressive exogenous model (NARX). Such structure consists basically in the feedback of the ANN using as part of its inputs the past outputs [104–107], as presented in Eq. (3):

$$y(t)=f\left(y(t-1),,,,,,y(t-2),,,,,,\dots,,,,,,y\left(t-n_{ty}\right),,,,,,U(t-1),,,,,,U(t-2),,,,,,U\left(t-n_{tu}\right)\right) \quad (3)$$

where, y is the output values, u is the process input values, n_{ty},n_{tu} the number of past values. The final structure of the ANN can thus be represented as shown in **Figure 7**.

After selecting the model structure, it is necessary for the overall architecture, which can be listed as: activation functions, number of hidden layers, and optimal number neurons. It is well known that one single layer is enough for a ANN model be able to approximate any function with relative precision [109]. The activation function is related to the dynamics of the systems being modeled, for example, pattern recognition case, where step functions are commonly used. To perform the training, usually the backpropagation method is employed [110–113]. The training is done until an acceptable MAPE is reached. The main point while identifying a ANN model is a careful selection of the optimal number of neurons, which is strictly correlated to the total number of parameters to be estimated. Thus, an excessive number of neurons might lead to a well-known problem, the overfitting. On the other hand, a small number might compromise the model prediction. In 1996, Schenker and Agarwal [114] proposed a method to identify the optimal number of neurons when few data are available, the dynamic cross-validation. The method consists in the usage of three data set, for example, set A, B, and C. The set A and B are employed in the training step, which will generate two different networks, for each neuron number. After the training, the network developed using

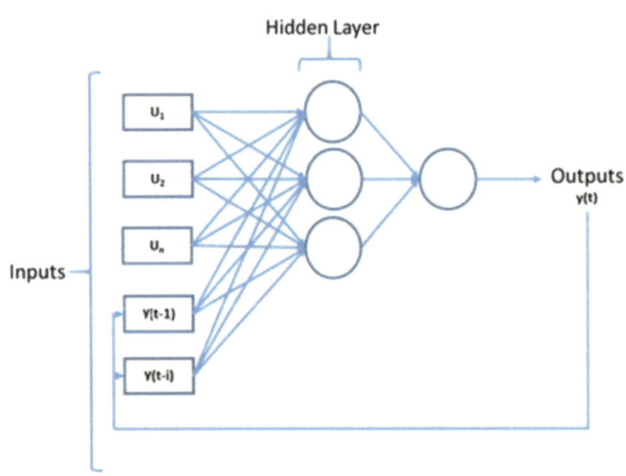

Figure 7.
Chosen structure for the neural network model: non-linear autoregressive exogenous model [108].

set A is validated using set B and the MAPE is calculated. The process continues up to a maximum number of neurons, which in the present work was 40 neurons. The optimal number of neurons is the one with lowest MAPE. The validation error is presented in **Figure 8** with its correspondent number of neurons. For the present case, the optimal number of neurons found was 5.

Finally, another network was trained using the optimal number of neurons. In order to avoid the overfitting, the early stopping criteria were employed [114–116]. This criterion consists in stop the training after a determined number of iteration where the validation error increased. The training of the final network was done with sets A and B, while the validation was done using set C. The general definitions of the final model are shown in **Table 2**.

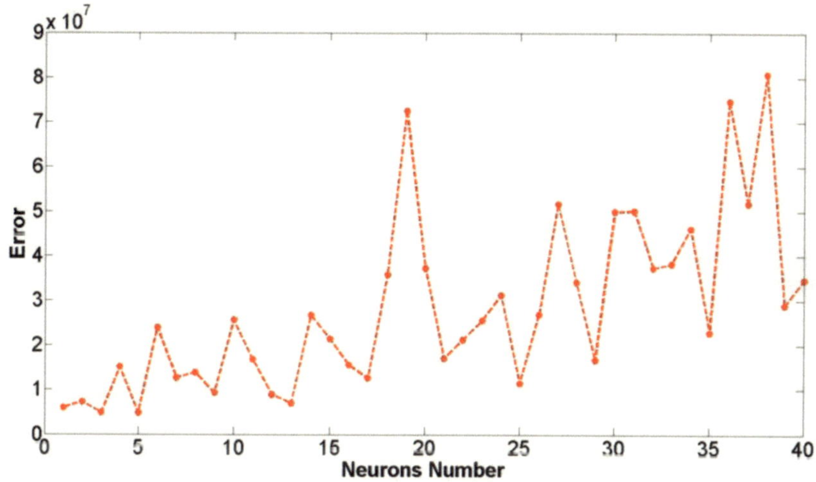

Figure 8.
Dynamic cross-validation for the selection of the optimal number of neurons of the hidden layer: validation errors for different number of neurons.

	ANN model parameters
Input	Database containing the energy consumption records of previous days
Output	Daily energy consumption
Total number of neurons evaluated	40
Total number of trainees done	40
Optimal number of neurons	5
Total iteration in training step	300
Minimum gradient	10^{-6}
Early stopping criteria	30
Transfer function in the first layer	Hyperbolic tangent sigmoid
Transfer function in the output layer	Linear function
Final model MAPE	6.54%

Table 2.
Characteristics of the proposed ANN model.

6. Results and discussion

6.1 Load demand model

In order to assess the generalization quality of the model, **Figure 9** shows the predicted data together with the validation data (real data).

As can be observed in the figure above, there are sudden variations in the daily consumption of energy, which repeat periodically in cycles of about 7 days. This refers to the load variation between workdays and weekends, with Saturdays presenting an intermediate value between a typical working day and the minimum consumption of Sundays. Overall, this type of curve can be taken as representative for an institutional building. Its variation depends directly on the occupation pattern of the University campus and, to a lesser extent, in the effect of temperature. The model developed using neural networks follows these consumption trends that were identified in **Figure 5** (working day versus non-working day) and **Figure 6** (seasonal variation of occupation and temperature).

The quality of the prediction was evaluated according to the MAPE, which was 6.54% for the final model. This means that through the proposed model, the campus managers can predict the electric consumption of any given day with an average error less than or equal to 6.54%. The average error is surprisingly similar to the ones achieved by different models for other university buildings (see the literature review in **Table 1**).

The error distribution, shown in **Figure 10**, revealed a slight trend of the model to underestimate the daily energy consumption.

The resulting set of errors showed a distribution with a high standard deviation. The standard deviation indicates how close the data points tend to be the mean of the set of errors. For the set of errors produced by this model, the standard deviation (sigma) is 20.75%. However, the model made some gross errors of up to -145% and + 85% at some points.

The CV depends on the standard deviation and on the mean of the forecast model data, as was shown in Eq. (2). Thus, the values calculated by the model showed a CV of 317%. This significant value of CV is due to the great variation between the load in working and in non-working days, typically between weekend and workweek. Together with the histogram of errors, **Figure 10** depicts the normal (or Gaussian) distribution of errors. This function is symmetric around the point

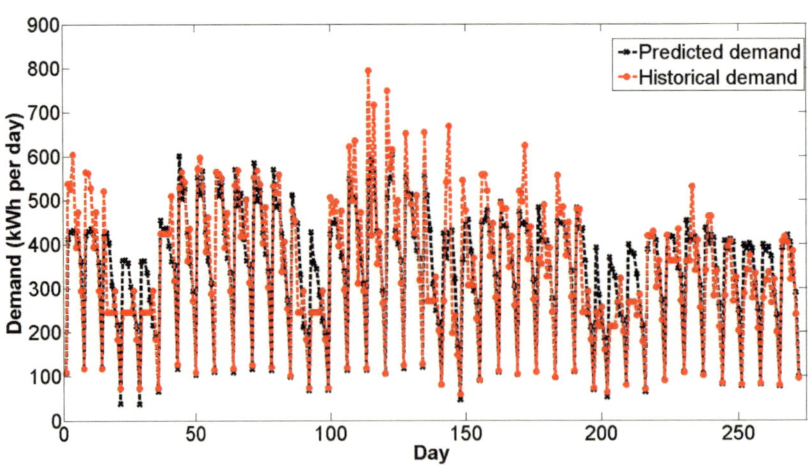

Figure 9.
Validation of the model with the demand data of the building from 300 consecutive days.

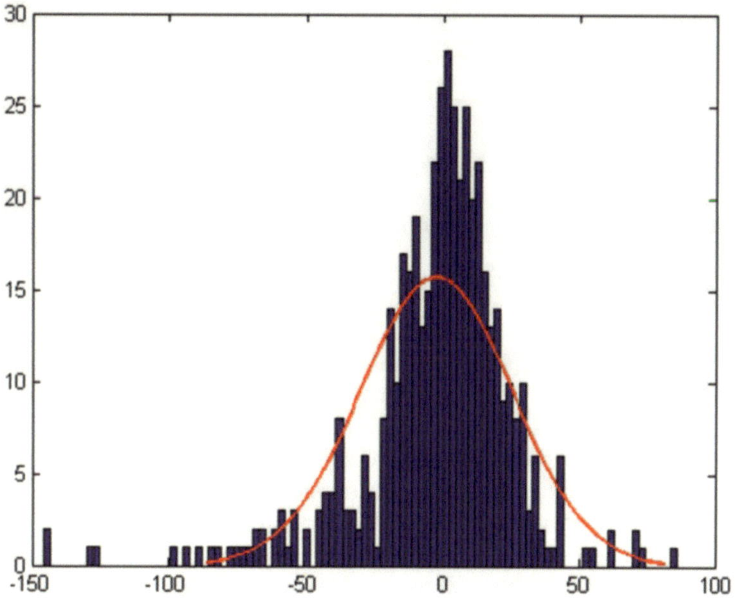

Figure 10.
Distribution of the errors made by the model.

−6.54 (mean value of the error). Within a normal distribution, the 3-sigma rule establishes that 68% of values are within one standard deviation away from the mean; about 95% of the values lie within two standard deviations; and about 99.7% are within three standard deviations. Therefore, it can be stated that by using the proposed ANN model, 68% of the forecasted values have an error of between −27.29 and +14.21% (MAPE ≤ 20.75%); 95% of the forecasted values have an error of between −48.04 and +34.96% (MAPE ≤ 41.50%); and about 99.7% of the forecasted values have an error of between −68.77 and +55.71% (MAPE ≤ 62.25%).

6.2 Correlation between the seasonal variation of load demand, solar, and wind energy availability

The proposed mathematical model can be taken as representative for the load profile of the campus where the building is inserted with an accuracy of 6.54%. This allows us to compare the load demand with the renewable energy availability in the campus. More precisely, allows the comparison of the seasonal variation of energy consumption versus the seasonal variation of the following meteorological parameters: wind speed and solar irradiation. There is a weather station in the campus that measures and records, among other parameters, global solar irradiation on a horizontal surface (MJ/m^2) and wind speed at 10 m height (m/s). The uncertainties of the measurements are ±5% for the solar pyranometer and ±1.5% for the wind anemometer [97]. Through this database, average values of wind speed and solar irradiation can be calculated for each day of the year, in order to build average curves that represent the seasonal variation of these two renewable sources. Then, these values can be compared with the load demand model, which yields the average value for energy consumption in the campus. To make this comparison, the Pearson product-moment correlation coefficient (hereafter Pearson correlation coefficient) will be used. This coefficient compares two sets of data and varies between −1 and 1. A value of 1 implies that a linear equation describes the relationship between the two compared variables perfectly, with all data points lying on a line for which

	Solar	Wind	Load demand
Solar	1	−0.008	0.803
Wind	−0.008	1	−0.505
Load demand	0.803	−0.505	1

Table 3.
Correlation (Pearson coefficient) between the seasonal variation of renewable energy resources and energy demand in the campus.

one increases as the other one increases. A value of −1 implies that all data points lie on a line for which one variable decreases as the other increases. A value of 0 for the coefficient implies that there is no linear correlation between the variables. The Pearson coefficient has proven to be useful in previous research in identifying which environmental variables (temperature and other weather conditions) correlate best (that is, have the greatest influence) in the energy consumption of buildings [117]. In our case, we are using the Pearson coefficient to assess the convenience of using some renewable energy sources by comparing its availability with the load of the campus. Three variables will be compared, namely "Solar," "Wind," and "Load demand." The Pearson correlation coefficient will indicate the strength of a linear relationship between them. As said, "Load demand" depends on the calendar but also on temperature, and thus may have some relationship with "Solar." "Solar" varies from a maximum in December to a minimum in August. "Wind" is the most intermittent and unpredictable, however tends to vary from a maximum in August to a minimum in March [97]. The Pearson correlation coefficient was calculated using the Statistical software Minitab® 16.2.1 and their resulting values are shown in **Table 3**.

Table 3 shows interesting results. "Solar" and "Wind" values show almost no relationship among them. When compared with the load demand of the campus, it was found that in the months were the load demand is higher the availability of wind resources tends to be lower and vice versa. The solar resource, meanwhile, showed a good correlation with the "Load demand." This is not surprising as the "Load demand" variable depends on temperature, which is related to solar irradiance. This correlation level means that in the months of high energy consumption, there is a higher availability of solar resource and vice versa. In other words, the variation of the solar resource matches very well the variation of the energy needs of the campus. When considering the daily variation of the load (as shown in **Figure 5**), the solar energy option gets reinforced, as most of the period with high load coincides with the peak of solar irradiation that occurs during the central hours of the day. Solar power is, therefore, the most convenient renewable energy source for this campus as is the one that best matches with the seasonal and daily variation of load demand.

7. Conclusion and future work

A reliable mathematical model was developed for the prediction of the electric load in a University campus. The neural network model was capable of forecasting the load with average error of 6.54%. The high standard deviation of the errors is the main weakness of this particular model. Load forecast models such as the one that is detailed in this article play an interesting role in the energy management of institutional buildings. First, as a powerful tool for the control of a smart grid that supplies either a single building or several of them grouped in a campus or a complex. Secondly, as a decision tool to assess the convenience of a set of renewable energy sources tend to vary seasonally. As was demonstrated in this study, statistical data that measure the availability of the local renewable sources can be compared with a load model in order to assess how well these energy sources match the variation of the energy needs of buildings. As future work the authors propose:

I. Applying calibration techniques to further reduce the error committed by the model;

II. Overcoming the high deviation of the errors by allowing the model to quickly recognize if a day is working-day or holiday;

III. Installing smart energy meters in the building with the aim to develop on-line building energy prediction using adaptive ANNs.

Acknowledgements

This work was financially supported by: Base Funding – UIDB/50020/2020 of the Associate Laboratory LSRE-LCM – funded by national funds through FCT/MCTES (PIDDAC).

Author details

Antonio Santos Sánchez[1]*, Maria João Regufe[2], Ana Mafalda Ribeiro[2] and Idelfonso B.R. Nogueira[2]*

1 Federal University of Ouro Preto, Brazil

2 Laboratory of Separation and Reaction Engineering, Associate Laboratory LSRE/LCM, Department of Chemical Engineering, Faculty of Engineering, University of Porto, Porto, Portugal

*Address all correspondence to: sanchezbahia@gmail.com and idelfonso@fe.up.pt

References

[1] Ruparathna R, Hewage K, Sadiq R. Improving the energy efficiency of the existing building stock: A critical review of commercial and institutional buildings. Renewable and Sustainable Energy Reviews. 2016;**53**:1032-1045. DOI: 10.1016/j.rser.2015.09.084

[2] Guillen-Garcia E, Zorita-Lamadrid AL, Duque-Perez O, Morales-Velazquez L, Osornio-Rios RA, Romero-Troncoso RDJ. Power consumption analysis of electrical installations at healthcare facility. Energies. 2017;**10**(1):64. DOI: 10.3390/en10010064

[3] Blanco ID, Vega AAC, López DP, González MD, Castro SA, Medrano MAP. Energy analytics in public buildings using interactive histograms. Energy and Buildings. 2017;**134**:94-104. DOI: 10.1016/j.enbuild.2016.10.026

[4] Kim YI, Ko JM, Choi SH. Methods for generating TLPs (typical load profiles) for smart grid-based energy programs. In: 2011 IEEE Symposium on Computational Intelligence Applications in Smart Grid (CIASG), April 11-15, 2011. 2011. pp. 1-6. DOI: 10.1109/CIASG.2011.5953331

[5] Yoldaş Y, Önen A, Muyeen SM, Vasilakos AV, Alan İ. Enhancing smart grid with microgrids: Challenges and opportunities. Renewable and Sustainable Energy Reviews. 2017;**72**:205-214. DOI: 10.1016/j.rser.2017.01.064

[6] Yang J, Rivard H, Zmeureanu R. On-line building energy prediction using adaptive artificial neural networks. Energy and Buildings. 2005;**37**(12):1250-1219. DOI: 10.1016/j.enbuild.2005.02.005

[7] Kim YS, Heidarinejad M, Dahlhausen M, Srebric J. Building energy model calibration with schedules derived from electricity use data. Applied Energy. 2017;**190**:997-1007. DOI: 10.1016/j.apenergy.2016.12.167

[8] Sánchez AS, Torres EA, Kalid RA. Renewable energy generation for the rural electrification of isolated communities in the Amazon region. Renewable and Sustainable Energy Reviews. 2015;**49**:278-290. DOI: 10.1016/j.rser.2015.04.075

[9] Gonzalez RM, van Goch TAJ, Aslam MF, Blanch A, Ribeiro PF. Microgrid design considerations for a smart-energy university campus. In: 2014 IEEE PES Innovative Smart Grid Technologies Conference Europe (ISGT-Europe), October 12-15, 2014. 2014. pp. 1-6. DOI: 10.1109/ISGTEurope.2014.7028743

[10] Ma Y, Borrelli F, Hencey B, Packard A, Bortoff S. Model predictive control of thermal energy storage in building cooling systems. In: Proceedings of the 48th IEEE Conference on Decision and Control, 2009 Held Jointly with the 2009 28th Chinese Control Conference, CDC/CCC 2009. IEEE; 2009. pp. 392-397. DOI: 10.1109/CDC.2009.5400677

[11] Shijue Z, Wanneng S, Guangdong C. A load balanced method based on campus grid. In: IEEE International Symposium on Communications and Information Technology. ISCIT; 2005. pp. 1516-1519. DOI: 10.1109/ISCIT.2005.1567160

[12] Lin Y, Barooah P, Mathieu JL. Ancillary services through demand scheduling and control of commercial buildings. IEEE Transactions on Power Systems. 2017;**32**(1):186-197. DOI: 10.1109/TPWRS.2016.2557764

[13] Kim W, Katipamula S. Development and validation of an intelligent

load control algorithm. Energy and Buildings. 2017;**135**:62-73. DOI: 10.1016/j.enbuild.2016.11.040

[14] Mahmood A, Javaid N, Razzaq S. A review of wireless communications for smart grid. Renewable and Sustainable Energy Reviews. 2015;**41**:248-260. DOI: 10.1016/j.rser.2014.08.036

[15] Yi P, Iwayemi A, Zhou C. Developing ZigBee deployment guideline under WiFi interference for smart grid applications. IEEE Smart Grid. 2011;**2**(1):110-120. DOI: 10.1109/TSG.2010.2091655

[16] Pipattanasomporn M, Feroze H, Rahman S. Multi-agent systems in a distributed smart grid: Design and implementation. In: Power Systems Conference and Exposition. PSCE'09; 2009. pp. 1-8. DOI: 10.1109/PSCE.2009.4840087

[17] Dong B, Cao C, Lee SE. Applying support vector machines to predict building energy consumption in tropical region. Energy and Buildings. 2005;**37**(5):545-553. DOI: 10.1016/j.enbuild.2004.09.009

[18] Yang C, Létourneau S, Guo H. Developing data-driven models to predict BEMS energy consumption for demand response systems. In: Modern Advances in Applied Intelligence (Proceedings of the 27th International Conference on Industrial Engineering and Other Applications of Applied Intelligent Systems, IEA/AIE 2014; 8481(Part I)):188-197. Springer International Publishing; 2014. DOI: 10.1007/978-3-319-07455-9_20

[19] Gerwig C. Short term load forecasting for residential buildings: An evaluation based on publicly available datasets. In: First International Conference on Smart Grid Inspired Future Technologies, SmartGIFT 2016, May 19-20 2016, Revised Selected Papers. Springer International Publishing; 2017. pp. 69-78. DOI: 10.1007/978-3-319-47729-9_8

[20] Alfares HK, Nazeeruddin M. Electric load forecasting: Literature survey and classification of methods. International Journal of Systems Science. 2002;**33**(1):23-34. DOI: 10.1080/00207720110067421

[21] Azadeh A, Ghaderi SF, Tarverdian S. Electrical energy consumption estimation by genetic algorithm. In: 2006 IEEE International Symposium on Industrial Electronics, July 9-13, 2006. Vol. 1. 9-13 July 2006. pp. 395-398. DOI: 10.1109/ISIE.2006.295626

[22] Huang CM, Huang CJ, Wang ML. A particle swarm optimization to identifying the ARMAX model for short-term load forecasting. IEEE Transactions on Power Systems. 2005;**20**(2):1126-1133. DOI: 10.1109/TPWRS.2005.846106

[23] Niu D, Li J, Li J, Liu D. Middle-long power load forecasting based on particle swarm optimization. Computers & Mathematcs with Applications. 2009;**57**(11-12):1883-1889. DOI: 10.1016/j.camwa.2008.10.044

[24] Niu D, Wang Y, Wu DD. Power load forecasting using support vector machine and ant colony optimization. Expert Systems with Applications. 2010;**37**(3):2531-2539. DOI: 10.1016/j.eswa.2009.08.019

[25] González-Romera E, Jaramillo-Morán MA, Carmona-Fernández D. Monthly electric energy demand forecasting with neural networks and Fourier series. Energy Conversion and Management. 2008;**49**(11):3135-3142. DOI: 10.1016/j.enconman.2008.06.004

[26] Massana J, Pous C, Burgas L, Melendez J, Colomer J. Short-term load forecasting in a non-residential building contrasting models and attributes.

Energy and Buildings. 2015;**92**:322-330. DOI: 10.1016/j.enbuild.2015.02.007

[27] Pai PF, Hong WC. Support vector machines with simulated annealing algorithms in electricity load forecasting. Energy Conversion and Management. 2005;**46**(17):2669-2688. DOI: 10.1016/j.enconman.2005.02.004

[28] Kavousi-Fard A, Samet H, Marzbani F. A new hybrid modified firefly algorithm and support vector regression model for accurate short term load forecasting. Expert Systems with Applications. 2014;**41**(13):6047-6056. DOI: 10.1016/j.eswa.2014.03.053

[29] Hong WC. Electric load forecasting by support vector model. Applied Mathematical Modelling. 2009;**33**(5):2444-2454. DOI: 10.1016/j.apm.2008.07.010

[30] Zhang F, Deb C, Lee SE, Yang J, Shah KW. Time series forecasting for building energy consumption using weighted support vector regression with differential evolution optimization technique. Energy and Buildings. 2016;**126**:94-103. DOI: 10.1016/j.enbuild.2016.05.028

[31] Paudel S, Elmitri M, Couturier S, Nguyen PH, Kamphuis R, Lacarrière B, et al. A relevant data selection method for energy consumption prediction of low energy building based on support vector machine. Energy and Buildings. 2017;**138**:240-256. DOI: 10.1016/j.enbuild.2016.11.009

[32] Yang HT, Huang CM, Huang CL. Identification of ARMAX model for short term load forecasting: An evolutionary programming approach. IEEE Transactions on Power Systems. 1996;**11**(1):403-408. DOI: 10.1109/PICA.1995.515202

[33] Darbellay GA, Slama M. Forecasting the short-term demand for electricity – Do neural networks stand a better chance? International Journal of Forecasting. 2000;**16**(1):71-83. DOI: 10.1016/S0169-2070(99)00045-X

[34] Taylor JW, Menezes LM, McSharry PE. A comparison of univariate methods for forecasting electricity demand up to a day ahead. International Journal of Forecasting. 2006;**22**(1):1-16. DOI: 10.1016/j.ijforecast.2005.06.006

[35] Newsham GR, Birt BJ. Building-level occupancy data to improve ARIMA-based electricity use forecasts. In: Proceedings of the 2nd ACM Workshop on Embedded Sensing Systems for Energy-Efficiency in Building. ACM; 2010. pp. 13-18. DOI: 10.1145/1878431.1878435

[36] Rahman S, Hazim O. A generalized knowledge-based short-term load-forecasting technique. IEEE Transactions on Power Systems. 1993;**8**(2):508-514. DOI: 10.1109/59.260833

[37] Safa M, Safa M, Allen J, Shahi A, Haas CT. Improving sustainable office building operation by using historical data and linear models to predict energy usage. Sustainable Cities and Society. 2017;**29**:107-117. DOI: 10.1016/j.scs.2016.12.001

[38] Li K, Su H, Chu J. Forecasting building energy consumption using neural networks and hybrid neuro-fuzzy system: A comparative study. Energy and Buildings. 2011;**43**:2893-2899. DOI: 10.1016/j.enbuild.2011.07.010

[39] Jurado S, Nebot À, Mugica F, Mihaylov M. Fuzzy inductive reasoning forecasting strategies able to cope with missing data: A smart grid application. Applied Soft Computing. 2017;**51**:225-238. DOI: 10.1016/j.asoc.2016.11.040

[40] Platon R, Dehkordi VR, Martel J. Hourly prediction of a building's electricity consumption using case-based reasoning, artificial neural

networks and principal component analysis. Energy and Buildings. 2015;**92**:10-18. DOI: 10.1016/j.enbuild.2015.01.047

[41] Tso GKF, Yau KKW. Predicting electricity energy consumption: A comparison of regression analysis decision tree and neural networks. Energy. 2007;**32**(9):1761-1768. DOI: 10.1016/j.energy.2006.11.010

[42] Boroojeni KG, Amini MH, Bahrami S, Iyengar SS, Sarwat AI, Karabasoglu O. A novel multi-time-scale modeling for electric power demand forecasting: From short-term to medium-term horizon. Electric Power Systems Research. 2017;**142**:58-73. DOI: 10.1016/j.epsr.2016.08.031

[43] Ortiz LE, Gonzalez JE, Gutierrez E, Arend M. Forecasting building energy demands with a coupled weather-building energy model in a dense urban environment. Journal of Solar Energy Engineering. 2017;**139**(1):1-8. DOI: 10.1115/1.4034909

[44] Sarwar R, Cho H, Cox SJ, Mago PJ, Luck R. Field validation study of a time and temperature indexed autoregressive with exogenous (ARX) model for building thermal load prediction. Energy. 2017;**119**:483-496. DOI: 10.1016/j.energy.2016.12.083

[45] He Y, Liu R, Li H, Wang S, Lu X. Short-term power load probability density forecasting method using kernel-based support vector quantile regression and copula theory. Applied Energy. 2017;**185**:254-266. DOI: 10.1016/j.apenergy.2016.10.079

[46] Cao J, Liu J, Man X. A united WRF/TRNSYS method for estimating the heating/cooling load for the thousand-meter scale megatall buildings. Applied Thermal Engineering. 2017;**114**:196-210. DOI: 10.1016/j.applthermaleng.2016.11.195

[47] Llanos J, Morales R, Núñez A, Sáez D, Lacalle M, Marín L, et al. Load estimation for microgrid planning based on a self-organizing map methodology. Applied Soft Computing. 2017;**53**:323-335. DOI: 10.1016/j.asoc.2016.12.054

[48] Muthalib MK, Nwankpa CO. Physically-based building load model for electric grid operation and planning. IEEE Smart Grid. 2017;**8**(1):169-177. DOI: 10.1109/TSG.2016.2595759

[49] Rodrigues F, Cardeira C, Calado JMF, Melício R. Family houses energy consumption forecast tools for smart grid management. In: Garrido P, Soares F, Moreira AP, editors. CONTROLO 2016. Springer International Publishing; 2017. pp. 691-699. DOI: 10.1007/978-3-319-43671-5_58

[50] Kalogirou SA. Applications of artificial neural-networks for energy systems. Applied Energy. 2000;**67**(1):17-35. DOI: 10.1016/S0306-2619(00)00005-2

[51] Taylor JW, McSharry PE. Short-term load forecasting methods: An evaluation based on European data. IEEE Transactions on Power Systems. 2007;**22**(4):2213-2219. DOI: 10.1109/TPWRS.2007.907583

[52] Espinoza M, Suykens JAK, De Moor B. Fixed-size least squares support vector machines: A large scale application in electrical load forecasting. Computational Management Science. 2006;**3**(2):113-129. DOI: 10.1007/s10287-005-0003-7

[53] Noh HY, Rajagopal R. Data-driven forecasting algorithms for building energy consumption. In: SPIE Proceedings Vol. 8692: Sensors and Smart Structures Technologies for Civil, Mechanical, and Aerospace Systems. 2013. DOI: 10.1117/12.2009894

[54] Raza MQ, Khosravi A. A review on artificial intelligence based load

demand forecasting techniques for smart grid and buildings. Renewable and Sustainable Energy Reviews. 2015;**50**:1352-1372. DOI: 10.1016/j.rser.2015.04.065

[55] Zhao HX, Magoulès F. A review on the prediction of building energy consumption. Renewable and Sustainable Energy Reviews. 2012;**16**(6):3586-3592. DOI: 10.1016/j.rser.2012.02.049

[56] Walker D, Creaco E, Vamvakeridou-Lyroudia L, Farmani R, Kapelan Z, Savić D. Forecasting domestic water consumption from smart meter readings using statistical methods and artificial neural networks. Procedia Engineering. 2015;**119**:1419-1428. DOI: 10.1016/j.proeng.2015.08.1002

[57] Rodger JA. A fuzzy nearest neighbor neural network statistical model for predicting demand for natural gas and energy cost savings in public buildings. Expert Systems with Applications. 2014;**41**(4, Part 2):1813-1829. DOI: 10.1016/j.eswa.2013.08.080

[58] Szoplik J. Forecasting of natural gas consumption with artificial neural networks. Energy. 2015;**85**:208-220. DOI: 10.1016/j.energy.2015.03.084

[59] Panapakidis IP, Dagoumas AS. Day-ahead natural gas demand forecasting based on the combination of wavelet transform and ANFIS/genetic algorithm/neural network model. Energy. 2017;**118**:231-245. DOI: 10.1016/j.energy.2016.12.033

[60] Kusiak A, Li M, Zhang Z. A data-driven approach for steam load prediction in buildings. Applied Energy. 2010;**87**(3):925-933. DOI: 10.1016/j.apenergy.2009.09.004

[61] Buratti C, Barbanera M, Palladino D. An original tool for checking energy performance and certification of buildings by means of artificial neural networks. Applied Energy. 2014;**120**:125-132. DOI: 10.1016/j.apenergy.2014.01.053

[62] Mohanraj M, Jayaraj S, Muraleedharan C. Applications of artificial neural networks for refrigeration, air-conditioning and heat pump systems—A review. Renewable and Sustainable Energy Reviews. 2012;**16**(2):1340-1358. DOI: 10.1016/j.rser.2011.10.015

[63] Yao Y, Yu Y. Combined forecasting models for air-conditioning load prediction. In: Modeling and Control in Air-Conditioning Systems. Berlin Heidelberg: Springer; 2017. pp. 297-327. DOI: 10.1007/978-3-662-53313-0_7

[64] Bou-Rabee M, Sulaiman SA, Saleh MS, Marafi S. Using artificial neural networks to estimate solar radiation in Kuwait. Renewable and Sustainable Energy Reviews. 2017;**72**:434-438. DOI: 10.1016/j.rser.2017.01.013

[65] Kashyap Y, Bansal A, Sao AK. Solar radiation forecasting with multiple parameters neural networks. Renewable and Sustainable Energy Reviews. 2015;**49**:825-835. DOI: 10.1016/j.rser.2015.04.077

[66] Ata R. Artificial neural networks applications in wind energy systems: A review. Renewable and Sustainable Energy Reviews. 2015;**49**:534-562. DOI: 10.1016/j.rser.2015.04.166

[67] Karabacak K, Cetin N. Artificial neural networks for controlling wind–PV power systems: A review. Renewable and Sustainable Energy Reviews. 2014;**29**:804-827. DOI: 10.1016/j.rser.2013.08.070

[68] Messalti S, Harrag A, Loukriz A. A new variable step size neural networks MPPT controller: Review, simulation and hardware implementation.

Renewable and Sustainable Energy Reviews. 2017;**68**(Part 1):221-233. DOI: 10.1016/j.rser.2016.09.131

[69] Almonacid F, Fernandez EF, Mellit A, Kalogirou S. Review of techniques based on artificial neural networks for the electrical characterization of concentrator photovoltaic technology. Renewable and Sustainable Energy Reviews. 2017;**75**:938-953. ISSN: 1364-0321. DOI: 10.1016/j.rser.2016.11.075

[70] Coughlin K, Piette MA, Goldman C, Kiliccote S. Statistical analysis of baseline load models for non-residential buildings. Energy and Buildings. 2009;**41**(4):374-381. DOI: 10.1016/j.enbuild.2008.11.002

[71] Kumar R, Aggarwal RK, Sharma JD. Energy analysis of a building using artificial neural network: A review. Energy and Buildings. 2013;**65**:352-358. DOI: 10.1016/j.enbuild.2013.06.007

[72] AlFuhaid AS, El-Sayed MA, Mahmoud MS. Cascaded artificial neural networks for short-term load forecasting. IEEE Transactions on Power Systems. 1997;**12**(4):1524-1529. DOI: 10.1109/59.627852

[73] González-Romera E, Jaramillo-Morán MA, Carmona-Fernández D. Forecasting of the electric energy demand trend and monthly fluctuation with neural networks. Computers and Industrial Engineering. 2007;**52**(3): 336-343. DOI: 10.1016/j.cie.2006.12.010

[74] Hashmi MU, Arora V, Priolkar JG. Hourly electric load forecasting using nonlinear AutoRegressive with eXogenous (NARX) based neural network for the state of Goa, India. In: IEEE International Conference on Industrial Instrumentation and Control (ICIC). 2015. pp. 1418-1423. DOI: 10.1109/IIC.2015.7150971

[75] Hu R, Wen S, Zeng Z, Huang T. A short-term power load forecasting model based on the generalized regression neural network with decreasing step fruit fly optimization algorithm. Neurocomputing. 2017;**221**:24-31. DOI: 10.1016/j.neucom.2016.09.027

[76] Khwaja AS, Zhang X, Anpalagan A, Venkatesh B. Boosted neural networks for improved short-term electric load forecasting. Electric Power Systems Research. 2017;**143**:431-437. DOI: 10.1016/j.epsr.2016.10.067

[77] Buitrago J, Asfour S. Short-term forecasting of electric loads using nonlinear autoregressive artificial neural networks with exogenous vector inputs. Energies. 2017;**10**(1):40. DOI: 10.3390/en10010040

[78] Yuce B, Mourshed M, Rezgui Y. An ANN-based energy forecasting framework for the district level smart grids. In: First International Conference on Smart Grid Inspired Future Technologies, SmartGIFT 2016, May 19-20, 2016, Revised Selected Papers. Springer International Publishing; 2017. pp. 107-117. DOI: 10.1007/978-3-319-47729-9_12

[79] Beccali M, Cellura M, Brano VL, Marvuglia A. Short-term prediction of household electricity consumption: Assessing weather sensitivity in a Mediterranean area. Renewable and Sustainable Energy Reviews. 2008;**12**:2040-2065. DOI: 10.1016/j.rser.2007.04.010

[80] Rodrigues F, Cardeira C, Calado JMF. The daily and hourly energy consumption and load forecasting using artificial neural network method: A case study using a set of 93 households in Portugal. Energy Procedia. 2014;**62**:220-229. DOI: 10.1016/j.egypro.2014.12.383

[81] Tascikaraoglu A, Boynuegri AR, Uzunoglu M. A demand side

management strategy based on forecasting of residential renewable sources: A smart home system in Turkey. Energy and Buildings. 2014;**80**:309-320. DOI: 10.1016/j.enbuild.2014.05.042

[82] Ekici BB, Aksoy UT. Prediction of building energy consumption by using artificial neural networks. Advances in Engineering Software. 2009;**40**(5):356-362. DOI: 10.1016/j.advengsoft.2008.05.003

[83] Burger EM, Moura SJ. Gated ensemble learning method for demand-side electricity load forecasting. Energy and Buildings. 2015;**109**:23-34. DOI: 10.1016/j.enbuild.2015.10.019

[84] Di Piazza MC, La Tona G, Luna M, Di Piazza A. A two-stage energy management system for smart buildings reducing the impact of demand uncertainty. Energy and Buildings. 2017;**139**:1-9. DOI: 10.1016/j.enbuild.2017.01.003

[85] Karatasou S, Santamouris M, Geros V. Modeling and predicting building's energy use with artificial neural networks: Methods and results. Energy and Buildings. 2006;**38**(8):949-958. DOI: 10.1016/j.enbuild.2005.11.005

[86] Chae YT, Horesh R, Hwang Y, Lee YM. Artificial neural network model for forecasting sub-hourly electricity usage in commercial buildings. Energy and Buildings. 2016;**111**:184-194. DOI: 10.1016/j.enbuild.2015.11.045

[87] Pino-Mejías R, Pérez-Fargallo A, Rubio-Bellido C, Pulido-Arcas JA. Comparison of linear regression and artificial neural networks models to predict heating and cooling energy demand, energy consumption and CO2 emissions. Energy. 2017;**118**:24-36. DOI: 10.1016/j.energy.2016.12.022

[88] González PA, Zamarreño JM. Prediction of hourly energy consumption in buildings based on a feedback artificial neural network. Energy and Buildings. 2005;**37**(6):595-601. DOI: 10.1016/j.enbuild.2004.09.006

[89] Jetcheva JG, Majidpour M, Chen WP. Neural network model ensembles for building-level electricity load forecasts. Energy and Buildings. 2014;**84**:214-223. DOI: 10.1016/j.enbuild.2014.08.004

[90] Escrivá-Escrivá G, Álvarez-Bel C, Roldán-Blay C, Alcázar-Ortega M. New artificial neural network prediction method for electrical consumption forecasting based on building end-uses. Energy and Buildings. 2011;**43**(11):3112-3119. DOI: 10.1016/j.enbuild.2011.08.008

[91] Roldán-Blay C, Escrivá-Escrivá G, Álvarez-Bel C, Roldán-Porta C, Rodríguez-García J. Upgrade of an artificial neural network prediction method for electrical consumption forecasting using an hourly temperature curve model. Energy and Buildings. 2013;**60**:38-46. DOI: 10.1016/j.enbuild.2012.12.009

[92] Deb C, Eang LS, Yang J, Santamouris M. Forecasting diurnal cooling energy load for institutional buildings using artificial neural networks. Energy and Buildings. 2016;**121**:284-297. DOI: 10.1016/j.enbuild.2015.12.050

[93] Mena R, Rodríguez F, Castilla M, Arahal MR. A prediction model based on neural networks for the energy consumption of a bioclimatic building. Energy and Buildings. 2014;**82**:142-155. DOI: 10.1016/j.enbuild.2014.06.052

[94] Yuan J, Farnham C, Azuma C, Emura K. Predictive artificial neural network models to forecast the seasonal hourly electricity consumption for a university campus. Sustainable Cities and Society. 2018;**42**:82-92. ISSN: 2210-6707. DOI: 10.1016/j.scs.2018.06.019

[95] Zhaoyang Y, Keun KM. Predicting electricity consumption in a building using an optimized backpropagation and Levenberg–Marquardt back-propagation neural network: Case study of a shopping mall in China. Sustainable Cities and Society. 2018;**42**:176-183. ISSN: 2210-6707. DOI: 10.1016/j.scs.2018.05.050

[96] Mocanu E, Nguyen PH, Gibescu M, Kling WL. Deep learning for estimating building energy consumption. Sustainable Energy, Grids and Networks. 2016:91-99. DOI: 10.1016/j.segan.2016.02.005

[97] de Jong P, Sánchez AS, Esquerre K, Kalid RA, Torres EA. Solar and wind energy production in relation to the electricity load curve and hydroelectricity in the northeast region of Brazil. Renewable and Sustainable Energy Reviews. 2013;**23**:526-535. DOI: 10.1016/j.rser.2013.01.050

[98] Sánchez AS, Nogueira IBR, Kalid RA. Uses of the reject brine from inland desalination for fish farming, Spirulina cultivation, and irrigation of forage shrub and crops. Desalination. 2015;**364**:96-107. DOI: 10.1016/j.desal.2015.01.034

[99] Sánchez AS, Cohim E, Kalid RA. A review on physicochemical and microbiological contamination of roof-harvested rainwater in urban areas. Sustainability Water Quality and Ecology. 2015;**6**:119-137. DOI: 10.1016/j.swaqe.2015.04.002

[100] Marinho M, Gonçalves MS, Kiperstok A. Water conservation as a tool to support sustainable practices in a Brazilian public university. Journal of Cleaner Production. 2014;**62**:98-106. DOI: 10.1016/j.jclepro.2013.06.053

[101] Galo JJM, Macedo MNQ, Almeida LAL, Lima ACC. Method for deployment of smart grids through the creation of a priority index. Renewable and Sustainable Energy Reviews. 2015;**51**:1421-1427. DOI: 10.1016/j.rser.2015.07.036

[102] Macedo MNQ, Galo JJM, Almeida LAL, Lima ACC. Demand side management using artificial neural networks in a smart grid environment. Renewable and Sustainable Energy Reviews. 2015;**41**:128-133. DOI: 10.1016/j.rser.2014.08.035

[103] Kialashaki A, Reisel JR. Modeling of the energy demand of the residential sector in the United States using regression models and artificial neural networks. Applied Energy. 2013;**108**:271-280. DOI: 10.1016/j.apenergy.2013.03.034

[104] Song JJ, Park S. Neural model predictive control for nonlinear chemical processes. Journal of Chemical Engineering of Japan. 1993;**26**(4):347-354. DOI: 10.1252/jcej.26.347

[105] Wang C, Klatt KU, Dünnebier G, Engell S, Hanisch F. Neural network-based identification of SMB chromatographic processes. Control Engineering Practice. 2003;**11**(8):949-959. DOI: 10.1016/S0967-0661(02)00212-5

[106] Costa MCB, Jardini AL, Maciel MRW, Embiruçu M, Filho RM. Empirical models for end-use properties prediction of LDPE: Application in the flexible plastic packaging industry. Materials Research. 2008;**11**(1):23-30. DOI: 10.1590/S1516-14392008000100005

[107] Noor RA, Ahmad Z, Don M, Uzir MH. Modelling and control of different types of polymerization processes using neural networks technique: A review. Canadian Journal of Chemical Engineering. 2010;**88**(6):1065-1084. DOI: 10.1002/cjce.20364

[108] Sánchez AS, Rodrigues DA, Fontes RM, Martins MF, Kalid RA,

Torres EA. Wave resource characterization through in-situ measurement followed by artificial neural networks' modeling. Renewable Energy. 2018;**115**:1055-1066. DOI: 10.1016/j.renene.2017.09.032

[109] Haykin S. Neural Network: A Comprehensive Foundation. 2nd ed. PTR Upper Saddle River, NJ, USA: Prentice Hall; 1999

[110] Nazari J, Ersoy OK. Implementation of back-propagation neural networks with MatLab. ECE Technical Reports, Paper 275; 1992. Available from: http://docs.lib.purdue.edu/ecetr/275

[111] Piuleac CG, Rodrigo MA, Cañizares P, Curteanu S, Sáez C. Ten steps modeling of electrolysis processes by using neural networks. Environmental Modelling and Software. 2010;**25**(1):74-81. DOI: 10.1016/j.envsoft.2009.07.012

[112] Mjalli FS, Ibrehem AS. Optimal hybrid modeling approach for polymerization reactors using parameter estimation techniques. Chemical Engineering Research and Design. 2011;**89**(7):1078-1087. DOI: 10.1016/j.cherd.2010.11.018

[113] Bessa I, Quito C, Pontes K. Artificial neural networks structure selection: The benefits of cross validation method. In: Eng Opt 2014: 4th International Conference on Engineering Optimization, September 8-11, 2014

[114] Schenker B, Agarwal M. Cross-validated structure selection for neural networks. Computers and Chemical Engineering. 1996;**20**(2):175-186. DOI: 10.1016/0098-1354(95)00013-R

[115] Bowden GJ, Maier HR, Dandy GC. Optimal division of data for neural network models in water resources applications. Water Resources Research. 2002;**38**(2):2-11. DOI: 10.1029/2001WR000266

[116] Hoque S, Farouk B, Haas CN. Development of metamodels for predicting aerosol dispersion in ventilated spaces. Atmospheric Environment. 2011;**45**(10):1876-1887. DOI: 10.1016/j.atmosenv.2010.12.046

[117] Kapetanakis DS, Mangina E, Finn DP. Input variable selection for thermal load predictive models of commercial buildings. Energy and Buildings. 2017;**137**:13-26. DOI: 10.1016/j.enbuild.2016.12.016